Protecting Endangered Species

Protecting Endangered Species

Editor

Chunwang Li

MDPI

Basel • Beijing • Wuhan • Barcelona • Belgrade • Novi Sad • Cluj • Manchester

Editor
Chunwang Li
Chinese Academy of Sciences
Beijing
China

Editorial Office
MDPI AG
Grosspeteranlage 5
4052 Basel, Switzerland

This is a reprint of articles from the Special Issue published online in the open access journal *Animals* (ISSN 2076-2615) (available at: https://www.mdpi.com/journal/animals/special_issues/BQ7T0I3X53).

For citation purposes, cite each article independently as indicated on the article page online and as indicated below:

Lastname, A.A.; Lastname, B.B. Article Title. *Journal Name* **Year**, *Volume Number*, Page Range.

ISBN 978-3-7258-2395-6 (Hbk)
ISBN 978-3-7258-2396-3 (PDF)
doi.org/10.3390/books978-3-7258-2396-3

Contents

About the Editor

Chunwang Li

Chunwang Li is an associate professor at the Institute of Zoology, Chinese Academy of Sciences and the University of Chinese Academy of Sciences. His research interests include ethology, behavioral ecology, behavioral endocrinology, and biodiversity conservation. In recent years, he has carried out many research projects and participated in joint research between China, Russia and the United States. He has also participated in many fieldwork surveys on wildlife diversity in many natural reserves and national parks. He has worked in the field for several years, gathering biological information about Père David's deer, Przewalski's gazelle, sand lizard and other animals. Many of his scientific achievements have been published in around 100 papers in journals in the categories of animal ecology and conservation biology. His recent research includes the behavioral ecology of desert lizards and endangered ungulates, as well as the investigation and assessment of biodiversity in protected areas.

animals MDPI

Editorial

Protecting Endangered Animal Species

Chunwang Li [1,2]

[1] Key Laboratory of Animal Ecology and Conservation Biology, Institute of Zoology, Chinese Academy of Sciences, Beijing 100101, China; licw@ioz.ac.cn
[2] College of Life Sciences, University of Chinese Academy of Sciences, Beijing 100049, China

Citation: Li, C. Protecting Endangered Animal Species. *Animals* **2024**, *14*, 2644. https://doi.org/10.3390/ani14182644

Received: 28 May 2024
Revised: 4 September 2024
Accepted: 11 September 2024
Published: 12 September 2024

Currently, global biodiversity loss is a growing problem, and more species are endangered and at risk of extinction. Due to the development of human society, the space left for wildlife has become more limited. Therefore, saving endangered species and conserving biodiversity is a matter of urgency. As of 2023, IUCN has assessed 89,856 animal species, 17,416 of which are listed as endangered, accounting for 19.38% of the total animal species assessed [1]. Among them, 11,195 vertebrate species are listed as endangered, accounting for 17.91% of the total vertebrate species assessed. In invertebrates, 6221 species were listed as endangered, accounting for 22.74% of the total invertebrate species assessed. Due to the vulnerability of endangered animals and their sensitivity to environmental degradation and human disturbance, they need to be given more attention. Therefore, the research on conservation of endangered species has naturally become a hotspot in the fields of ecology and conservation biology.

In this Special Issue on Protecting Endangered Species, we collected 17 papers that related to the conservation of endangered species and threatened animals. These publications cover some important aspects of endangered species conservation, such as ecological adaptation of animals, habitat alteration, climate change, population dynamics of endangered animals, conservation translocation, conservation genetics, pollution effect on animals, and construction of protected areas.

Studies have shown that animals adapt to the ambient environment through their physiological regulation and behavioral strategies [2]. Animals with strong ecological adaptability are able to resist uncertain changes in the environment. However, in the context of dramatic changes in the environment, those species that are less adaptable require more attention.

The gut microbiota plays an important role in regulating the physiological function of wild animals [3], while dietary factors and environmental changes could affect the gut microbiota of animals [4]. In the golden snub-nosed monkey (*Rhinopithecus roxellana*), the different fiber intake results in a significant difference in the gut microbiota between the wild and the captive populations [5]. In contrast to wild populations, captive golden snub-nosed monkeys have less beneficial bacteria and more potentially pathogenic bacteria [5]. In captive *Cyprinus chilia*, the gut microbial community structure is significantly changed after release, along with the trend in initially decreasing and then increasing gut fullness [6]. This suggests that the difference in gut microbiota between captive and released animals should be considered in ex situ conservation.

In sympatric animals, the food niche is another issue of concern. Yue et al. (2023) found that there was a significant difference in the food composition of the Tibetan macaque (*Macaca thibetana*) and the gray snub-nosed monkey (*Rhinopithecus brelichi*) [7]. However, the food niches of these two monkey species almost entirely overlapped in winter. Therefore, although the differentiation of dietary habits can reduce niche overlap and interspecific competition, this will be reversed in different seasons.

For animals, habitats are shelters to survive and keep their populations growing. However, many endangered animals are experiencing global changes that lead to habitat degradation or habitat loss, especially climate change [8]. Thus, predicting potential suitable habitats is essential for endangered species, especially in the context of current global climate change.

Among all the vertebrates, amphibians and reptiles are the most vulnerable to climate change. Tao et al. (2024) found that the currently suitable habitats for Wushan salamander (*Liua shihi*) are concentrated in the Daba Mountains, in southwestern China [9]. An optimistic estimate is that under future climate conditions, the area of suitable habitats will increase. Similarly, the main environmental factors influencing the distribution of Szechwan rat snake (*Euprepiophis perlacea*) include the distance from streams and the slope degree, and their potential habitat will not decrease in the context of future climate scenarios [10].

For mammals, climate change also affects the survival of individuals and the habitat of their population. The marbled polecat (*Vormela peregusna*) is a mustelid mammal endemic to Eurasia and listed as a vulnerable (VU) by IUCN due to its low population size and increasing human threats. It is reported that the marbled polecat has a wide range of distribution under current conditions [11]. However, climate change is predicted to severely affect the distribution of the marbled polecat and substantially lead to a significant reduction in the area of suitable habitats in the future [11]. As with terrestrial mammals, the negative effects of climate change are widely found in marine mammals (especially ice-breeding marine mammals). e.g., spotted seals in the North Pacific may face the loss of breeding habitats caused by climate change [12].

In addition to climate change, human activities such as poaching, bycatch, deforestation, overgrazing, urbanization, road killing, and pollution affect endangered animals negatively and directly. It is reported that global extinctions in the marine environment were mainly attributed to overexploitation, followed by invasive species, habitat destruction, trophic cascades, and pollution [13]. Zuo et al. (2023) found that bycatch and stranding incidents occurred widely across the Shandong Peninsula throughout all seasons from 2000 to 2018. Meanwhile, the widespread use of fishing gear was the principal cause of death and injury to finless porpoises during that period [14]. Chilvers and McClelland (2023) reviewed information from pre-emptive captures and translocations of threatened wildlife undertaken during past oil spills and island pest eradications and listed a number of incidents in which these processes have affected animals [15]. They suggested that in order to reduce the negative impact of pollution on endangered animals, wildlife can be captured and transferred before foreseeable contamination occurs [15].

For protecting endangered animals, in situ conservation and ex situ conservation are the two main conservation strategies that are attracting increasing attention, especially in the face of climate change [16]. The most commonly used measure in in situ conservation is the establishment of protected areas. While the methods in ex situ conservation usually include introduction, reintroduction, translocation, and assisted migrations.

The giant panda (*Ailuropoda melanoleuca*) is the flagship species of animal conservation worldwide. In order to effectively protect the giant panda, its habitat, and the entire ecosystem of the giant panda's range, China has established a number of nature reserves over the past few decades that have developed into the Giant Panda National Park, which covers an area of 27,134 square kilometers. Despite the success of in situ conservation, some small populations of giant pandas are still facing the problems of being non-self-sustaining [17,18]. While the successful conservation of Przewalski's gazelle (*Procapra przewalskii*) has led to population increase, it has caused them to face another problem, that of excessive density in their current range [19]. To promote the further conservation of these two species, the conservation translocation method based on ecological models and GAP analysis is/will be used to solve the different problems of conservation in the giant panda and Przewalski's gazelle [17,18,20].

In this Special Issue, some ecological models (e.g., InVEST-HQ model, MaxEnt model, GAP analysis) are reported being used in the study of many endangered species such as Wushan salamander [9], Szechwan rat snake [10], Przewalski's gazelle [20], spotted seal [12], yellow-throated martens (*Martes flavigula*), and leopard cats (*Prionailurus bengalensis*) [21], to evaluate and predict suitable habitats and potential distribution areas for endangered animals. These ecological models are also used to determine gaps in existing protected areas, identify priority conservation areas [21] or predict potential corridors between habitats [22].

In addition to macro-ecological methods, other methods, such as the conservation genetics method [23,24], remote sensing of biodiversity [25], and passive acoustics techniques [26], have also been used for inventory and monitoring of endangered animals. This suggests that there is a growing number of new technologies being developed and used in the study and the conservation of endangered species, rather than just the traditional ecological methods. Moreover, an expert opinion survey suggests that even in the basic activities of museum-based biological collection, the new scientific methods are relied upon to improve the effectiveness of biocollections for biodiversity conservation [27].

In sum, the contributions cover studies on the ecological adaptation of endangered animals, the effects of climate change and human activities on endangered animals, and the approaches and methods of animal conservation. Although there are fewer than twenty papers and reviews in this Special Issue, they point out some serious problems endangered animals are facing and reflect the research trends in the conservation of endangered species. There is still a long way to go to protect endangered species, and the problems faced by endangered animals need to be addressed on the basis of in-depth research on ecology and conservation biology.

Acknowledgments: The author thanks all colleagues who contributed their work to this Special Issue.

Conflicts of Interest: The authors declare no conflicts of interest.

References

1. IUCN. The IUCN Red List of Threatened Species. Version 2024-1. 2024. Available online: https://www.iucnredlist.org (accessed on 15 May 2024).
2. Bernhardt, J.R.; O'Connor, M.I.; Sunday, J.M.; Gonzalez, A. Life in fluctuating environments. *Phil. Trans. R. Soc. B.* **2020**, *375*, 20190454. [CrossRef] [PubMed]
3. Dallas, J.W.; Warne, R.W. Captivity and animal microbiomes: Potential roles of microbiota for influencing animal conservation. *Microb. Ecol.* **2023**, *85*, 820–838. [CrossRef] [PubMed]
4. Nova, E.; Gómez-Martinez, S.; González-Soltero, R. The Influence of Dietary Factors on the Gut Microbiota. *Microorganisms* **2022**, *10*, 1368. [CrossRef]
5. Wang, Y.; Yang, X.; Zhang, M.; Pan, H. Comparative analysis of gut microbiota between wild and captive golden snub-nosed monkeys. *Animals* **2023**, *13*, 1625. [CrossRef]
6. Zhu, T.; Yang, D.; Gong, J.; Wang, C.; Miao, X.; Liang, Y.; Li, X. Initial post-release performance of cultured *Cyprinus chilia* juveniles in a shallow lake in southwestern China. *Animals* **2023**, *13*, 3196. [CrossRef]
7. Yue, L.; Wang, C.; Meng, B.; Xie, B.; Cao, H.; Su, H.; Zhang, M. The food niche overlap and interspecific relationship between the sympatric Tibetan macaque and grey snub-nosed monkey. *Animals* **2023**, *13*, 2536. [CrossRef]
8. Priya, A.K.; Muruganandam, M.; Rajamanickam, S.; Sivarethinamohan, S.; Gaddam, M.K.R.; Velusamy, P.; Gomathi, R.; Ravindiran, G.; Gurugubelli, T.R.; Muniasamy, S.K. Impact of climate change and anthropogenic activities on aquatic ecosystem—A review. *Environ. Res.* **2023**, *238*, 117233.
9. Tao, J.; Hu, Y.; Jiang, J.; Yang, W.; Zhao, T.; Su, S. Prediction of potential suitable distribution areas for an endangered salamander in China. *Animals* **2024**, *14*, 1390. [CrossRef] [PubMed]
10. Song, X.; Jiang, Y.; Zhao, L.; Jin, L.; Yan, C.; Liao, W. Predicting the potential distribution of the Szechwan rat snake (*Euprepiophis perlacea*) and its response to climate change in the Yingjing area of the Giant Panda National Park. *Animals* **2023**, *13*, 3828. [CrossRef]
11. Cheng, X.; Han, Y.; Lin, J.; Jiang, F.; Cai, Q.; Shi, Y.; Cui, D.; Wen, X. Time to step up conservation: Climate change will further reduce the suitable habitats for the vulnerable species marbled polecat (*Vormela peregusna*). *Animals* **2023**, *13*, 2341. [CrossRef]
12. Yang, L.; Zhuang, H.; Liu, S.; Cong, B.; Huang, W.; Li, T.; Liu, K.; Zhao, L. Estimating the spatial distribution and future conservation requirements of the spotted seal in the North Pacific. *Animals* **2023**, *13*, 3260. [CrossRef] [PubMed]
13. Nikolaou, A.; Katsanevakis, S. Marine extinctions and their drivers. *Reg. Environ. Chang.* **2023**, *23*, 88. [CrossRef]
14. Zuo, T.; Li, Y.; Cheng, Z.; Wang, J.; Sun, J.; Yuan, W.; Niu, M. Finless porpoise bycatch and stranding along the Shandong Peninsula, China, based on public reports from 2000 to 2018. *Animals* **2023**, *13*, 3868. [CrossRef] [PubMed]
15. Chilvers, B.L.; McClelland, P.J. Lessons learned for pre-emptive capture management as a tool for wildlife conservation during oil spills and eradication events. *Animals* **2023**, *13*, 833. [CrossRef] [PubMed]

16. Moir, M.L.; Vesk, P.A.; Brennan, K.E.; Poulin, R.; Hughes, L.; Keith, D.A.; McCarthy, M.A.; Coates, D.J. Considering extinction of dependent species during translocation, ex situ conservation, and assisted migration of threatened hosts. *Conserv. Biol.* **2012**, *26*, 199–207. [CrossRef]
17. Wang, Y.; Wei, W.; Yuan, F.; Cao, D.; Zhang, Z. The science underlying giant panda conservation translocations. *Animals* **2023**, *13*, 3332. [CrossRef]
18. Zhang, Y.; Liu, J.; Yu, J.; Li, C.; Zhao, X.; Mo, L.; Wu, W.; Gai, Y.; Xu, Q.; Ni, J.; et al. Enhancing the viability of a small giant panda population through individual introduction from a larger conspecific group: A scientific simulation study. *Animals* **2024**, *14*, 2345. [CrossRef]
19. Liang, D.; Liu, Y.; Ping, X.; Jiang, Z.; Li, C. Ensuring recovery for the Przewalski's gazelle. *Science* **2021**, *374*, 163. [CrossRef]
20. Liang, D.; Li, C. Habitat suitability, distribution modelling and gap analysis of Przewalski's gazelle conservation. *Animals* **2024**, *14*, 149. [CrossRef]
21. Jeong, A.; Kim, M.; Lee, S. Analysis for priority conservation areas using Habitat Quality models and MaxEnt models. *Animals* **2024**, *14*, 1680. [CrossRef]
22. Roshani; Rahaman, M.H.; Masroor, M.; Sajjad, H.; Saha, T.K. Assessment of habitat suitability and potential corridors for Bengal tiger (*Panthera tigris tigris*) in Valmiki Tiger Reserve, India, using MaxEnt model and Least-cost modeling approach. *Environ. Model Assess* **2024**, *29*, 405–422. [CrossRef]
23. Vella, A.; Vella, N. Conservation genetics of the loggerhead sea turtle, *Caretta*, from the Central Mediterranean: An insight into the species' reproductive behaviour in Maltese waters. *Animals* **2024**, *14*, 137. [CrossRef]
24. Gómez-Lobo, D.A.; Monteoliva, A.P.; Fernandez, A.; Arbelo, M.; de la Fuente, J.; Pérez-Gil, M.; Varo-Cruz, N.; Servidio, A.; Pérez-Gil, E.; Borrell, Y.J.; et al. Mitochondrial variation of bottlenose dolphins (*Tursiops truncatus*) from the Canary Islands suggests a key population for conservation with high connectivity within the North-East Atlantic Ocean. *Animals* **2024**, *14*, 901. [CrossRef]
25. Cavender-Bares, J.; Schneider, F.D.; Santos, M.J.; Armstrong, A.; Carnaval, A.; Dahlin, K.M.; Fatoyinbo, L.; Hurtt, G.C.; Schimel, D.; Townsend, P.A.; et al. Integrating remote sensing with ecology and evolution to advance biodiversity conservation. *Nat. Ecol. Evol.* **2022**, *6*, 506–519. [CrossRef]
26. Cheng, Z.; Li, Y.; Pine, M.K.; Wan, X.; Zuo, T.; Niu, M.; Wang, J. Acoustic presence of cetaceans in the Miaodao Archipelago, China. *Animals* **2023**, *13*, 1306. [CrossRef]
27. Alpeeva, E.V.; Sharova, N.P.; Sharov, K.S.; Vorotelyak, E.A. Russian biodiversity collections: A professional opinion survey. *Animals* **2023**, *13*, 3777. [CrossRef]

animals

MDPI

Article

Enhancing the Viability of a Small Giant Panda Population Through Individual Introduction From a Larger Conspecific Group: A Scientific Simulation Study

Yuzhen Zhang [1,2], Jiabin Liu [2], Jiaojiao Yu [2], Cheng Li [2], Xing Zhao [1], Li Mo [2], Wei Wu [2], Yulin Gai [1], Qiang Xu [3], Jiubin Ni [4], Limin Shen [5], Haijun Gu [6], Jindong Zhang [1,*], Dunwu Qi [2,*] and Xiaodong Gu [6]

1 Key Laboratory of Southwest China Wildlife Resources Conservation Ministry of Education, China West Normal University, Nanchong 637001, China; zhangyz9711@163.com (Y.Z.); zukoxi98@163.com (X.Z.); yulingai0818@163.com (Y.G.)
2 Sichuan Key Laboratory of Conservation Biology for Endangered Wildlife, Chengdu Research Base of Giant Panda Breeding, Chengdu 610081, China; jiabin_liu2013@126.com (J.L.); yujiaojiao2016@126.com (J.Y.); lslicheng@126.com (C.L.); jasmineae@126.com (L.M.); happyeffiewu@163.com (W.W.)
3 World Wide Fund for Nature, China Office, Beijing 100006, China; qxu@wwfchina.org
4 The Nature Conservancy (USA) Beijing Representative Office, Beijing 100600, China; nijiubin@163.com
5 Tangjiahe National Nature Reserve, Guangyuan 628100, China; tjhshenlm@163.com
6 Sichuan Forestry and Grassland Bureau, Chengdu 610082, China; ghyciom@163.com (H.G.); gu1751@cn.com (X.G.)
* Correspondence: zhangjd224@cwnu.edu.cn (J.Z.); qidunwu@163.com (D.Q.)

Simple Summary: Individual quantity is the guarantee of the long-term survival of a population. Currently, approximately 70% of the 33 local populations of wild giant pandas (*Aliuropoda melanoleuca*) face the risk of extinction. We studied four typical populations, from Tangjiahe, Wanglang, Liziping, and Daxiangling, and attempted to translocate individuals from large populations to enhance the viability of small populations. Our results validated a classic conclusion that the introduction of individuals from Tangjiahe and Wanglang significantly improved the genetic diversity in Liziping and Daxiangling. To ensure the long-term survival of two small populations over the next 100 years, we have devised multiple specific strategies for individual introduction. Our research has certain value for the rejuvenation of small populations of giant pandas, and we hope to provide a reference for the protection of endangered animals such as giant pandas.

Abstract: Currently, nearly 70% of giant panda populations are facing survival challenges. The introduction of wild individuals can bring vitality to them. To explore this possibility, we hypothetically introduced giant pandas from Tangjiahe and Wanglang into Liziping and Daxiangling Nature Reserves. We collected feces from these areas and analyzed the genetic diversity and population viability before and after introduction using nine microsatellite loci. The results showed the genetic level and viability of the large populations were better than the small populations. We investigated the effects of time intervals (2a, 5a, and 10a; year: a) and gender combinations (female: F; male: M) on the rejuvenation of small populations. Finally, five introduction plans (1F/2a, 2F/5a, 1F1M/5a, 3F/10a, and 2F1M/10a) were obtained to make Liziping meet the long-term survival standard after 100 years, and six plans (1F/2a, 2F/5a, 1F1M/5a, 4F/10a, 3F1M/10a, and 2F2M/10a) were obtained in Daxiangling. The more females were introduced, the greater the impact on the large populations. After introducing individuals, the number of alleles and expected heterozygosity of the Liziping population are at least 6.667 and 0.688, and for the Daxiangling population, they are 7.111 and 0.734, respectively. Our study provides theoretical support for the translocation of giant pandas, a reference for the restoration of other endangered species worldwide.

Keywords: conservation of giant pandas (*Aliuropoda melanoleuca*); genetic diversity; population rejuvenation; population viability

Citation: Zhang, Y.; Liu, J.; Yu, J.; Li, C.; Zhao, X.; Mo, L.; Wu, W.; Gai, Y.; Xu, Q.; Ni, J.; et al. Enhancing the Viability of a Small Giant Panda Population Through Individual Introduction From a Larger Conspecific Group: A Scientific Simulation Study. *Animals* **2024**, *14*, 2345. https://doi.org/10.3390/ani14162345

Academic Editor: Miho Inoue-Murayama

Received: 23 June 2024
Revised: 10 August 2024
Accepted: 12 August 2024
Published: 14 August 2024

1. Introduction

More than 157,100 species are currently recorded in the IUCN Red List, of which over 44,000 species are facing the threat of extinction [1]. Biodiversity is crucial for nature conservation [2]. The preservation of endangered species and the achievement of sustainable species development have long been the central focus and primary research direction for conservation biologists. As a rare relict species in China, giant pandas (*Aliuropoda melanoleuca*) serve as flagship and umbrella species [3]. The conservation efforts dedicated to giant pandas not only offer protection to other species inhabiting the same region but also provide valuable insights for safeguarding other endangered species worldwide [4]. In recent years, there has been an overall increase in the population of giant pandas, leading to a decrease in their threatened status from endangered to vulnerable [5]. However, the survival prospects for small populations are not very optimistic due to habitat fragmentation and increased human activities [6,7].

In order to reduce the damage of human activities and habitat fragmentation to the giant panda population and better protect them in situ, China has established 67 giant panda nature reserves. However, despite these measures, nearly 33% of the population still remains unprotected [8]. Over the past decade or so, increasing the population size through individual releases has been a major focus of researchers [9]. Looking back at the releases, the survival rates of about 75% for wild individuals released, the survival rates of released captive individuals with wild lineage is about 50%, and no clear evidence of survival after the release of captive individuals of captive parentage have obtained. Of this total, about 66.7% of the individuals had wild individuals as their parents. As the most successful case of wild individual release, "Luxin" is the only panda found to have given birth to a cub in the wild [10–12]. Individual introduction can reduce the risk of extinction of small populations [13], and historical data suggests that ex situ releases of wild individuals have the highest success rates. In contrast to individuals which often led to the development of behavioral defects due to the captive environment [14], researchers would not have to devote considerable time and effort to develop the survival skills of wild individuals [15]. Although the establishment of ecological corridors can also provide assistance [16], the implementation of these measures may face constraints in terms of financial resources and time availability, and there is no guarantee that they will be fully utilized by pandas [17]. Therefore, directly using wild individuals as a source of introduction is an ideal approach for the genetic rescue of small populations [18], and this will also be a new perspective on small population restoration.

With the idea that the giant pandas have reached an evolutionary dead end refuted, it is widely accepted that the giant panda is a species with a medium or high genetic diversity [19–21]; the evolutionary potential is not low compared to species in the same family [22]. However, looking at the six mountains, the Xiangling areas were comparatively lower [23]. Under natural conditions, populations with lower genetic levels often face a greater risk of extinction and show a decreased resistance to random factors [24,25]. Population viability analysis, through simulating population dynamics, concretizes the challenges that giant pandas may face at a specific point in the future [26]. Although population parameters vary slightly among similar studies, one consistent finding is that smaller populations tend to exhibit weaker viability [27–29]. From past research, most studies have focused on the population of giant pandas in entire mountains or counties [22,30,31]. Due to the vast study areas, these investigations may fail to fully capture the genetic dynamics and survival status of the smaller, localized areas. The Tangjiahe and Wanglang populations belong to the large population in the Minshan Mountains with good habitat connectivity, while the Liziping and Daxiangling populations are distributed in the Xiangling Mountains with broken habitats [32]. Our research selected the four populations as the study objects and conducted a comprehensive assessment of the genetic status and viability of giant panda populations. Utilizing the findings from the studies, we simulated the introduction of individuals from a population with a high genetic diversity to address the deficiencies observed in the small populations. This is the first theoretical study on the introduction of wild individuals. Our plan aims to mitigate the risks associated with genetic bottlenecking and enhance the long-term survival prospects of these populations. Furthermore, it is our aspiration that this study can provide valuable insights and guidance for the conservation efforts dedicated to giant pandas, as well as other endangered species worldwide.

2. Materials and Methods

2.1. Study Area and Sample Collection

Our study area includes four nature reserves: Tangjiahe (TJH), Wanglang (WL), Liziping (LZP), and Daxiangling (DXL), which are respectively located in Qingchuan, Pingwu, Shimian, and Yingjing Counties in Sichuan Province, Southwest China. In the fourth survey conducted from 2011 to 2014, 39 wild giant pandas were found in TJH, 28 in WL, 22 in LZP, and seven in DXL. From 2019 to 2023, we investigated the number of giant pandas in these four regions again and 156 fecal samples were collected from TJH, 126 from WL, 92 from LZP, and 40 from DXL (Figure 1). All samples were collected with sterile gloves and stored in more than 95% anhydrous ethanol.

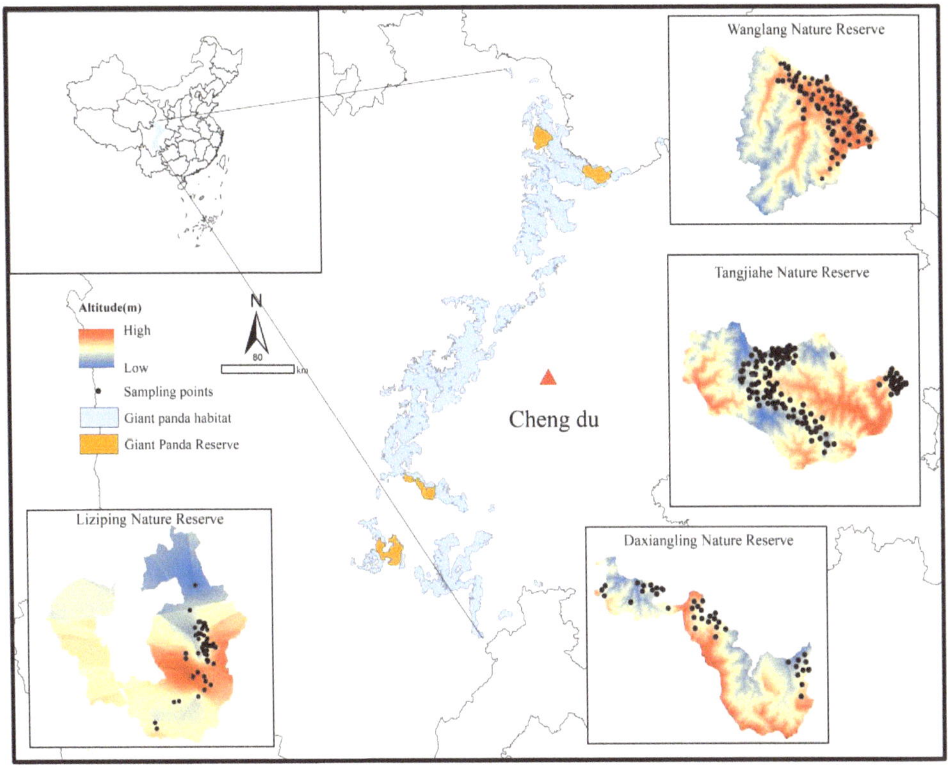

Figure 1. Sampling distribution points of four giant panda nature reserves.

2.2. DNA Extraction and Microsatellite Amplification

The total DNA from fecal samples was extracted using the QIAamp Fast DNA Stool Mini Kit (Hilden, Germany) following the manufacturer's instructions and stored at −20 °C. In this study, nine microsatellite markers with high polymorphic information content and stability were utilized (Table 1). The upstream primer was fluorescently labeled with FAM, HEX, and TAM at the 5′ end. The reaction system was 20 uL: Taq PCR Mastermix 10 μL, upper and downstream primers 0.5 μL each, DNA 2 μL, BSA 1 μL, and ddH_2O was added to 20 μL. The reaction conditions: 95 °C for 15 min; 94 °C for 30 s, 48 °C~60 °C for

90 s, 72 °C for 60 s, for a total of 30 cycles; 60 °C for 30 min; and stored at 4 °C. The PCR products were detected through electrophoresis on a 2% agarose gel. Primer synthesis and microsatellite genotyping were conducted by Sangon Biotech Co., Ltd. (Shanghai, China).

Table 1. Information of nine microsatellite loci.

Loci	Repeat Motif	Primer Sequence (5′–3′)	Size (bp)
Panda-22	(CAA)12	F: AGGGGAGAGAACATTGCTCG R: GAAGCCAGCCCAACTTTTCC	177–186
Ame-μ26	(CA)11	F: TTTTCAGGCCTCCGAAAAC R: ATTCCCAATAAAGCAAATCAGA	114–120
GPL-60	(TCTT)12	F: TGCCGGAAAGTTCTAAGCAT R: TTTCTCTCCCTCTCCCCTTC	218–238
Ame-μ13	(CA)18	F: GGAAGCATTAAGGAAAACATGC R: AATGATGACCATTTCAAACGC	142–171
Ame-μ11	(CA)12	F: TATGCCACCTGCCCAGAC R: GATGGAAAGAGTAGAGCCAAGG	228–236
Ame-μ10	(CA)16	F: ACCGTGCTCTTAATCCCCTT R: CCCATGCTTATGAGAAACAGG	138–160
GPZ-6	(AAAG)11	F: CCTGGCAGGGCAAAGTATT R: CCCCGTGAAAACATCAAGAC	194–222
GPZ-47	(AATG)20	F: GACCTCAGTGTACGCCCAGT R: CTGGACAGGCAGGTAGAAGC	174–210
GPL-47	(TCTA)20	F: TCCCCCTCTATGGTAAAAGG R: CCATGTTGGGTGTAGGGATT	140–172

2.3. *Data Analysis*

2.3.1. Individual Identification

The Micro-Checker 2.2.3 was utilized to identify any missing or invalid alleles [33]. The Gimlet 1.3.3 software was employed to calculate the *p*-value for the joint differentiation rate among the 9 loci [34]. Subsequently, the Microsatellite Toolkit was employed to discern individual similarities [35]. Samples were considered to belong to the same individual if either all alleles across the nine loci were identical or only one allele within a single locus differed.

2.3.2. Genetic Diversity and Genetic Structure

GenAlEx 6.5 was used to calculate the number of alleles (Na), observed heterozygosity (Ho), expected heterozygosity (He), and Shannon–Wiener Index (I) [36]; the Hardy–Weinberg equilibrium was tested by Genepop 4.7 [37].

Structure 2.3.4 was used to analyze the source of individuals in a population by the Bayesian clustering method [38]. The value of K was set to range from one to eight, and ten independent operations were carried out. Before 1,000,000 formal calculations were repeated, 100,000 preliminary experiments were conducted. The calculation results were estimated using the Structure Harvester online tool to determine the optimal K value [39].

2.3.3. Population Viability

The extinction rates of four populations in the next 100 years were calculated using Vortex 10.6 [40], and each simulation was repeated 1000 times. Parameter setting: The initial breeding age of female and male wild giant pandas is seven and eight years old, respectively, the maximum breeding age is 20 years old, and the maximum lifespan is 26 years old [41,42]. The sex ratio is 1:1, and the annual reproductive rate is 62.5%, with a single birth rate of 58.33% and a twin birth rate of 4.17%, and all males have the ability to engage in reproduction [41]. Additionally, the mortality rates at different age stages can be found in Table S1 of Supplementary Materials [41]. Natural disasters include bamboo flowering and forest fire, the frequency of bamboo flowering is 1/60, and the frequency of forest fire is 1/30; their impact on survival and reproduction is both 10% [43,44]. The lethal equivalent of inbreeding depression is 3.14 based on a study of 40 captive mammals in North American zoos [45]. The criteria for long-term survival are an extinction rate of less than 2% and gene diversity greater than 0.9 [46].

The number of individuals in the reserve obtained from this study represents the initial population size. The maximum population density of giant pandas was 3.03/km^2 [47,48]. The area of the reserves and the coverage rate of edible bamboo for giant pandas are sourced from the fourth survey report [32]. The maximum environmental capacity can be found in Table 2.

Table 2. Maximum environmental capacity of giant pandas in four nature reserves.

Reserve	Area (km^2)	Suitable and Sub-Suitable Habitats	Coverage Rate of Edible Bamboo	Maximum Capacity
TJH	400.00	51.00% [49]	83.94%	519
WL	322.97	-	41.81%	409
DXL	284.50	-	98.58%	850
LZP	479.40	26.58% [50]	69.90%	270

3. Results

3.1. Genotyping and Individual Identification

A total of 227 reliable genotypes were identified from the samples. Micro-checker analysis indicated that the amplification results were not affected by null alleles or allele dropout. The joint differentiation rate of nine loci was high, with a PID value of 1.73×10^{-9}, and the probability P (sib) of misjudgment caused by twins was 4.06×10^{-4}. Even if the locus with the highest polymorphism (GPL-47) fails to amplify, the probability of misjudgment caused by twins only increases to 1.15×10^{-3}, which is much smaller than 0.01 [51]; this satisfies the requirement for population size evaluation. Therefore, this study retained genotypes at eight loci and obtained a total of 139 unique genotypes. Among them, there are 56 genotypes from TJH, 45 from WL, 25 from LZP, and 13 from DXL.

3.2. Genetic Diversity and Genetic Structure

A total of 83 alleles were detected from the 139 unique genotypes. There were 17 private alleles that were present only in a single population, and six of these occurred in the TJH, five in the DXL, and three in the WL and LZP, respectively. Only 66 alleles were shared among all four populations.

The average allele numbers were 7.556 and 6.333 for TJH and WL populations, respectively, while LZP and DXL populations had average allele numbers of 5.667 and 5.778, respectively. The expected heterozygosity (He) ranked from high to low were 0.746 (WL), 0.743 (DXL), 0.725 (TJH), and 0.654 (LZP). The highest observed heterozygosity was 0.578 (WL), followed by 0.536 (TJH) and 0.530 (LZP), and the lowest was 0.430 (DXL). The Shannon–Wiener values ranked from high to low were 1.534 (TJH), 1.531 (DXL), 1.528 (WL), and 1.316 (LZP). The comprehensive genetic parameters indicated that the genetic level of TJH and WL populations was higher, followed by DXL and LZP populations. The Hardy–Weinberg equilibrium test revealed that there were seven loci with deviations from equilibrium in TJH, six in WL and DXL, and four in LZP ($p < 0.01$) (Table 3).

Then, we analyzed the structure of the four populations individually. The TJH and WL populations both exhibit three gene clusters, suggesting that they originate from three subpopulations. The DXL and LZP populations are composed of two gene clusters, indicating that they originate from two subpopulations (Figures 2 and 3).

Table 3. Genetic diversity of four populations based on nine loci.

	TJH (n = 56)					WL (n = 45)					LZP (n = 25)					DXL (n = 13)				
	Na	**I**	**Ho**	**He**	**P**	**Na**	**I**	**Ho**	**He**	**P**	**Na**	**I**	**Ho**	**He**	**P**	**Na**	**I**	**Ho**	**He**	**P**
Panda-22	4.000	1.249	0.547	0.682	0.0303	5.000	1.005	0.778	0.575	0.9990	4.000	1.204	0.250	0.666	0.0001	5.000	1.413	0.769	0.722	0.0486
Ame-μ26	7.000	1.554	0.518	0.748	0	4.000	1.276	0.778	0.704	0.8261	5.000	1.201	0.560	0.594	0.0146	5.000	1.487	0.231	0.749	0
GPL-60	5.000	1.396	0.554	0.715	0.0564	5.000	1.534	0.659	0.767	0.0713	4.000	0.961	0.480	0.554	0.2801	5.000	1.380	0.385	0.710	0.0023
Ame-μ13	11.000	1.589	0.607	0.705	0	7.000	1.347	0.452	0.684	0	6.000	1.220	0.400	0.610	0.0026	9.000	2.034	0.308	0.849	0
Ame-μ11	6.000	1.315	0.518	0.672	0.0005	4.000	1.354	0.682	0.735	0.0899	5.000	1.026	0.640	0.538	0.6727	5.000	1.445	0.333	0.743	0.0007
Ame-μ10	8.000	1.681	0.547	0.771	0	7.000	1.675	0.300	0.787	0	8.000	1.770	0.800	0.801	0.3761	6.000	1.662	0.692	0.790	0.0120
GPZ-6	8.000	1.438	0.464	0.645	0	6.000	1.655	0.659	0.787	0.0057	5.000	1.127	0.320	0.570	0.0045	4.000	1.029	0.308	0.559	0.0086
GPL-47	10.000	1.880	0.618	0.816	0.0001	9.000	1.953	0.548	0.838	0	6.000	1.455	0.800	0.729	0.7751	7.000	1.654	0.462	0.769	0.0270
GPZ-47	9.000	1.704	0.446	0.770	0	10.000	1.955	0.350	0.838	0	8.000	1.880	0.520	0.827	0	6.000	1.667	0.385	0.796	0
Average	7.556	1.534	0.536	0.725	—	6.333	1.528	0.578	0.746	—	5.667	1.316	0.530	0.654	—	5.778	1.531	0.430	0.743	—

Note: The number of individuals (n); the number of alleles (Na); the observed heterozygosity (Ho); the expected heterozygosity (He); Shannon–Wiener index (I); Hardy–Weinberg equilibrium (P).

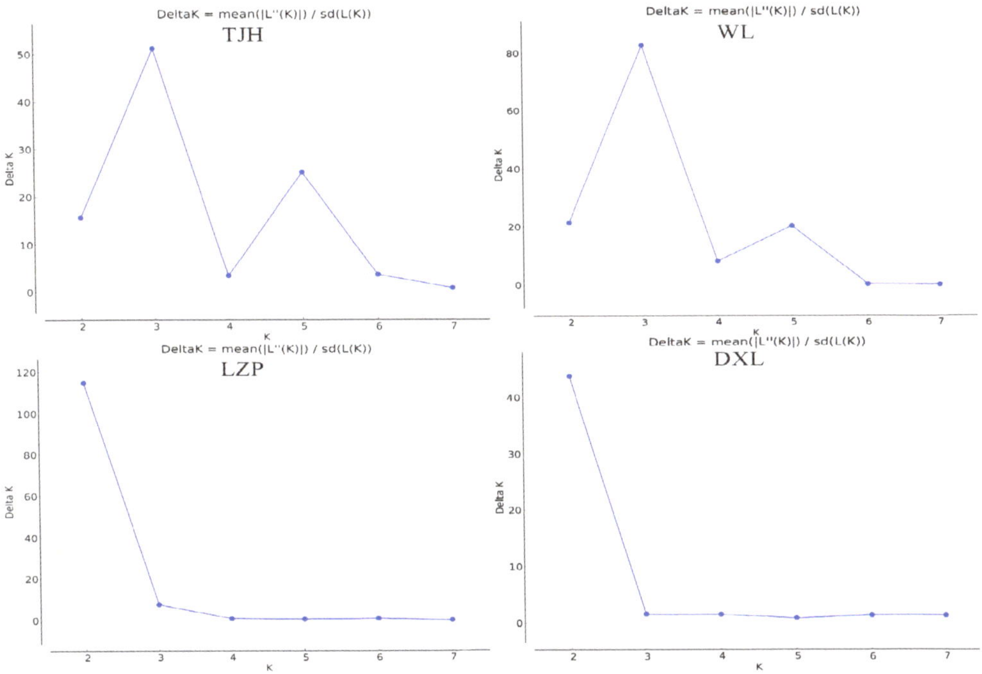

Figure 2. Optimal K values for the analysis of the genetic structure of the four populations.

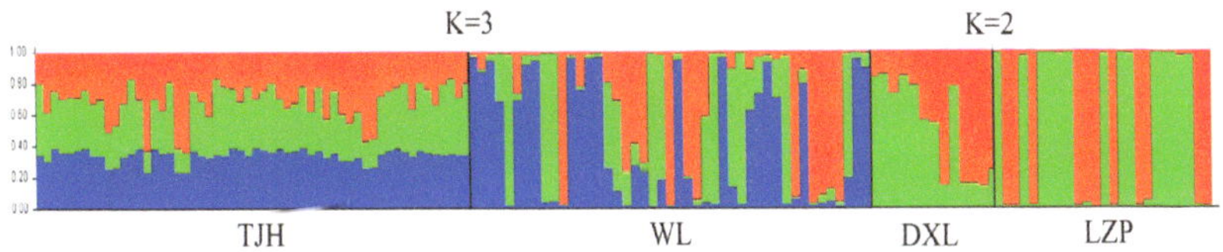

Figure 3. The map of the gene cluster of four giant panda populations. Note: Red, green, and blue represent different gene clusters.

3.3. *Population Viability*

In an ideal state without inbreeding depression and natural disasters such as bamboo flowering and forest fires, the extinction rates of the four populations were 2.4% (TJH), 4.0% (WL), 19.6% (LZP), and 52.5% (DXL), respectively, and none of them meet the requirements for long-term survival. The initial number of individuals is synergistically related to genetic diversity and inversely proportional to the cumulative extinction rate. The TJH (N = 56) population has the strongest viability, while the DXL (N = 13) population has the highest risk of extinction. Under the separate effects of inbreeding depression or natural disasters, the former has a stronger negative impact on the LZP and DXL populations compared to the latter. If the two kinds of effects coexist, the extinction rate of the four populations will increase significantly, and the extinction rate of small populations is higher than 50%, which is about three~six times that of large populations. It is important to note that there is the migration of exotic individuals among the TJH and WL populations. Therefore, it is possible that the defined extinction risk is relatively high based on a set of static values. The two small populations on the brink of extinction are relatively isolated, and if no assistance is provided by humans, it is highly likely that they will vanish within the next 100 years (Figure 4).

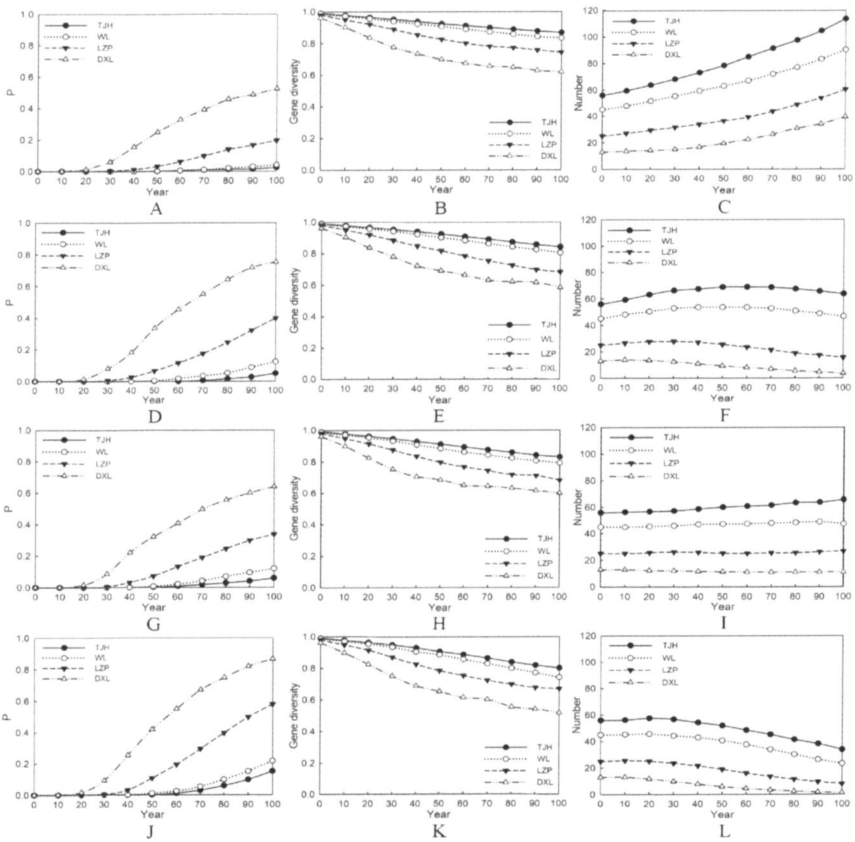

Figure 4. Variation of viability under different conditions in the next 100 years. Note: (**A–C**) are unaffected by inbreeding depression and disasters; (**D–F**) are affected by inbreeding depression only; (**G–I**) are affected by disasters only; (**J–L**) are affected by both inbreeding depression and disasters; P is the extinction rate, same as below.

3.4. Rejuvenation of Small Populations

3.4.1. Viability of Small Populations after Introducing Individuals

This study explored the effects of introducing individuals with different gender combinations (female: F; male: M) every two, five, and ten years (year: a) on the viability of small populations under ideal conditions. When introducing individuals, start increasing from one only until long-term survival conditions are met (Table S2 of Supplementary Materials).

We designed 16 plans to rejuvenate LZP and ultimately found that five plans were the most suitable (Figure 5). The extinction rate of all the applicable plans was zero, so they are not shown in the figure. Based on the extinction rate, genetic diversity, and the expected number of individuals in the next 100 years, it is most beneficial for the survival of LZP to introduce 1F/2a, then 2F/5a and 1F1M/5a, and finally 3F/10a and 2F1M/10a. More females means a higher birth rate, which can produce more new individuals. Therefore, the growth rate of individual numbers in the LZP population was the highest when 1F/2a was introduced, while the growth rate was lowest when 1F1M/5a or 2F1M/10a were introduced.

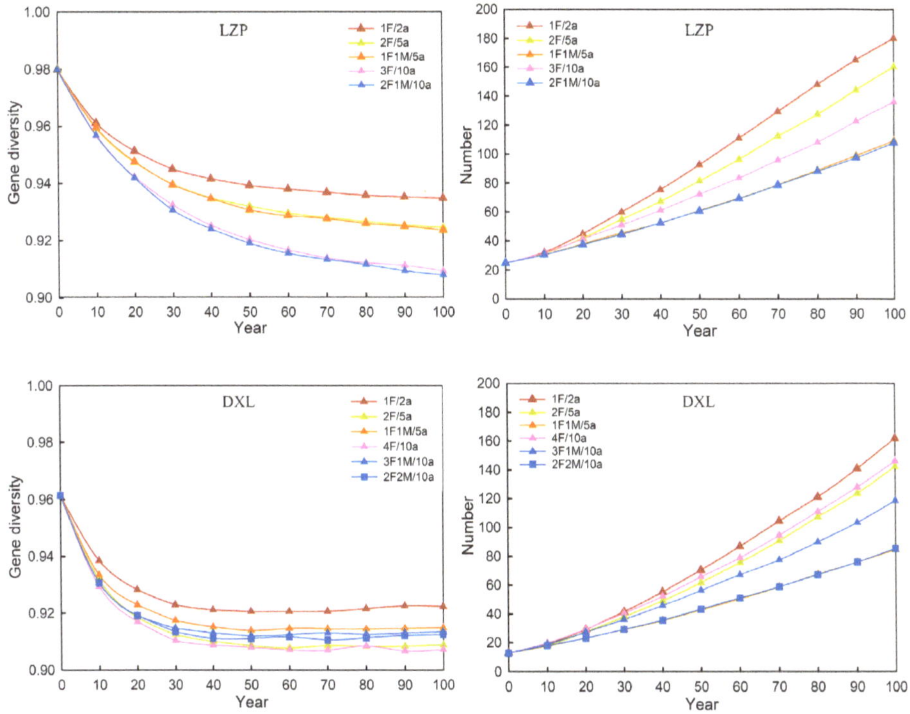

Figure 5. Introduction plans for small populations to meet long-term survival standards. Note: "Year" is represented by "a"; "female" and "male" are represented by "F" and "M", respectively, same as below.

For the DXL with the smallest initial number of individuals, we explored 21 plans, of which only six plans were applicable (Figure 5), and the extinction rate was zero. The most significant way to expand the population of DXL is still to replenish 1F/2a. Then, there are 1F1M/5a, 3F1M/10a, and 2F2M/10a, which have similar effects. Finally, there are 2F/5a and 4F/10a, which have weaker enhancement effects on the viability of the DXL population compared to the first four plans. As for the population size of DXL after 100 years, it still depends on the number of females that were introduced.

3.4.2. Viability of Large Populations after Introducing Individuals

We further investigated the impact of applicable introduction plans on two large populations (Figure 6; Table S3 of Supplementary Materials). Among the five plans applicable to LZP, 1F/2a has the greatest impact, which leads to the highest extinction rate of TJH and WL in the next 100 years, and the lowest gene diversity and individual number. Then, the damage to the large populations in descending order is 2F/5a, 3F/10a, 1F1M/5a, and 2F1M/10a. The more females were introduced, the smaller the population size of TJH and WL in 100 years.

Among the six plans applicable to DXL (Figure 6), 1F/2a is still the most harmful introduction plan for TJH and WL. The impact of 2F/5a and 4F/10a on the large population is slightly lower than that of 1F/2a. Introducing 3F1M/10a can obviously show that the gap between the large populations and their ideal state has narrowed. In addition, 1F1M/5a and 2F2M/10a are the least harmful to large populations.

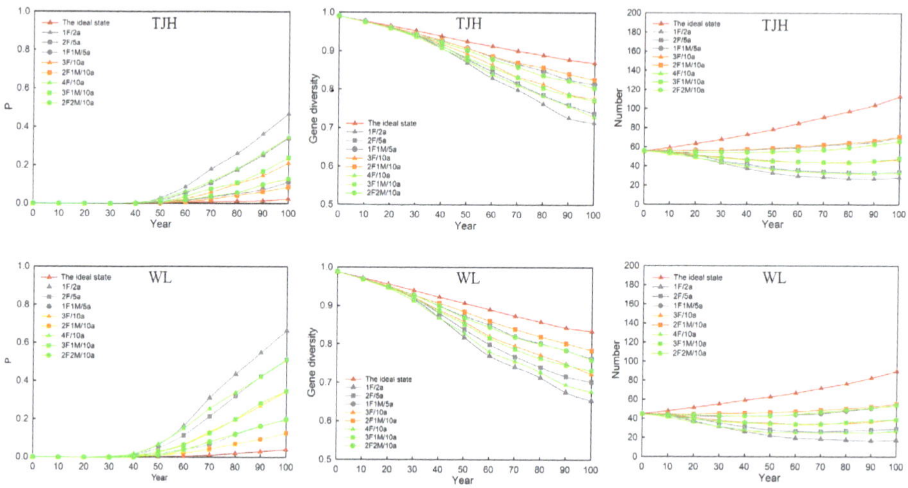

Figure 6. Effect of suitable introduction plans on large populations. Note: The gray is suitable for DXL and LZP, the orange is only suitable for LZP, and the green is only suitable for DXL; the red is the viability of large populations before introduction.

3.4.3. Genetic Level of Small Populations after Introducing Individuals

Estimation of Genetic Level of LZP

In an ideal state, the number of giant pandas in TJH after 100 years is 113 (Table S3 of Supplementary Materials). After introducing individuals to LZP according to five introduction plans, The number of remaining individuals in TJH was 28 (introduce 1F/2a)~71 (introduce 2F1M/10a), with the introduced rate of 37% [(113 − 71)/113]~75% [(113 − 28)/113]. By introducing 21 (56 × 37%)~42 (56 × 75%) individuals from 56 TJH individuals to LZP, the genetic level of LZP after 100 years was estimated (Figure 7A). Selecting the 42 individuals of TJH with the highest genotype similarity to LZP individuals, when introducing the top 21 individuals, the number of alleles and expected heterozygosity were 7.111 and 0.688, respectively. When all were introduced, the number of alleles and expected heterozygosity were 7.667 and 0.703, respectively. Therefore, it is speculated that the number of alleles of LZP after introducing individuals from TJH will be 7.111 to 7.667, and the expected heterozygosity will be 0.688 to 0.703.

The number of giant pandas in WL after 100 years is 90 in an ideal state. After revitalizing LZP, the number of remaining individuals in WL was about 17 (introduce 1F/2a)~56 (introduce 2F1M/10a), and the introduced rate was 38% [(90 − 56)/90]~81% [(90 − 17)/90]. The genetic status of LZP in the future can be estimated by introducing 17 (45 × 38%)~36 (45 × 81%) individuals from the current 45 WL individuals (Figure 7B). The number of alleles and expected heterozygosity of LZP were 6.667 and 0.706, respectively, when 17 individuals of WL with the highest similarity to LZP individuals were selected. When 36 individuals were selected, the number of alleles and expected heterozygosity were 7.000 and 0.741, respectively. Therefore, it is speculated that when introducing individuals from WL to LZP, the number of alleles in LZP will be 6.667~7.000, with an expected heterozygosity of 0.706~0.741.

Estimation of Genetic Level of DXL

Similarly, for the rejuvenation of DXL (Table S3 of Supplementary Materials), the introduced rate of TJH is 38% (introduce 1F1M/5a; [(113 − 70)/113])~75% (introduce 1F/2a; [(113 − 28)/113]), and it is estimated that in the future, the genetic level of DXL will need to introduce 21 (56 × 38%)~42 (56 × 75%) individuals from TJH (Figure 7A). The number of alleles and expected heterozygosity were 7.556 and 0.734 when introducing the top 21 individuals, respectively. When introducing 42 individuals, the number of alleles and expected heterozygosity were 7.889 and 0.740, respectively. Therefore, it is speculated that the number of alleles in DXL will be 7.556~7.889 and the expected heterozygosity will be 0.734~0.740 when individuals are introduced from TJH to DXL.

For rejuvenating DXL, the introduced rate of WL is 39% (introduce 1F1M/5a; [(90 − 55)/90])~81% (introduce 1F/2a; [(90 − 17)/90]). According to this ratio, WL needs to introduce 18 (45 × 39%)~36 (45 × 81%) individuals to speculate the genetic level of DXL (Figure 7B). When the first 18 individuals with the highest similarity to DXL were introduced, the number of alleles and expected heterozygosity were 7.111 and 0.758, respectively. When 36 individuals were introduced, the maximum number of alleles and expected heterozygosity were 7.333 and 0.764, respectively. Therefore, it is speculated that the number of alleles of DXL will be 7.111~7.333, and the expected heterozygosity will be 0.758~0.764 when individuals are introduced from WL to DXL.

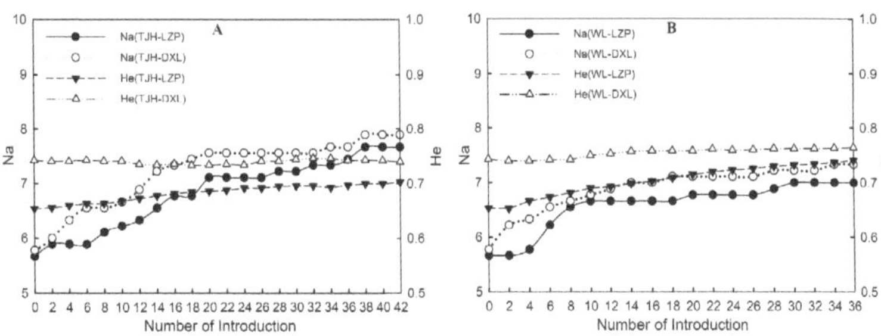

Figure 7. The future genetic level of LZP and DXL populations. Note: (**A**) shows the genetic level changes of two small populations after introducing individuals from TJH; (**B**) shows the genetic level changes of two small populations after introducing individuals from WL.

4. Discussion

Reserves are highly effective tools for biodiversity conservation, and their establishment and systematic planning represent a valuable legacy that can be passed down through generations in the field of conservation [52]. TJH and WL Nature Reserves are part of population A, situated in the northern region of the Minshan Mountains, where the largest known population of wild giant pandas currently exists [53]. The population of giant pandas in TJH inhabits Qingchuan County and maintains communication with the wild populations in Wen County and Pingwu County [54]. Since the fourth survey, our research has found that the population of TJH has increased to 56 individuals, a growth of approximately 43.6% [32]. As a core part of the Jiuzhai–Baima local population, the population of WL has increased by nearly 60.7% compared to the previous survey, with a current count of around 45 individuals [32], due to the healthy gene exchange among large populations. Of the existing wild populations in the six mountains, the Xiaoxiangling populations were geographically separated by the presence of a railway and the Dadu River, impeding gene exchange with other mountain populations. Consequently, this isolation poses an extremely high risk of extinction [55,56]. In particular, the LZP Nature Reserve in Shimian County, which is divided into two parts by National Highway 108, serves as a prime example of isolated small populations. Daxiangling is the mountain with the lowest population density of giant pandas, except for Xiaoxiangling [8]. The distribution of individuals is also quite scattered, with the majority living in Yingjing County [57]. In recent years, the population growth in DXL Nature Reserve has been less than ten individuals [32], and the severity of inbreeding depression is a cause for concern [58]. In macro perspective, the habitat research of giant pandas has been quite comprehensive [59], while the scientific questions at the micro level still intrigue us. Specifically, we are curious about the genetic differences and survival abilities among wild populations of different scales, as well as the most favorable approach for individual introduction to rejuvenate small populations. The research objects we have chosen are relatively representative and can provide us with an objective answer.

Genetic variation reflects the evolutionary potential of species [60]. This study employed nine microsatellite markers, with polymorphic information content ranging from 0.648 (Panda-22) to 0.810 (GPL-47), all of which were higher than 0.5, this indicates that these markers can provide ample genetic information for our research [61]. The TJH and WL populations have a high number of alleles and heterozygosity, indicating that their genetic diversity is at a high level, which is consistent with previous research findings [62,63]. In addition to the geographical advantage of the reserves, the efforts made by the government in management and protection should not be ignored [18]. The small populations have relatively lower levels of genetic diversity and genetic structure analysis showed that they also have fewer ancestral components. We further adopted population viability analysis and used previous wild giant panda parameters to evaluate the long-term survival ability of the four populations [41,44,46,64], and obtained the same results. In an ideal state, the extinction rate of DXL is as high as 52.5%, while LZP is 19.6%. Their situation is much more dangerous compared to TJH's 2.4% and WL's 4.0%. It is also not difficult to observe from the results that small populations are more susceptible to the effects of inbreeding depression. If inbreeding depression and natural disasters occur together, the extinction rates of LZP and DXL will be increased to 57.9% and 86.7%, respectively, which is about three~six times that of the two large populations at this time. Essentially, this is a matter of population size [65]. The greater the initial number of individuals, the higher the probability of survival [66]. If the number of individuals in the four populations is doubled in the ideal state, the two large populations far exceed the long-term survival standard, and the extinction rate of DXL is reduced by about three times and LZP by about ten times. Compared to the rejuvenation of isolated small populations, increasing the number of large populations requires less effort. We believe that it is expected to realize the self-maintenance of wild populations in the near future by making large populations stronger first and then providing provenances for small populations.

Based on this, we explored the number and sex of introductions needed to meet the long-term survival criteria for the small populations at intervals of two, five, and ten years in an ideal state. Introducing 1F/2a, 2F/5a, and 1F1M/5a is sufficient for DXL and LZP. If replenishment is made every ten years, the DXL with fewer initial individuals needs four individuals to meet the standard, while the LZP needs only three individuals. Obviously, populations with fewer individuals always require more attention and assistance [67]. Moreover, an interesting finding is that in the DXL with the smallest population size, the extinction rates are all zero, and the gene diversity obtained from 1F1M/5a, 3F1M/10a, and 2F2M/10a are slightly higher than that obtained from 2F/5a and 4F/10a. Aligning with the perspectives of Yang et al. [68], the female-biased sex combinations provide greater benefits to population survival compared to introducing only one sex or having more males. Certainly, the giant panda is a species with female-biased dispersal, it has a polygynous mating mechanism [69], and the population litter size is determined by the number of females [54]. Therefore, as far as the plans we have designed are concerned, the more females were introduced to DXL and LZP, the larger the estimated population size will be in 100 years [70]. On the contrary, when these individuals come from TJH or WL, the more females were introduced, the lower the survival rate of TJH and WL. In particular, the most effective introduction plan 1F/2a for LZP and DXL made the extinction rate of TJH soar from 2.4% to 46.7%, and WL soared from 4.0% to 66.4%. The plans that not only meet the long-term survival of small populations but also have less harm to large populations are 1F1M/5a, 2F1M/10a, and 2F2M/10a. Thus, fewer males can be taken into account in our selection of individuals.

Previous research has shown that after the introduction of individuals, there are new gene frequencies and alleles in this small group [68,71]. We compared the expected number in large populations after 100 years in an ideal state with the number of large populations after introducing individuals into small populations, and we estimated the number of TJH and WL that need to be introduced by reducing the proportion of individuals, so that the genetic level that the small population may reach after 100 years can be inferred. Consistent with previous studies, the genetic level of LZP and DXL will be significantly improved in the future. No matter which large population is selected, the number of alleles of LZP and DXL can reach at least 6.667 and 7.111, respectively, and the expected heterozygosity can reach at least 0.688 and 0.734, respectively. Compared with the genetic level of giant pandas in five mountains except Daxiangling studied by Zhang et al. [22], DXL and LZP populations after rejuvenation have richer genetic diversity.

We cannot deny that wild individual translocation is the method with the fastest and the highest success rate in saving small populations [18,72]. From a short-term perspective, the introduction of new genetic resources can quickly improve the genetic diversity of small populations [18,71]. From a long-term perspective, migrating individuals from other populations to establish heterogeneous populations reduces the risk of extinction and alters the developmental trajectory of extinction at a certain point in the future [73]. However, as it stands, it is not easy to implement. The direct introduction of wild individuals is not realistic [74], we can only carry out it indirectly, and a better introduction plan may need to be studied in the future. Here are some indirect individual introduction plans, which may provide ideas for the emergence of better schemes. Firstly, the endangerment of giant pandas is linked to the environmental pressures they are under [75]. The Xiangling Mountains present a more challenging environment for the survival of giant pandas [76], it is separated from Qionglai Mountains and Liangshan Mountains by the influence of rivers, railways, and national highways [32]. It is possible to establish ecological corridors to introduce individuals from the large populations of Lewu and Baishahe into the Xiangling Mountains. Secondly, like "Luxin", individuals rescued from other mountains can be reintroduced into the Xiangling Mountains, thus achieving long-distance individual translocation. In addition, this would allow the offspring of wild individuals rescued to be born and raised in semi-wild conditions before being released into small populations, similar to "Zhangxiang" [70]. It should be noted that based on our existing data, it may not be appropriate for TJH and WL to assist small populations in the current situation where they cannot sustain themselves in the long term. However, it is not difficult to achieve large population growth by strengthening genetic exchange with surrounding populations and protecting habitats [70,77]. A planned individual introduction would be a win–win for both large and small populations at this time. In the future, it is hoped that self-sustainability can be achieved by introducing wild individuals, thereby reducing the pressure on the wild release of captive individuals. Of course, in further research, we need to consider more factors to ensure the successful implementation of the rejuvenation plan, such as the topography of the habitat, the types and areas of edible bamboo, and the interrelations between intruders and residents.

5. Conclusions

Currently, the rejuvenation of small populations mainly relies on the release of captive individuals into the wild, which requires significant human, material, and financial resources. Our research indicated that the genetic diversity of large wild populations was higher than that of small isolated populations. It can help small populations overcome their challenges by introducing individuals. Enhancing gene flow between local populations is crucial. It is necessary for us to continue conducting in-depth research on the specific plans for implementing individual introduction in the future.

Supplementary Materials: The following supporting information can be downloaded at https://www.mdpi.com/article/10.3390/ani14162345/s1.

Author Contributions: Conceptualization: D.Q.; data curation: J.Y., Y.Z., L.S., Q.X., J.N., H.G. and X.G.; formal analysis: J.Z., D.Q., J.L., Y.Z. and Y.G.; funding acquisition: J.Z. and D.Q.; investigation: J.Y., C.L., X.Z., L.M., W.W., Q.X., J.N., L.S., H.G. and X.G.; methodology: J.Y., C.L., X.Z., L.M., W.W., Y.G., J.Z., D.Q., J.L. and Y.Z.; project administration: W.W.; software: J.Y., C.L., X.Z. and L.M.;

writing—original draft: Y.Z.; writing—review and editing: J.Z., D.Q., J.L. and Y.G. All authors have read and agreed to the published version of the manuscript.

Funding: This study was financially supported by the National Natural Science Foundation of China (No. U21A20193; 42071279), The key project of the State Forestry and Grassland Administration "Study on Key Technologies for Conservation of Wild Giant Panda Populations and Its Habitats within Giant Panda National Park System" (No. CGF2024001); Sichuan Science and Technology Program (No. 2023NSFSC1156; 2024NSFSC0032), Chengdu Science and Technology Program (No. 2023-YF09-00017-SN) and Chengdu Research Base of Giant Panda Breeding (No. 2024CPB-P01; 2024CPB-B06; 2022CPB-P04; 2022CPB-C06; 2025CPB-C13).

Institutional Review Board Statement: This study does not require ethical approval.

Informed Consent Statement: Not applicable.

Data Availability Statement: The data that have been used are confidential.

Acknowledgments: We are deeply grateful for the support and assistance provided by the "Southwest Mountain Region Ecological and Sustainable Development Exchange Center" in this research.

Conflicts of Interest: This manuscript fully complies with all the rules of your esteemed journal. It has not been submitted to any other journal and there are no conflicts of interest. All authors have agreed to the submission of this manuscript.

References

1. IUCN. The IUCN Red List of Threatened Species. Version 2024-1. Available online: https://www.iucnredlist.org (accessed on 5 August 2024).
2. Corlett, R.T. Safeguarding our future by protecting biodiversity. *Plant Divers.* **2020**, *42*, 221–228. [CrossRef] [PubMed]
3. Li, S.; McShea, W.J.; Wang, D.; Gu, X.; Zhang, X.; Zhang, L.; Shen, X. Retreat of large carnivores across the giant panda distribution range. *Nat. Ecol. Evol.* **2020**, *4*, 1327–1331. [CrossRef] [PubMed]
4. Li, B.V.; Pimm, S.L. China's endemic vertebrates sheltering under the protective umbrella of the giant panda. *Conserv. Biol.* **2015**, *30*, 329–339. [CrossRef] [PubMed]
5. Swaisgood, R.R.; Wang, D.; Wei, F. Panda Downlisted but not Out of the Woods. *Conserv. Lett.* **2017**, *11*, e12355. [CrossRef]
6. Li, C.; Connor, T.; Bai, W.; Yang, H.; Zhang, J.; Qi, D.; Zhou, C. Dynamics of the giant panda habitat suitability in response to changing anthropogenic disturbance in the Liangshan Mountains. *Biol. Conserv.* **2019**, *237*, 445–455. [CrossRef]
7. Zhang, J.; Hull, V.; Ouyang, Z.; Li, R.; Connor, T.; Yang, H.; Zhang, Z.; Silet, B.; Zhang, H.; Liu, J. Divergent responses of sympatric species to livestock encroachment at fine spatiotemporal scales. *Biol. Conserv.* **2017**, *209*, 119–129. [CrossRef]
8. National Forestry and Grassland Administration. *The 4th National Survey Report on Giant Panda in China*; Science Press: Beijing, China, 2021.
9. Swaisgood, R.R.; Martin-Wintle, M.S.; Owen, M.A.; Zhou, X.; Zhang, H. Developmental stability of foraging behavior: Evaluating suitability of captive giant pandas for translocation. *Anim. Conserv.* **2018**, *21*, 474–482. [CrossRef]
10. He, K.; Dai, Q.; Foss-Grant, A.; Gurarie, E.; Fagan, W.F.; Lewis, M.A.; Qing, J.; Huang, F.; Yang, X.; Gu, X.; et al. Movement and activity of reintroduced giant pandas. *Ursus* **2019**, *29*, 163–174. [CrossRef]
11. Hong, M.; Wei, W.; Zhou, H.; Tang, J.; Han, H.; Zhang, Z. Creative conservation in China: Releasing captive giant pandas into the wild. *Environ. Sci. Pollut. Res.* **2019**, *26*, 31548–31549. [CrossRef]
12. Lei, M.; Yuan, S.; Yang, Z.; Hong, M.; Yang, X.; Gu, X.; Huang, F.; Zhang, Z. Comparison of microhabitats and foraging strategies between the captive-born Zhangxiang and wild giant pandas: Implications for future reintroduction. *Environ. Sci. Pollut. Res.* **2015**, *22*, 15089–15096. [CrossRef] [PubMed]
13. Fang, S.G.; Wan, Q.H.; Fujihara, N. Genetic Diversity of the Giant Panda(*Ailuropoda melanoleuca*) between Big and Small Populations. *J. Appl. Anim. Res.* **2002**, *21*, 65–74. [CrossRef]
14. Mathews, F.; Orros, M.; McLaren, G.; Gelling, M.; Foster, R. Keeping fit on the ark: Assessing the suitability of captive-bred animals for release. *Biol. Conserv.* **2005**, *121*, 569–577. [CrossRef]
15. Peng, J.; Jiang, Z.; Hu, J. Status and conservation of giant panda (*Ailuropoda melanoleuca*): A review. *Folia Zool.* **2001**, *50*, 81–88.
16. Qi, D.; Zhang, S.; Zhang, Z.; Hu, Y.; Yang, X.; Wang, H.; Wei, F. Measures of giant panda habitat selection across multiple spatial scales for species conservation. *J. Wildl. Manag.* **2012**, *76*, 1092–1100. [CrossRef]
17. Yang, Z.; Hu, J.; Liu, N. The influence of dispersal on the metapopulation viability of Giant Panda (*Aliuropoda melanoleuca*) in the Minshan Mountains. *Acta Zool. Acad. Sci. Hung.* **2007**, *53*, 169–184.
18. Wei, F.; Swaisgood, R.; Hu, Y.; Nie, Y.; Yan, L.; Zhang, Z.; Qi, D.; Zhu, L. Progress in the ecology and conservation of giant pandas. *Conserv. Biol.* **2015**, *29*, 1497–1507. [CrossRef]
19. Hu, Y.; Qi, D.; Wang, H.; Wei, F. Genetic evidence of recent population contraction in the southernmost population of giant pandas. *Genetica* **2010**, *138*, 1297–1306. [CrossRef]

20. Li, W.; Zhou, C.; Cheng, M.; Tu, H.; Wang, G.; Mao, Y.; Huang, Y.; Chen, M.; Price, M.; Meng, Y.; et al. Large-scale genetic surveys for main extant population of wild giant panda (*Ailuropoda melanoleuca*) reveals an urgent need of human management. *Evol. Appl.* **2023**, *16*, 738–749. [CrossRef]
21. Qiao, M.; Connor, T.; Shi, X.; Huang, J.; Huang, Y.; Zhang, H.; Ran, J. Population genetics reveals high connectivity of giant panda populations across human disturbance features in key nature reserve. *Ecol. Evol.* **2019**, *9*, 1809–1819. [CrossRef] [PubMed]
22. Zhang, B.; Li, M.; Zhang, Z.; Goossens, B.; Zhu, L.; Zhang, S.; Hu, J.; Bruford, M.W.; Wei, F. Genetic Viability and Population History of the Giant Panda, Putting an End to the "Evolutionary Dead End"? *Mol. Biol. Evol.* **2007**, *24*, 1801–1810. [CrossRef]
23. Yang, B.; Yang, C.; Tu, F.; Du, Y.; Liu, Y.; Li, D.; Zhang, H.; Huang, Y. The Genetics Analysis of Captive Giant Panda Reintroduction. *Sichuan J. Zool.* **2013**, *32*, 149–155. [CrossRef]
24. Allendorf, F.W. Genetic drift and the loss of alleles versus heterozygosity. *Zoo Biol.* **1986**, *5*, 181–190. [CrossRef]
25. Frankham, R.; Briscoe, D.A.; Ballou, J.D. *Introduction to Conservation Genetics*; Cambridge University Press: Cambridge, UK, 2002.
26. Boyce, M.S. Population Viability Analysis. *Annu. Rev. Ecol. Syst.* **1992**, *23*, 481–497. [CrossRef]
27. Gong, M.; Song, Y.; Yang, Z.; Lin, C. Important population viability analysis parameters for giant pandas(*Aliuropoda melanoleuca*). *Zool. Res.* **2012**, *33*, 18–24. [CrossRef]
28. Guo, J.; Chen, Y.; Hu, J. Population viability analysis of giant pandas in the Yele Nature Reserve. *J. Nat. Conserv.* **2002**, *10*, 35–40. [CrossRef]
29. Zhou, Z.; Pan, W. Analysis of the Viability of a Giant Panda Population. *J. Appl. Ecol.* **1997**, *34*, 363–374. [CrossRef]
30. Lu, Z.; Johnson, W.E.; Menotti-Raymond, M.; Yuhki, N.; Martenson, J.S.; Mainka, S.; Shi-Qiang, H.; Zhihe, Z.; Li, G.; Pan, W.; et al. Patterns of Genetic Diversity in Remaining Giant Panda Populations. *Conserv. Biol.* **2002**, *15*, 1596–1607. [CrossRef]
31. Ma, T.; Hu, Y.; Russo, I.R.M.; Nie, Y.; Yang, T.; Xiong, L.; Ma, S.; Meng, T.; Han, H.; Zhang, X.; et al. Walking in a heterogeneous landscape: Dispersal, gene flow and conservation implications for the giant panda in the Qinling Mountains. *Evol. Appl.* **2018**, *11*, 1859–1872. [CrossRef] [PubMed]
32. Sichuan Provincial Forestry Department. *Giant Panda in Sichuan-Report on the Fourth Giant Panda Survey in Sichuan Province*; Sichuan Science and Technology Press: Chengdu, China, 2015.
33. Van Oosterhout, C.; Hutchinson, W.F.; Wills, D.P.M.; Shipley, P. micro-checker: Software for identifying and correcting genotyping errors in microsatellite data. *Mol. Ecol. Notes.* **2004**, *4*, 535–538. [CrossRef]
34. Valière, N. gimlet: A computer program for analysing genetic individual identification data. *Mol. Ecol. Notes.* **2002**, *2*, 377–379. [CrossRef]
35. Park, S.D.E. Trypanotolerance in West African Cattle and the Population Genetic Effects of Selection. Ph.D. Thesis, University of Dublin, Dublin, Ireland, 2001.
36. Peakall, R.O.D.; Smouse, P.E. genalex 6: Genetic analysis in Excel. Population genetic software for teaching and research. *Mol. Ecol. Notes.* **2005**, *6*, 288–295. [CrossRef]
37. Raymond, M.; Rousset, F. GENEPOP (Version 1.2): Population Genetics Software for Exact Tests and Ecumenicism. *J. Hered.* **1995**, *86*, 248–249. [CrossRef]
38. Evanno, G.; Regnaut, S.; Goudet, J. Detecting the number of clusters of individuals using the software structure: A simulation study. *Mol. Ecol.* **2005**, *14*, 2611–2620. [CrossRef]
39. Mukesh; Kumar, V.P.; Sharma, L.K.; Shukla, M.; Sathyakumar, S. Pragmatic Perspective on Conservation Genetics and Demographic History of the Last Surviving Population of Kashmir Red Deer (*Cervus elaphus hanglu*) in India. *PLoS ONE* **2015**, *10*, e0117069. [CrossRef] [PubMed]
40. Lacy, R.C. VORTEX: A computer simulation model for population viability analysis. *Wildl. Res.* **1993**, *20*, 45–65. [CrossRef]
41. Wei, F.; Hu, J. Study on Breeding of Wild Giant Panda in Wolong Nature Reserve. *Acta Theriol. Sin.* **1994**, *14*, 243–248. [CrossRef]
42. Wei, F.; Hu, J.; Xu, G.; Jiang, M.; Deng, Q.; Zhong, Z. Preliminary compilation of life table of wild giant panda. *Acta Theriol. Sin.* **1989**, *9*, 81–86. [CrossRef]
43. Jiang, H.; Hu, J. Population Viability Analysis for the Giant Panda in Baoxing County, Sichuan. *Sichuan J. Zool.* **2010**, *29*, 161–165.
44. Zhu, L.; Wu, P.; Zhang, H.; Hu, J. Viability analysis of giant panda population in Xiaoxiangling mountains. *J. China West Norm. Univ. (Nat. Sci.)* **2008**, *29*, 112–116.
45. Ralls, K.; Ballou, J.D.; Templeton, A. Estimates of Lethal Equivalents and the Cost of Inbreeding in Mammals. *Conserv. Biol.* **1988**, *2*, 185–193. [CrossRef]
46. Wei, F.; Fgeng, Z.; Hu, J. Population Viability Analysis Computer Model of Giant Panda Population in Wuyipeng, Wolong Natural Reserve, China. *Bears Their Biol. Manag.* **1997**, *9*, 19–23. [CrossRef]
47. Hu, J.; Reid, D.G.; Dong, S.; Wang, W.; Huang, Y. Foraging behavior and carrying capacity of giant pandas after bamboo flowering. *J. China West Norm. Univ. (Nat. Sci.)* **1990**, *11*, 103–113. [CrossRef]
48. Hu, J.; Xia, L. *The Giant Panda in Wolong*; Sichuan Science and Technology Press: Chengdu, China, 1985.
49. Rong, Z.; Zhou, H.; Wei, W.; Zhang, J.; Shen, L.; Zhang, Z. Giant panda habitat suitability assessment in Tangjiahe Nature Reserve based on MAXENT model. *J. Lanzhou Univ. Nat. Sci.* **2017**, *53*, 269–273. [CrossRef]
50. Ruan, T.; Han, H.; Wei, W.; Qiu, L.; Hong, M.; Tang, J.; Zhou, H.; Zhang, Z. Habitat suitability evaluation for giant panda in Liziping National Nature Reserve, Sichuan Province. *Glob. Ecol. Conserv.* **2021**, *30*, e01780. [CrossRef]
51. Waits, L.P.; Luikart, G.; Taberlet, P. Estimating the probability of identity among genotypes in natural populations: Cautions and guidelines. *Mol. Ecol.* **2001**, *10*, 249–256. [CrossRef]

52. Fuller, R.A.; McDonald-Madden, E.; Wilson, K.A.; Carwardine, J.; Grantham, H.S.; Watson, J.E.M.; Klein, C.J.; Green, D.C.; Possingham, H.P. Replacing underperforming protected areas achieves better conservation outcomes. *Nature* **2010**, *466*, 365–367. [CrossRef]
53. Li, Y. Research on the habitat status of a population of Giant Pandas in the Minshan Mountain before and after the Wenchuan earthquake. *J. Sichuan For. Sci. Technol.* **2009**, *30*, 43–47. [CrossRef]
54. Zhang, Z.; Hu, J.; Wu, H.; Hou, W. An analysis on population viability for Giant Panda in Tangjiahe. *Acta Ecol. Sin.* **2002**, *22*, 990–998.
55. Ran, J.; Liu, S.; Wang, H.; Zeng, Z.; Sun, Z.; Liu, S. A survey of disturbance of Giant Panda habitat in the Xiaoxiangling Mountains of Sichuan Province. *Acta Theriol. Sin.* **2004**, *24*, 277–281. [CrossRef]
56. Zhu, L.; Zhan, X.; Wu, H.; Zhang, S.; Meng, T.; Bruford, M.W.; Wei, F. Conservation Implications of Drastic Reductions in the Smallest and Most Isolated Populations of Giant Pandas. *Conserv. Biol.* **2010**, *24*, 1299–1306. [CrossRef]
57. Ran, J.; Zeng, Z.; Liu, S.; Wang, H. A survey of the Giant Panda population and habitats in the Daxiangling Mountains. *J. Sichuan Univ. (Nat. Sci. Ed.)* **2006**, *43*, 889–893.
58. Xu, W.; Ouyang, Z.; Jiang, Z.; Zhen, H.; Liu, J. Assessment of giant panda habitat in the Daxiangling Mountain Range, Sichuan, China. *Biodivers. Sci.* **2006**, *14*, 223–231.
59. Zhao, C.; Yue, B.; Ran, J.; Moermond, T.; Hou, N.; Yang, X.; Gu, X. Relationship between human disturbance and Endangered giant panda Ailuropoda melanoleuca habitat use in the Daxiangling Mountains. *Oryx* **2016**, *51*, 146–152. [CrossRef]
60. Chen, Y.; Zhu, Y.; Wan, Q.; Lou, J.; Li, W.; Ge, Y.; Fang, S. Patterns of Adaptive and Neutral Diversity Identify the Xiaoxiangling Mountains as a Refuge for the Giant Panda. *PLoS ONE* **2013**, *8*, e70229. [CrossRef]
61. Botstein, D.; White, R.L.; Skolnick, M.; Davis, R.W. Construction of a Genetic Linkage Map in Man Using Restriction Fragment Length Polymorphisms. *Am. J. Hum. Genet.* **1980**, *32*, 314–331. [PubMed]
62. Shen, F.; Zhang, Z.; He, W.; Yue, B.; Zhang, A.; Zhang, L.; Hou, R.; Wang, C.; Watanabe, T. Microsatellite variability reveals the necessity for genetic input from wild giant pandas (*Ailuropoda melanoleuca*) into the captive population. *Mol. Ecol.* **2009**, *18*, 1061–1070. [CrossRef]
63. Yang, J.; Zhang, Z.; Shen, F.; Yang, X.; Zhang, L.; Chen, L.; Zhang, W.; Zhu, Q.; Hou, R. Microsatellite variability reveals high genetic diversity and low genetic differentiation in a critical giant panda population. *Curr. Zool.* **2011**, *57*, 717–724. [CrossRef]
64. Lindenmayer, D.B.; Clark, T.W.; Lacy, R.C.; Thomas, V.C. Population viability analysis as a tool in wildlife conservation policy: With reference to Australia. *Environ. Manag.* **1993**, *17*, 745–758. [CrossRef]
65. Schenkman, L. Hope for Wild Pandas. *Sci. China Ser. C Life Sci.* **2010**, *328*, 553. [CrossRef]
66. Soulé, M.; Gilpin, M.; Conway, W.; Foose, T. The millenium ark: How long a voyage, how many staterooms, how many passengers? *Zoo Biol.* **2005**, *5*, 101–113. [CrossRef]
67. Ma, B.; Lei, S.; Qing, Q.; Wen, Y. Should the endangered status of the Giant panda really be reduced? The case of Giant panda conservation in Sichuan, China. *Animals* **2018**, *8*, 69. [CrossRef] [PubMed]
68. Yang, Z.; Gu, X.; Nie, Y.; Huang, F.; Huang, Y.; Dai, Q.; Hu, Y.; Yang, Y.; Zhou, X.; Zhang, H.; et al. Reintroduction of the giant panda into the wild: A good start suggests a bright future. *Biol. Conserv.* **2018**, *217*, 181–186. [CrossRef]
69. Zhan, X.J.; Zhang, Z.J.; Wu, H.; Goossens, B.; Li, M.; Jiang, S.W.; Bruford, M.W.; Wei, F.W. Molecular analysis of dispersal in giant pandas. *Mol. Ecol.* **2007**, *16*, 3792–3800. [CrossRef] [PubMed]
70. Wang, Y.; Wei, W.; Yuan, F.; Cao, D.; Zhang, Z. The Science Underlying Giant Panda Conservation Translocations. *Animals.* **2023**, *13*, 3332. [CrossRef]
71. Chen, Z.; Huang, F.; He, K.; Feng, F.; Gu, X.; Yang, X.; Dai, Q.; Qi, D.; Yang, Z. The genetic structure of small Giant Panda population in the Liziping Nature Reserve after reintroduction. *J. Hechi Univ.* **2016**, *36*, 1–7.
72. Rummel, L.; Martínez–Abraín, A.; Mayol, J.; Ruiz–Olmo, J.; Mañas, F.; Jiménez, J.; Gómez, J.A.; Oro, D. Use of wild–caught individuals as a key factor for success in vertebrate translocations. *Anim. Biodivers. Conserv.* **2016**, *39*, 207–219. [CrossRef]
73. Fox, G.A. Extinction Risk of Heterogeneous Populations. *Ecology* **2005**, *86*, 1191–1198. [CrossRef]
74. Eillisl, S.; Pan, W.; Xie, Z.; Wildt, D. The giant panda as a social, biological and conservation phenomenon. In *Giant Pandas: Biology, Veterinary Medicine and Management*; Wildt, D.E., Zhang, A., Zhang, H., Janssen, D.L., Ellis, S., Eds.; Cambridge University Press: Cambridge, UK, 2006.
75. Wei, F.; Hu, Y.; Yan, L.; Nie, Y.; Wu, Q.; Zhang, Z. Giant Pandas Are Not an Evolutionary cul-de-sac: Evidence from Multidisciplinary Research. *Mol. Biol. Evol.* **2014**, *32*, 4–12. [CrossRef] [PubMed]
76. Zhu, L.; Zhang, S.; Gu, X.; Wei, F. Significant genetic boundaries and spatial dynamics of giant pandas occupying fragmented habitat across southwest China. *Mol. Ecol.* **2011**, *20*, 1122–1132. [CrossRef]
77. Tallmon, D.; Luikart, G.; Waples, R. The alluring simplicity and complex reality of genetic rescue. *Trends Ecol. Evol.* **2004**, *19*, 489–496. [CrossRef]

animals

MDPI

Article

Analysis of Priority Conservation Areas Using Habitat Quality Models and MaxEnt Models

Ahmee Jeong, Minkyung Kim and Sangdon Lee *

Department of Environmental Science and Engineering, Ewha Womans University, Seoul 03760, Republic of Korea; worldami@ewhain.net (A.J.); enviecol@ewha.ac.kr (M.K.)
* Correspondence: lsd@ewha.ac.kr

Simple Summary: This study investigated core habitat areas for yellow-throated martens and leopard cats, two endangered forest species sensitive to habitat fragmentation in Korea. Overlaying the InVEST-HQ and MaxEnt models, priority conservation areas were identified by analyzing gaps in currently protected areas. The core area (14.7%) was mainly distributed in forests such as the Baekdudaegan Mountains Reserve and 12.9% was outside protected areas, and only 1.8% was protected. Conservation priority areas were identified as those with more than 95% forest cover, offering an appropriate habitat for the two species. These findings can be used to identify priority conservation areas through objective habitat analysis and as a basis for protected area designation and assessment of endangered species habitat conservation, thereby contributing to biodiversity and ecosystem conservation.

Abstract: This study investigated core habitat areas for yellow-throated martens (*Martes flavigula*) and leopard cats (*Prionailurus bengalensis*), two endangered forest species sensitive to habitat fragmentation in Korea. Overlaying the InVEST-HQ and MaxEnt models, priority conservation areas were identified by analyzing gaps in currently protected areas. The InVEST-HQ model showed that habitat quality ranged from 0 to 0.86 on a scale from 0 to 1, and the majority of the most suitable areas on the Environmental Conservation Value Assessment Map, designated as grade 1, were derived correctly. The MaxEnt model analysis accurately captured the ecological characteristics of the yellow-throated marten and the leopard cat and identified probable regions of occurrence. We analyzed the most suitable yellow-throated marten and leopard cat habitats by superimposing the two results. Gap analysis determined gaps in existing protected areas and identified priority conservation areas. The core area (14.7%) was mainly distributed in forests such as the Baekdudaegan Mountains Reserve in regions such as Gyeongbuk, Gyeongnam, and Gangwon; 12.9% was outside protected areas, and only 1.8% was protected. The overlap results between protected and non-protected areas were compared with different land use types. Conservation priority areas were identified as those with more than 95% forest cover, offering an appropriate habitat for the two species. These findings can be used to identify priority conservation areas through objective habitat analysis and as a basis for protected area designation and assessment of endangered species habitat conservation, thereby contributing to biodiversity and ecosystem conservation.

Keywords: habitat quality; InVEST; MaxEnt; protected areas; gap analysis; endangered species

Citation: Jeong, A.; Kim, M.; Lee, S. Analysis of Priority Conservation Areas Using Habitat Quality Models and MaxEnt Models. *Animals* **2024**, *14*, 1680. https://doi.org/10.3390/ani14111680

Academic Editor: Chunwang Li

Received: 19 April 2024
Revised: 16 May 2024
Accepted: 27 May 2024
Published: 4 June 2024

1. Introduction

Biodiversity is a critical component of Earth's ecosystems, providing multiple resources and playing an essential role in maintaining the stability and functioning of ecosystems. This biodiversity is intimately connected to our lives and well-being and its conservation is of great value. The loss of biodiversity due to ecosystem degradation also affects the services ecosystems provide humans [1]. The most effective way to conserve ecosystems and biodiversity is to protect habitats and designate areas of high ecological value as protected areas [2]. In Korea, top predators such as tigers (*Panthera tigris altaica*), leopards (*Panthera pardus orientalis*), and wolves (*Canis lupus coreanus*) have become extinct or ecologically irrelevant over the past century, increasing the importance of the remaining predators such as yellow-throated martens (*Martes flavigula*) and medium to large mammals such as leopard cats (*Prionailurus bengalensis*). Yellow-throated martens and leopard cats, both listed as Class II endangered species by the Ministry of Environment, exhibit a wide range of behavior and are highly vulnerable to the effects of habitat fragmentation because of their small population size [3]. Specifically, 63% of South Korea's territory is covered in forests [4]. Endangered forest-dependent species like the yellow-throated marten and leopard cat are important subjects for studies on forest landscape conservation and the designation of protected areas [5,6]. In addition, yellow-throated martens and leopard cats are flagship species for the Korean ecosystem, and targeting them will increase the likelihood of protecting other populations, which will help conserve biodiversity.

Current domestic wildlife reserves are small in area and have vague designation criteria; there is a need for objective assessment and specific protections for endangered species [4]. Many countries have historically prioritized and managed conservation solely by assessing areas rich in biodiversity, which may underestimate the potential value of species with broader influence ranges. Therefore, numerous recent studies have assessed the diverse functions and values of ecosystems in order to address these limitations [3,7].

Ensembling results by combining multiple models instead of using only a single model has become popular in recent research. This is to overcome the limitations of a single model and produce more reliable results [8,9]. In this study, we followed this approach and combined the results of two or more models. In particular, the InVEST-HQ model has the advantage of being able to consider land use change, but has limitations in its application to endangered species habitats. On the other hand, the MaxEnt model provides more realistic results because it bases its predictions on actual occurrence data of species. Therefore, in this study, we combined the results of these two models, a technique that has been adopted by other studies recently [10–13].

This study analyzed priority areas for habitat conservation for endangered forest species, such as the yellow-throated marten and leopard cat, that are sensitive to habitat fragmentation. To analyze the habitat of these two species, we (1) evaluated protected areas designated and managed for wildlife habitat protection; (2) habitat quality in Korea was assessed using the InVEST Habitat Quality Model, which evaluates habitat quality in the area using land cover; and (3) used the MaxEnt model to predict the probability of species occurrence, allowing us to identify suitable habitats for endangered species with small populations. Finally, we combined the results of the two models to identify the most suitable core area likely to be used by the yellow-throated marten and leopard cat and conducted a gap analysis with existing protected areas to identify priority conservation areas. The findings provide a basis for habitat conservation for the yellow-throated marten and leopard cat, as well as assist in the designation of new protected areas and evaluation of existing ones.

2. Materials and Methods

2.1. Study Flow

The overall research flow is depicted in Figure 1. After overlaying the results of the InVEST Habitat Quality and MaxEnt models, we conducted a gap analysis with existing protected areas to identify priority areas for conservation.

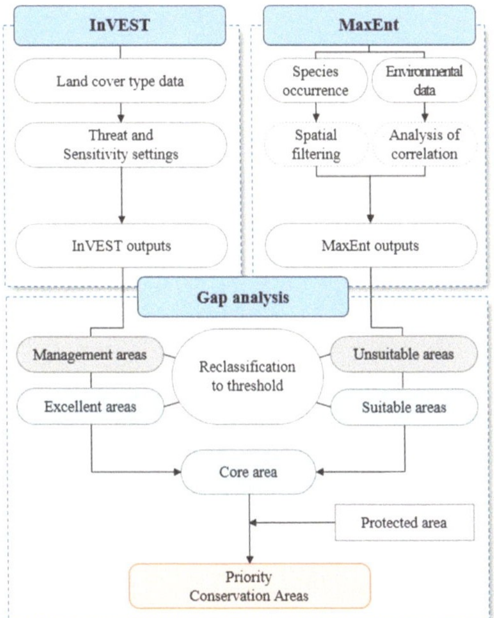

Figure 1. Study flow.

2.2. *Study Species and Site*

The yellow-throated marten and leopard cat are classified as least concern by the IUCN Red List of Threatened Species and are listed Class II endangered species by the Ministry of Environment and in the vulnerable category on the National Biodiversity Red List in South Korea. (Figure 2). The yellow-throated marten is found in mountainous areas throughout Asia in dense forests and forested valleys near streams. In contrast, the leopard cat is found throughout Korea in forests and fields, except on Jeju Island and some other islands. While both species have experienced a dramatic decline in their populations in recent years due to deforestation, lack of habitat, and poaching [14], there has been insufficient research on conserving their habitats in Korea. The yellow-throated marten and leopard cat are the flagship species of the Korean ecosystem, and we chose them as targets because of their common use of forests. If we can identify critical habitats through habitat analysis, we can inform the selection of additional protected areas.

Figure 2. Study species: (**a**) yellow-throated marten (*Martes flavigula*), and (**b**) leopard cat (*Prionailurus bengalensis*) [15].

The study was conducted across South Korea as follows, and each administrative region was divided into nine regions for area comparison (Figure 3b). South Korea is located between China to the west and Japan to the east (Figure 3a), between 33~43° latitude and 124~132° longitude. It is a peninsula bordered on three sides by the sea and consists of 63% forests. The Baekdudaegan Mountains Reserve stretches from north to south, forming a high plateau in the east and a low one in the west. There are also four distinct seasons, so the vegetation varies with different types of forests depending on the region.

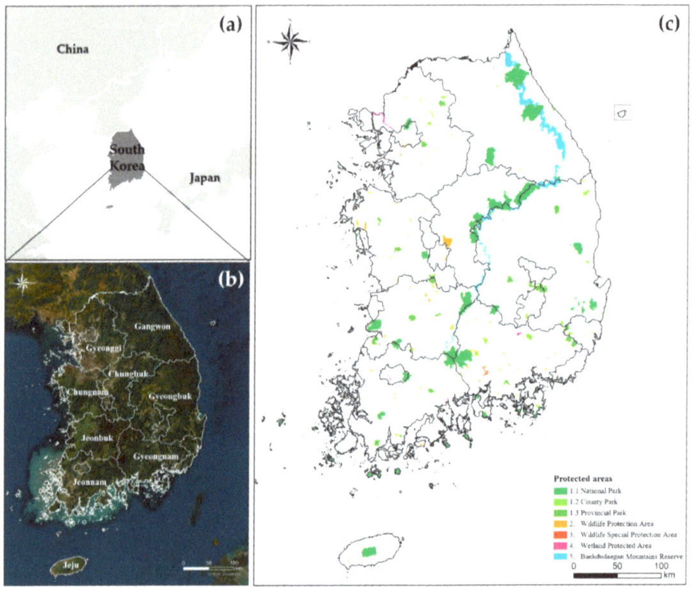

Figure 3. (**a**) Study site; (**b**) provinces in Republic of Korea; and (**c**) protected areas selected in the study.

2.3. Selection of Protected Areas

This study selected protected areas closely associated with forest wildlife habitats, with yellow-throated marten and leopard cat as the target species: the protected areas designated under the Fourth Basic Plan for Wildlife (2021–2025) and the Baekdudaegan Mountains Reserve, excluding Special Islands that are not habitats for the yellow-throated marten and leopard cat [2]. Finally, a total of five protected areas were extracted from the existing protected areas and built into a 1 km^2 resolution raster (Figure 3c): the Wildlife Protection Area, the Wildlife Special Protection Area, the Wetland Protected Area, the Nature Park (including National Park, County Park, and Provincial Park), and the Baekdudaegan Mountains Reserve [16].

2.4. Analyzing Habitat Quality with the InVEST-HQ Model

2.4.1. InVEST-HQ Model

To analyze the habitat quality for yellow-throated martens and leopard cats, it is crucial to employ multiple metrics and models that can quantitatively assess the diverse benefits of an ecosystem. The InVEST (Integrated Valuation and Environmental Services and Tradeoffs) model used in this study consists of various ecosystem service valuation items that can be valued using relevant variables based on land cover. It is open source and easily accessible, and the analysis results can be visualized on a map [17]. Therefore, it has recently been used as a decision-support model for ecosystem services [13,18].

One of the InVEST models, Habitat Quality (HQ), utilizes land cover to assess an area's habitat quality. This model can be used as an indicator of biodiversity [17]. The InVEST-HQ model can easily overlap with the species distribution model [13].

2.4.2. Creating Threats and Sensitivity Variables

The threats in the InVEST-HQ model were prioritized based on the threats presented in the ecological papers of the yellow-throated marten and leopard cat, except poaching, which is challenging to incorporate into raster data [19–22]. Nine threats were selected: residential areas, industrial areas, commercial areas, recreational areas, roads, public utilities, agricultural land (excluding paddy and dry fields), paddy fields, and dry fields. Among these, roads were divided into traffic areas in the land cover map and further classified into three levels based on the maximum speed (MAX_SPD), taken from national standard node-link data (Table 1).

Table 1. Threat and sensitivity values for the InVEST Habitat Quality model input for yellow-throated marten and leopard cat.

	Description	HS [1]	Sensitivity to Threat [2]										
			Res	**Ind**	**Com**	**Rec**	**Road 1**	**Road 2**	**Road 3**	**Pub**	**Agri**	**Pad**	**Dry**
Threat	**Weight**	-	0.58	0.88	0.88	0.88	0.59	0.40	0.20	0.88	0.57	0.67	0.57
	Maximum Distance	-	3.8	5.9	5.9	5.9	2.4	2.0	1.6	5.9	3.4	4.7	4.7
	Decay [3]	-	Expo	Line	Line	Line	Expo	Expo	Expo	Line	Line	Line	Line
Sensitivity	Residential area	0.05	0	0	0	0	0	0	0	0	0	0	0
	Industrial area	0.05	0	0	0	0	0	0	0	0	0	0	0
	Commercial area	0.05	0	0	0	0	0	0	0	0	0	0	0
	Recreational area	0.05	0	0	0	0	0	0	0	0	0	0	0
	Road	0.05	0	0	0	0	0	0	0	0	0	0	0
	Public utility area	0.05	0	0	0	0	0	0	0	0	0	0	0
	Paddy field	0.25	0.33	0.41	0.41	0.41	0.25	0.15	0.05	0.41	0.31	0	0.10
	Dry field	0.30	0.33	0.41	0.41	0.41	0.25	0.15	0.05	0.41	0.31	0.16	0
	Facility plantation	0.25	0.33	0.41	0.41	0.41	0.25	0.15	0.05	0.41	0.31	0.16	0
	Orchard	0.30	0.33	0.41	0.41	0.41	0.25	0.15	0.05	0.41	0.31	0.16	0
	Other plantations	0.40	0.33	0.41	0.41	0.41	0.25	0.15	0.05	0.41	0.31	0.16	0
	Broadleaved forest	0.86	0.60	0.75	0.75	0.75	0.52	0.42	0.32	0.75	0.66	0.61	0.51
	Coniferous forest	0.86	0.60	0.75	0.75	0.75	0.52	0.42	0.32	0.75	0.66	0.61	0.51
	Mixed forest	0.86	0.60	0.75	0.75	0.75	0.52	0.42	0.32	0.75	0.66	0.61	0.51
	Natural grassland	0.50	0.36	0.45	0.45	0.45	0.33	0.23	0.13	0.45	0.46	0.41	0.41
	Artificial grassland	0.34	0.36	0.45	0.45	0.45	0.33	0.23	0.13	0.45	0.46	0.41	0.41
	Inland wetland	0.70	0.56	0.70	0.70	0.70	0.55	0.45	0.35	0.70	0.75	0.85	0.75
	Coastal wetland	0.70	0.56	0.70	0.70	0.70	0.55	0.45	0.35	0.70	0.75	0.85	0.75
	Bare ground	0.08	0.11	0.14	0.14	0.14	0.05	0.04	0.03	0.14	0.15	0.15	0.15
	Artificial ground	0.08	0.11	0.14	0.14	0.14	0.05	0.04	0.03	0.14	0.15	0.15	0.15
	Inland water	0.65	0.58	0.73	0.73	0.73	0.55	0.45	0.35	0.73	0.65	0.75	0.65
	Marine water	0.65	0.58	0.73	0.73	0.73	0.55	0.45	0.35	0.73	0.65	0.75	0.65

[1] HS: habitat suitability; [2] Res: residential area, Ind: industrial area, Com: commercial area, Rec: recreational area, Road 1: ~60 km/h, Road 2: 60~80 km/h, Road 3: 80 km~/h, Agri: agricultural land, Pub: public utility, Pad: paddy field, Dry: dry field; [3] Decay: Expo: exponential, Line: linear.

Next, we set the suitability and sensitivity values for the threat factors using the values verified in South Korean research. Habitat suitability and sensitivity can range from 0 to 1, with values closer to 1 indicating higher suitability and sensitivity. Values set by experts in [18] were preferentially used for domestic conditions, and residential values that were not included in their report were derived from other previous studies [23,24]. The proportions of roads and agricultural land were set to match those in previous studies, considering the available information from studies that assessed forested mammals and forests [23,25,26]. All threats except roads (national standard node-link, 2023) were extracted from the land cover map (Environmental Geographic Information Service, 2022). The finalized threat and sensitivity table is presented in Table 1.

This study used the InVEST (v3.14.0) model (https://naturalcapitalproject.stanford.edu/software/invest (accessed on 1 October 2023)) for analysis. In addition, the habitat quality results were compared to the Environmental Conservation Value Assessment Map (ECVAM) for validation, and the habitat excellence areas were examined based on ECVAM ratings [27].

2.5. Analyzing Potential Habitats with the MaxEnt Model

2.5.1. MaxEnt Model

There is an increasing need for a species distribution model that can predict the probability of species occurrence in the case of endangered wildlife, where identifying the habitat is challenging due to the small population. The MaxEnt (Maximum Entropy) model is one of the species distribution models that can predict the probability of distribution by utilizing actual species occurrence data and environmental variables [28]. As the data surveyed in Korea only include occurrence data, the MaxEnt model is more suitable than other models [29,30]. For this study, the MaxEnt model was selected to predict potential habitats for the yellow-throated marten and leopard cat.

2.5.2. Creating Occurrence Data and Environmental Variables

The occurrence data of yellow-throated martens and leopard cats were obtained from the 4th National Natural Environment Survey. Spatial autocorrelation (SAC), a measure of the spatial dependence of the data, was determined using the average nearest neighbor index in R (v.4.3.1). Coordinates were clustered to some extent, and to avoid overfitting due to spatial autocorrelation we used the spThin package in R to ensure that each point was at least 1 km^2 apart and adjusted accordingly. In total, we used 446 occurrence points for the yellow-throated marten and 2379 points for the leopard cat.

The environmental variables used in the MaxEnt model were topography (elevation, gradient, aspect, and topographic wetness), distance (distance from residential areas, used areas, roads, and agricultural land), climate, vegetation (normalized difference vegetation index, diameter, age, and density), and land cover, based on previous studies of yellow-throated martens and leopard cats [6,20,25,31–34] (Table 2).

Table 2. Environmental variables used for MaxEnt modeling.

Classification	Code	Variables	Type	Source
Topography	DEM	Elevation	Continuous	Digital Elevation Model (National Geographic Information Institute, 2014)
	Slope	Gradient		
	Aspect	Aspect		
	TWI	Topographic Wetness Index		Topographic Wetness Index (Korea Institute of Geoscience and Mineral Resources, 2019)
Distance	Res	Distance from residential area		Land cover map (Environmental Geographic Information Service, 2022)
	Used	Distance from used area		
	Road	Distance from road		
	Agri	Distance from agricultural land		
	Water	Distance from water		
Climate	Bio3	Isothermality (Bio2/Bio7) (×100)		WorldClim (1970~2000)
	Bio7	Temperature annual range (Bio5-Bio6)		
	Bio12	Annual precipitation		
	Bio13	Precipitation of wettest month		
Vegetation	NDVI	Normalized Difference Vegetation Index		Normalized Difference Vegetation Index (Korea Institute of Geoscience and Mineral Resources, 2022)
	DMCLS	Diameter of forest	Categorical	1: 5000 Forest type map (Korea Forest Service, 2022)
	AGCLS	Age of forest		
	DNST	Density of forest		
Land cover	LULC	Land cover type		Land cover map (Environmental Geographic Information Service, 2022)

All environmental variables were utilized at a spatial resolution of 1 km^2, and Pearson's correlation coefficient (r) was computed to prevent multicollinearity. Most of the highly correlated variables ($|r| > 0.7$) were climate variables. After removing a proportion of the variables, 18 environmental variables were constructed (Table 2).

For the MaxEnt (v.3.4.4) model (https://biodiversityinformatics.amnh.org/open_source/maxent/ (accessed on 1 October 2023)), we used 70% of the occurrence points as training data for model development and the remaining 30% as test data to verify the model results. The training and test data were selected five times using the cross-validation method. In the model, the background points were set to 10,000, the regularization multipliers were set to 1, and the auto features included linear, quadratic, product, and hinge types. The output from the model was formatted as logistic. This is because the effect on the appearance of a species can be assessed with a value between 0 and 1 through the logistic setting, and an appropriate threshold can be set to generate a binary map marked with suitable and unsuitable habitats [35].

The model's predictive power was assessed by measuring the AUC (area under cover) of the ROC (receiver operating characteristic) curve. The AUC value of the ROC curve, which can be used to evaluate the accuracy of the MaxEnt model, is close to 1, indicating that the model has a high predictive probability (the explanatory power of the model is considered meaningful when it is around 0.7 or higher) [28]. The jackknife test was also conducted to determine the relative importance of environmental variables in model generation, along with the response curve.

2.6. Gap Analysis

After selecting core areas for the yellow-throated martens and leopard cats, we conducted a gap analysis to identify priority conservation areas by comparing the core areas derived from the nested InVEST-HQ and MaxEnt models with existing protected areas. Gap analysis is an analytical method that identifies gaps in the status of different components of wildlife habitat and conservation and determines suitable habitats [7,8].

The habitat quality results of the InVEST-HQ model were used for the gap analysis to extract the habitat quality of yellow-throated marten and leopard cat occurrence points, with the mean value set as the threshold. Areas above the threshold were classified as "excellent" habitat areas, while areas below the threshold were classified as habitat "management" areas [18]. The "maximum test sensitivity plus specificity" threshold from the logistic output of the MaxEnt model was used to identify potential habitat areas. Areas that exceeded the average thresholds for both species were classified as "suitable" areas, while areas below the threshold were classified as "unsuitable" [35].

The converted "excellent" habitat areas and "suitable" areas were overlaid, and the "core area" was selected as habitats expected to be excellent for yellow-throated martens and leopard cats and with a high probability of occurrence. In addition, by overlaying the previously extracted protected areas, we compared the excellent habitat areas, suitable areas, and core habitats outside and inside the protected areas. Finally, we selected priority areas for conservation with Table 3.

Table 3. Determining the ranking of conservation priorities and explaining the overlay map.

Ranking	Overlay Map			Explain
	InVEST-HQ	MaxEnt	PA [1]	
1	O	O		Core area outside PA
2	O	O	O	Core area inside PA
3		O		Suitable area outside PA
4		O	O	Suitable area inside PA
5	O			Excellent area outside PA
6	O		O	Excellent area inside PA
-			O	Over-PA
-				Unsuitable habitat area

[1] PA: protected area.

3. Results and Discussion

3.1. Habitat Quality Analysis Results

The results of the habitat quality analysis for yellow-throated martens and leopard cats using the InVEST-HQ model are shown in Figure 4a. Habitat quality ranged from 0 to 0.86 on a scale of 0 (low quality) to 1 (high quality), averaging 0.53 ± 0.27. The redder the color of the map, the higher the habitat quality. We found that the highest habitat quality was centered around significant mountain ranges such as Gangwon and parts of Gyeongbuk, where there is significant forest cover outside of city centers. Areas with high habitat quality have a higher likelihood of biodiversity sustainability [13], and active protection and management of these areas are needed to conserve them.

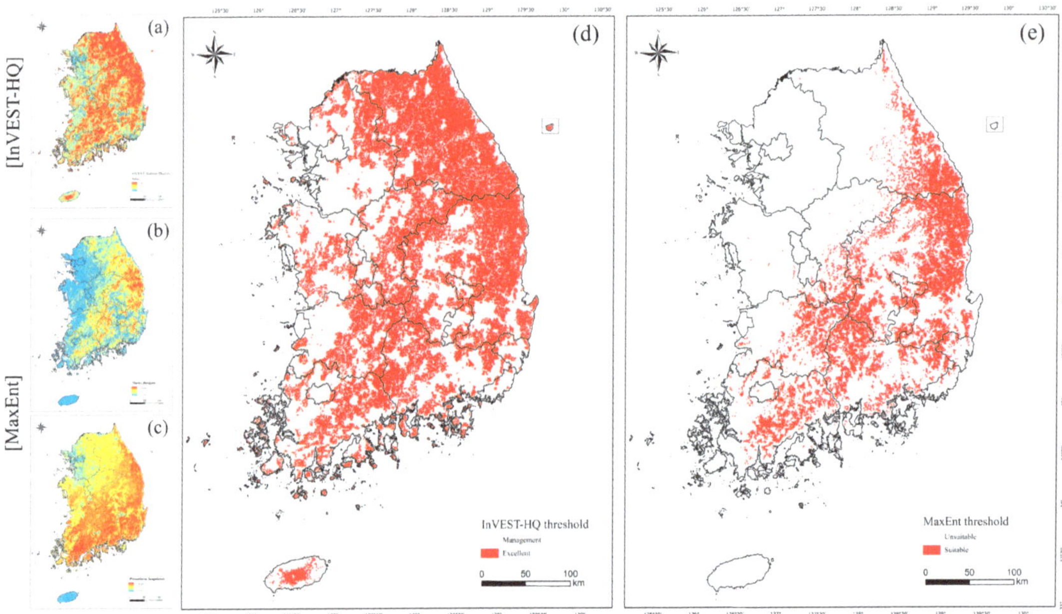

Figure 4. (**a**) InVEST habitat quality map; (**b**) potential distribution map of the yellow-throated marten; (**c**) potential distribution map of the leopard cat; (**d**) excellent/management areas of habitat quality map; and (**e**) suitable/unsuitable areas of potential distribution map.

The average habitat quality value of the yellow-throated marten and leopard cat was calculated to establish the threshold. The final threshold was determined to be 0.67, with 44,021 km^2 (43.9%) of excellent habitat areas above this value and 56,178 km^2 (56.1%) of habitat management areas below it (Figure 4d). To evaluate the validity of the habitat quality results, we compared them with the ECVAM [27]. We found that most of the excellent habitat areas for yellow-throated martens and leopard cats were rated as grade 1. This confirms that the excellent habitat areas were appropriately identified based on Korean conditions.

3.2. Potential Habitat Analysis Results

In our analysis, the AUC values for the yellow-throated marten and leopard cat were 0.810 and 0.645, respectively (Table 4). In the case of leopard cats, it is difficult to exceed 0.7 on average because the occurrence coordinates are widely distributed across the country, similar to general species, and like previous studies that ran MaxEnt models for leopard cats at the national level, we found that the AUC values were relatively low (0.561, 0.629, and 0.761) [25,32,36]. Therefore, although the AUC value does not indicate a very high accuracy, it can be evaluated as relatively reliable because it aligns with relatively low values in previous studies [25].

Table 4. AUC values, relative importance of environmental variables, and jackknife test by MaxEnt model.

Species	AUC	Percent Contribution	Jackknife
Martes flavigula	0.810	**DEM** [1] > Bio3 > Bio7 > **Used** > **Slope**	**DEM** > **Slope** > **Used** > DMCLS > AGCLS
Prionailurus bengalensis	0.645	**Bio7** > **Slope** > DEM > **LULC** > Used	**Bio7** > Bio3 > **Slope** > Bio12 > **LULC**

[1] Bold variables indicate those common to both percent contribution and jackknife tests.

In addition, although we have addressed spatial clustering in our study, there is still some degree of SAC remaining. This clustering could be due to differences in survey methods or specific behavioral patterns of the species. Due to time and data constraints, adjustments to the SAC were limited, but future studies will address this issue in more depth with spatial econometric models [37].

The results of the jackknife test to assess the contribution of variables to the distribution of potential habitats for the yellow-throated marten and leopard cat and the relative importance of each variable are shown in Table 4. Of the environmental variables, elevation, slope, and distance from used areas were the most important variables affecting the distribution of yellow-throated martens. Bio7 (temperature annual range), slope, and land cover were the most critical variables for leopard cats. The response curves indicated that yellow-throated martens were more likely to be located at higher elevations, slopes, and distances from used areas, which is similar to the findings of previous studies that have shown that yellow-throated martens are more likely to occur in forests with higher elevations and a more complex canopy structure, especially in broadleaf forests with plant fruits and rodents that can serve as food sources [20,33]. In the case of leopard cats, we found that they were more likely to be located in forests and wetlands with annual temperature differences between 30 °C and 33 °C and low slopes, which aligns with previous studies that showed that leopard cat prefers relatively low elevation, medium-hardwood forests with gentle slopes, and abundant streamside grasslands [19,38]. Therefore, this study's MaxEnt model of yellow-throated martens and leopard cats supports the results of these previous studies.

The results of predicting potential habitats for yellow-throated martens and leopard cats using the MaxEnt model are shown in Figure 4b,c. The predicted values range from 0 to 1, with redder colors on the map indicating a higher distribution probability. In the case of the yellow-throated marten, the distribution probability was concentrated in areas with high mountain ranges, centered on the Baekdudaegan Mountains Reserve. In the case of leopard cats, the distribution probability was high throughout the country except for some urban centers. Since the leopard cat uses forests as its primary habitat but has ecological characteristics that allow it to live throughout Korea, it is appropriate to establish protected areas centered on major forest areas where leopard cats can live when considering long-term biodiversity conservation [14].

For thresholding, we used 0.42 ± 0.06, the average of the values for both species where the sum of sensitivity and specificity was maximized as the final threshold. The area of possible occurrences with a threshold of 0.42 or suitable areas was 17,943 km^2 (17.9%), while the area of unsuitable areas with a threshold of 0.42 or lower was 82,256 km^2 (82.1%) (Figure 4e).

3.3. Core Area Analysis Results

To identify the core habitat, we overlaid the habitat results from the InVEST-HQ and MaxEnt analysis, focusing on forests shared by yellow-throated martens and leopard cats (Figure 5). The core area was selected by overlaying the best and most likely habitat areas divided by the thresholds of the InVEST-HQ and MaxEnt models. The core area (red) covered 14,718 km^2, or 14.7% of the total area; MaxEnt results only (green), 3225 km^2 (3.2%), InVEST-HQ results only (blue), 29,303 km^2 (29.2%), and unsuitable habitat areas (gray) covered 52,953 km^2 (52.9%) (Figure 5).

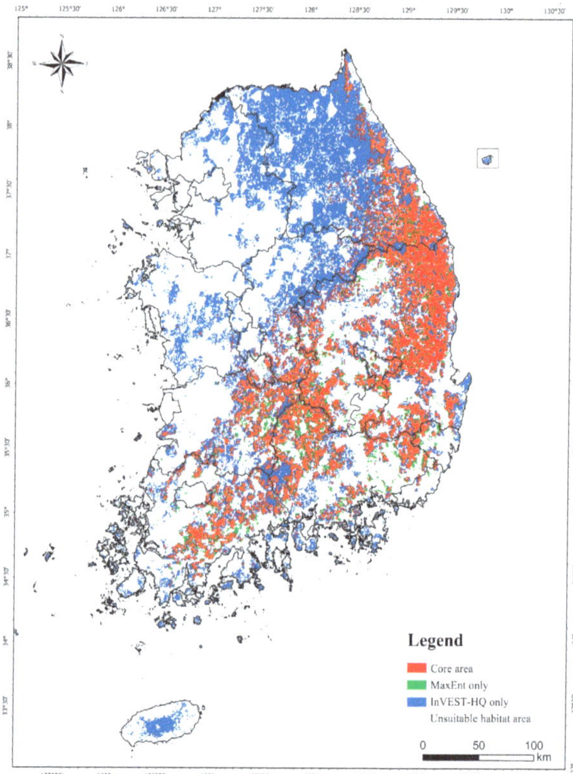

Figure 5. Mapping overlay of the core area map by MaxEnt and InVEST-HQ.

The difference between the InVEST-HQ and MaxEnt models is visible on the map because the habitat quality value of forests was set to be high to match the ecological characteristics of yellow-throated martens and leopard cats in the input data. The InVEST-HQ model evaluates habitat quality by considering actual land use changes, and it is necessary to refine the habitat quality evaluation system further in conjunction with other models [13]. In this study, MaxEnt results using actual yellow-throated marten and leopard cat occurrence coordinates yielded similar predictions to the ecological characteristics of each species, supporting the utility of MaxEnt models for supporting decision making at the site selection stage in the management plan of protected areas in conjunction with the InVEST-HQ model [11,12].

3.4. Priority Areas for Conservation

Finally, we analyzed the gap between protected areas and our analysis-derived core area to identify priority areas for conservation. For this, we overlaid the maps of protected areas on the maps of wildlife habitats and the core area of yellow-throated martens and leopard cats, where they are likely to occur. The overlapping maps identified priority areas for conservation, as shown in Figure 6.

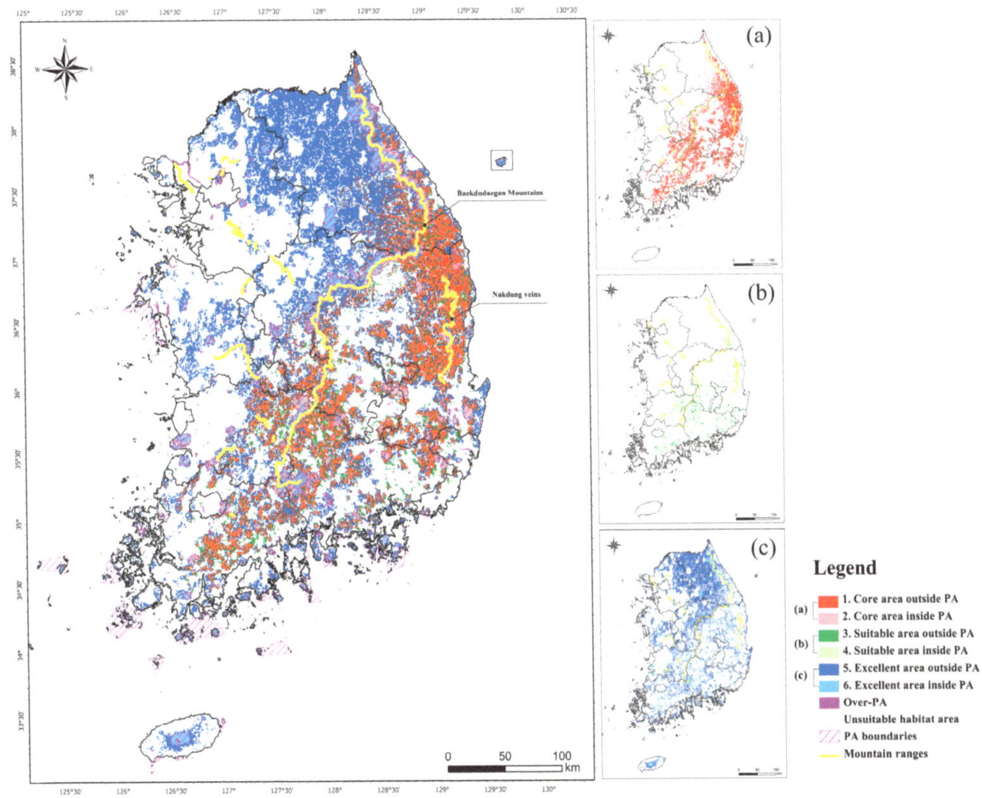

Figure 6. Gap analysis results for conservation priority area: (**a**) core area outside/inside PA, (**b**) suitable area outside/inside PA, and (**c**) excellent area outside/inside PA.

On the conservation priority areas map, the red color scheme indicates core areas in and outside protected areas, the green color scheme indicates suitable areas in and outside protected areas, and the blue color scheme indicates excellent habitat areas in and outside protected areas. Protected areas that do not overlap with the core areas are shown in purple as overprotected areas, and the boundaries of protected areas are also shown in purple. Otherwise, gray indicates unsuitable habitat areas, and yellow indicates the Baekdudaegan Mountains Reserve and prominent forest veins (Figure 6).

The gap analysis sorted administrative regions in order of the largest area of core area outside of the highest priority protected areas for conservation (Figure 7) and identified the proportion of this area in each region (Table 5). The results showed that the highest priority areas for conservation were Gyeongbuk (5932 km^2), Gyeongnam (2130 km^2), and Gangwon (1845 km^2), while Gyeonggi and Jeju had small areas (Figure 6).

The total area of core areas outside protected areas was 12,914 km^2 (12.9%), while core area inside protected areas was only 1804 km^2 (1.8%) (Table 5). Thus, we found that most of the core area for yellow-throated martens and leopard cats in this study was not included in protected areas, suggesting that current protected areas could benefit from a reassessment.

The comparison of land use type and overlap inside and outside protected areas showed that more than 95% of the externally predicted area is forested (Figure 8). This indicates that these areas have a high potential for new habitats for conserving the study's target species, the yellow-throated marten and leopard cat. In particular, core areas in Gyeongbuk (5932 km^2), Gyeongnam (2130 km^2), and Gangwon (1845 km^2), which are currently unprotected, were found to be centered on forest veins such as the Baekdudaegan

Mountains Reserve and Nakdong veins. The Baekdudaegan Mountains Reserve is an important migration corridor and habitat for critical plants and animals, including endangered species [4]. New habitat candidates centered on forests should be considered for designation as protected areas, reflecting the ecological characteristics of yellow-throated martens and leopard cats, which use forests as their primary habitat.

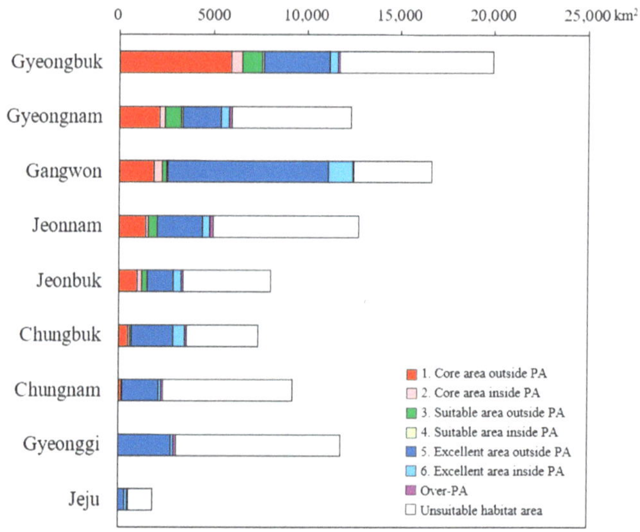

Figure 7. Mapping overlay of core area map by MaxEnt and InVEST-HQ.

Table 5. Percentage and conservation priority ranking area (km^2) by administrative district.

Ranking		1	2	3	4	5	6	-	-
Map Color (Figures 6 and 7)									
District		**Core Area Outside PA**	**Core Area Inside PA**	**Suitable Area Outside PA**	**Suitable Area Inside PA**	**Excellent Area Outside PA**	**Excellent Area Inside PA**	**Over-PA**	**Unsuitable Habitat Area**
Gyeongbuk	19,914	5932 (29.79%)	570 (2.86%)	1054 (5.29%)	72 (0.36%)	3559 (17.87%)	432 (2.17%)	92 (0.46%)	8203 (41.19%)
Gyeongnam	12,373	2130 (17.21%)	293 (2.37%)	877 (7.09%)	62 (0.50%)	2051 (16.58%)	399 (3.22%)	140 (1.13%)	6421 (51.90%)
Gangwon	16,664	1845 (11.07%)	425 (2.55%)	239 (1.43%)	25 (0.15%)	8640 (51.85%)	1259 (7.56%)	82 (0.49%)	4149 (24.90%)
Jeonnam	12,790	1407 (11.00%)	143 (1.12%)	463 (3.62%)	21 (0.16%)	2415 (18.88%)	368 (2.88%)	195 (1.52%)	7778 (60.81%)
Jeonbuk	8087	952 (11.77%)	260 (3.22%)	269 (3.33%)	25 (0.31%)	1408 (17.41%)	398 (4.92%)	107 (1.32%)	4668 (57.72%)
Chungbuk	7397	505 (6.83%)	86 (1.16%)	83 (1.12%)	2 (0.03%)	2238 (30.26%)	614 (8.30%)	80 (1.08%)	3789 (51.22%)
Chungnam	9263	142 (1.53%)	27 (0.29%)	32 (0.35%)	1 (0.01%)	1936 (20.90%)	126 (1.36%)	102 (1.10%)	6897 (74.46%)
Gyeonggi	11,861	1 (0.01%)	0 (0.00%)	0 (0.00%)	0 (0.00%)	2801 (23.62%)	156 (1.32%)	122 (1.03%)	8781 (74.03%)
Jeju	1850	0 (0.00%)	0 (0.00%)	0 (0.00%)	0 (0.00%)	356 (19.24%)	147 (7.95%)	16 (0.86%)	1331 (71.95%)
Total	100,199	12,914 (12.89%)	1804 (1.80%)	3017 (3.01%)	208 (0.21%)	25,404 (25.35%)	3899 (3.89%)	936 (0.93%)	52,017 (51.91%)

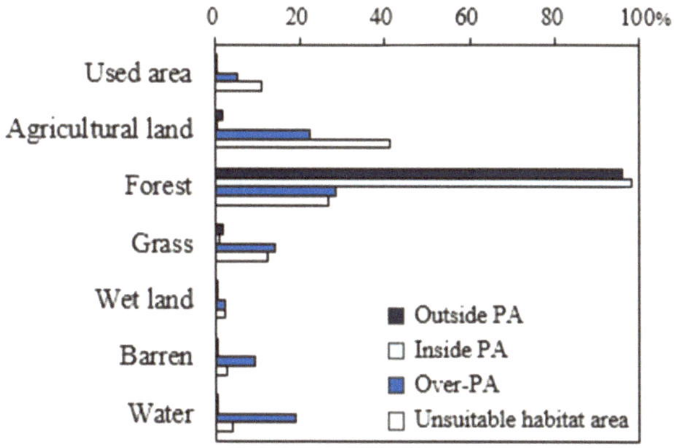

Figure 8. Land use type for gap analysis result areas.

4. Conclusions

This study focused on the yellow-throated marten and leopard cat—two endangered forest species that are sensitive to habitat fragmentation—by superimposing the results of the InVEST-HQ and MaxEnt models to analyze core areas with high habitat quality and high occurrence potential. A gap analysis was conducted to identify gaps in existing protected areas and priority areas for conservation.

The "core area" of the yellow-throated marten and leopard cat, which is the overlap between excellent areas and suitable areas, was analyzed to be 14,718 km^2, or 14.7% of the national area. Of this, "core area outside protected areas", which does not overlap with protected areas, was 12,914 km^2 (12.9%), and "core area inside protected areas" was only 1804 km^2 (1.8%). This suggests that much of the core habitat is not included in protected areas and that protected areas need to be adequately assessed.

These findings suggest that among the "core area outside protected areas", the highest priority areas for conservation are located in areas such as Gyeongbuk (5932 km^2), Gyeongnam (2130 km^2), and Gangwon (1845 km^2) and are centered on forests, especially the Baekdudaegan Mountains Reserve and Nakdong veins. As the area of forests that serve as the primary habitat for wildlife in Korea continues to decline, a comprehensive ecosystem conservation plan is needed to protect existing forests and designate new protected areas. In particular, when comparing the overlap of land use types and protected areas in this study, areas with more than 95% forest cover were identified as forest areas, appropriately designated as yellow-throated marten and leopard cat protected areas.

The results of this study provide an objective habitat analysis of yellow-throated martens and leopard cats to identify priority areas for conservation, which can be used as a basis for the evaluation and designation of protected areas to conserve habitats for threatened species for biodiversity and ecosystem conservation. Future research should consider various additional taxonomic groups to inform the selection of more effective protected areas. Additionally, it is necessary to further develop the habitat assessment method by supplementing the inputs of the InVEST-HQ and MaxEnt models and comparing them with other ecosystem service models or species distribution models.

Author Contributions: Formal analysis, A.J. and M.K.; Investigation, M.K.; Data curation, A.J.; Supervision, S.L.; Project administration, S.L.; Funding acquisition, S.L. All authors have read and agreed to the published version of the manuscript.

Funding: This research was supported by Korea Environmental Industry & Technology Institute (KEITI) funded by Korea Ministry of Environment (MOE) (grant number 2020002990006, 2021003360002, RS-2022-KE002025) and the National Research Foundation of Korea (NRF-2021R1A2C1011213, 2020R1I1A1A01058327).

Institutional Review Board Statement: Not applicable.

Informed Consent Statement: Not applicable.

Data Availability Statement: The data that support the findings of this study are available from the corresponding author, upon reasonable request.

Conflicts of Interest: The authors declare no conflicts of interest.

References

1. Costanza, R.; d'Arge, R.; De Groot, R.; Farber, S.; Grasso, M.; Hannon, B.; Limburg, K.; Naeem, S.; O'Neill, R.V.; Paruelo, J.; et al. The value of the world's ecosystem services and natural capital. *Nature* **1997**, *387*, 253–260. [CrossRef]
2. Pullin, A.S.; Bangpan, M.; Dalrymple, S.; Dickson, K.; Haddaway, N.R.; Hauari, H.; Oliver, S. Human well-being impacts of terrestrial protected areas. *Environ. Evid.* **2013**, *2*, 19. [CrossRef]
3. Lyra-Jorge, M.C.; Ciocheti, G.; Pivello, V.R. Carnivore mammals in a fragmented landscape in northeast of São Paulo State, Brazil. *Biodivers. Conserv.* **2008**, *17*, 1573–1580. [CrossRef]
4. ME (Ministry of Environment). *Endangered Wildlife Conservation Master Plan ('18~'27)*; Ministry of Environment: Sejong, Republic of Korea, 2018; pp. 7–92.
5. Mergey, M.; Helder, R.; Roeder, J.J. Effect of forest fragmentation on space-use patterns in the European pine marten (*Martes martes*). *J. Mammal.* **2011**, *92*, 328–335. [CrossRef]
6. Lee, S.R.; Lee, S.M.; Song, W.K.; Lee, M.J. Habitat Potential Mapping of Marten (*Martes flavigula*) and Leopard Cat (*Prionailurus bengalensis*) in South Korea Using Artificial Neural Network Machine Learning. *Appl. Sci.* **2017**, *7*, 912. [CrossRef]
7. Scott, J.M.; Davis, F.; Csuti, B.; Noss, R.; Butterfield, B.; Groves, C.; Anderson, H.; Caicco, S.; D'Erchia, F.; Edwards, T.C.; et al. Gap Analysis: A Geographic Approach to Protection of Biological Diversity. *Wildl. Monogr.* **1993**, *123*, 3–41. Available online: https://www.jstor.org/stable/3830788 (accessed on 1 October 2023).
8. Sher, A.A.; Primack, R.B.; Lee, S.D.; Kim, Y.H.; Kim, J.G.; Nam, D.H.; Park, J.H.; Bae, Y.J.; Seo, J.E.; Lee, G.S. *Conservation Biology*; Worldscience: Seoul, Republic of Korea, 2021; pp. 288–292.
9. Araújo, M.B.; New, M. Ensemble forecasting of species distributions. *Trends Ecol. Evol.* **2007**, *22*, 42–47. [CrossRef] [PubMed]
10. Marmion, M.; Parviainen, M.; Luoto, M.; Heikkinen, R.K.; Thuiller, W. Evaluation of consensus methods in predictive species distribution modelling. *Divers. Distrib.* **2009**, *15*, 59–69. [CrossRef]
11. Menon, S.; Pontius Jr, R.G.; Rose, J.; Khan, M.L.; Bawa, K.S. Identifying conservation-priority areas in the tropics: A land-use change modeling approach. *Conserv. Biol.* **2001**, *15*, 501–512. [CrossRef]
12. Seto, K.C.; Güneralp, B.; Hutyra, L.R. Global forecasts of urban expansion to 2030 and direct impacts on biodiversity and carbon pools. *Proc. Natl. Acad. Sci. USA* **2012**, *109*, 16083–16088. [CrossRef]
13. Huang, Z.; Qian, L.; Cao, W. Developing a novel approach integrating ecosystem services and biodiversity for identifying priority ecological reserves. *Resour. Conserv. Recycl.* **2022**, *179*, 106128. [CrossRef]
14. Lee, S.D.; Miller-Rushing, A.J. Degradation, urbanization, and restoration: A review of the challenges and future of conservation on the Korean Peninsula. *Biol. Conserv.* **2014**, *176*, 262–276. [CrossRef]
15. NIE (National Institute of Ecology). 2023. Available online: https://www.nie.re.kr/ (accessed on 1 October 2023).
16. KDPA (Korea Database on Protected Areas). 2022. Available online: http://www.kdpa.kr/ (accessed on 1 October 2023).
17. Natural Capital Project. InVEST 3.14.0. Stanford University, University of Minnesota, Chinese Academy of Sciences, The Nature Conservancy, World Wildlife Fund, Stockholm Resilience Centre, and the Royal Swedish Academy of Sciences. 2023. Available online: https://naturalcapitalproject.stanford.edu/software/invest (accessed on 1 October 2023).
18. KEI (Korea Environment Institute). *Development of Decision Supporting Framework to Enhance Natural Capital Sustainability: Focusing on Ecosystem Service Analysis*; Korea Environment Institute: Sejong, Republic of Korea, 2015; pp. 63–104. [CrossRef]
19. Choi, T.Y.; Kwon, H.S.; Woo, D.G.; Park, C.H. Habitat Selection and Management of the Leopard Cat (*Prionailurus bengalensis*) in a Rural Area of Korea. *Korean J. Environ. Ecol.* **2012**, *26*, 322–332.
20. Woo, D.G.; Choi, T.Y.; Kwon, H.S.; Lee, S.G.; Lee, J.C. The Food Habits and Habitat Use of Yellow-Throated Martens (*Martes flavigula*) by Snow Tracking in Korean Temperate Forest During the Winter. *J. Environ. Impact Assess.* **2015**, *24*, 532–548. [CrossRef]
21. Chen, M.T.; Liang, Y.J.; Kuo, C.C.; Pei, K.J.C. Home ranges, movements and activity patterns of leopard cats (*Prionailurus bengalensis*) and threats to them in Taiwan. *Mammal Study* **2016**, *41*, 77–86. [CrossRef]
22. Oleynikov, A.Y.; Yudin, G.V.; Salkina, G.P.; Sedash, G.A. Current population status of yellow-throated marten *Martes (Charronia) flavigula* in Russia. *Russ. J. Theriol.* **2022**, *21*, 63–69. [CrossRef]
23. Kim, T.Y.; Song, C.H.; Lee, W.K.; Kim, M.I.; Lim, C.H.; Jeon, S.W.; Kim, J.S. Habitat Quality Valuation Using InVEST Model in Jeju Island. *J. Korean Soc. Environ. Restor. Technol.* **2015**, *18*, 1–11. [CrossRef]
24. Seo, S.B. Analysis of Habitat Quality in the Nam-Han River Upstream Watershed Using InVEST Model. Master's Thesis, Ewha Womans University, Seoul, Republic of Korea, 2017.
25. Ryu, J.E.; Choi, Y.Y.; Jeon, S.W.; Sung, H.C. Evaluation of Habitat Function of National Park Based on Biodiversity and Habitat Value. *Korea Soc. Environ. Restor. Technol.* **2018**, *21*, 39–60.
26. KNPRI (Korea National Park Research Institute). *National Parks Ecosystem Services Valuation Study*; Korea National Park Research Institute: Gangwon, Republic of Korea, 2019.

27. Jeong, S.H.; Sung, K.J. Assessing Habitat Quality and Risk of Coastal Areas in Busan. *J. Environ. Impact Assess.* **2022**, *31*, 95–105. [CrossRef]
28. Phillips, S.J.; Anderson, R.P.; Schapire, R.E. Maximum entropy modeling of species geographic distributions. *Ecol. Model.* **2006**, *190*, 231–259. [CrossRef]
29. Kim, J.Y.; Seo, C.W.; Kwon, H.S.; Ryu, J.E.; Kim, M.J. A Study on the Species Distribution Modeling using National Ecosystem Survey Data. *J. Environ. Impact Assess.* **2012**, *21*, 593–607. [CrossRef]
30. Kwon, H.S. Applying Ensemble Model for Identifying Uncertainty in the Species Distribution Models. *J. Korean Soc. Geospat. Inf. Sci.* **2014**, *22*, 47–52. [CrossRef]
31. Grassman, L.I.; Tewes, M.E.; Silvy, N.J. Ranging, habitat use and activity patterns of binturong Arctictis binturong and yellow-throated marten *Martes flavigula* in north-central Thailand. *Wildl. Biol.* **2005**, *11*, 49–57. [CrossRef]
32. Jeon, S.W.; Kim, J.U.; Jung, H.C.; Lee, W.K.; Kim, J.S. Species Distribution Modeling of Endangered Mammals for Ecosystem Services Valuation. *J. Korea Soc. Environ. Restor. Technol.* **2014**, *17*, 111–122. [CrossRef]
33. Lim, C.H.; Lee, C.S.; Jung, S.H.; Park, Y.S. A Study on the Trail Management in National Park Using Habitat Suitability Assessment: A Case Study of Yellow-throated Marten Habitats in Mt. Mudeung National Park. *J. Korea Soc. Environ. Restor. Technol.* **2017**, *20*, 63–75.
34. Yoo, D.Y.; Lim, T.Y.; Kim, W.M.; Song, W.K. A Comparative Study on HSI and MaxEnt Habitat Prediction Models: About *Prionailurus bengalensis*. *J. Korea Soc. Environ. Restor. Technol.* **2021**, *24*, 1–14. [CrossRef]
35. Liu, C.; Berry, P.M.; Dawson, T.P.; Pearson, R.G. Selecting Thresholds of Occurrence in the Prediction of Species Distributions. *Ecography* **2005**, *28*, 385–393. [CrossRef]
36. Kim, J.Y.; Kwon, H.S.; Seo, C.W.; Kim, M.J. A nationwide analysis of mammalian biodiversity hotspots in South Korea. *J. Environ. Impact Assess.* **2014**, *23*, 453–465. [CrossRef]
37. Sillero, N.; Barbosa, A.M. Common mistakes in ecological niche models. *Int. J. Geogr. Inf. Sci.* **2021**, *35*, 213–226. [CrossRef]
38. Lim, S.J.; Kim, J.Y.; Park, Y.C. Analysis of habitat characteristics of leopard cat (*Prionailurus bengalensis*) in Odaesan National Park. *J. Agric. Life Sci.* **2015**, *49*, 99–111. [CrossRef]

animals

MDPI

Article

Prediction of Potential Suitable Distribution Areas for an Endangered Salamander in China

Jiacheng Tao [1], Yifeng Hu [1], Jianping Jiang [2], Wanji Yang [3], Tian Zhao [1,*] and Shengqi Su [1,*]

1 College of Fisheries, Southwest University, Chongqing 400715, China; tcheng632@163.com (J.T.); hyf990312@163.com (Y.H.)
2 CAS Key Laboratory of Mountain Ecological Restoration and Bioresource Utilization & Ecological Restoration Biodiversity Conservation Key Laboratory of Sichuan Province, Chengdu Institute of Biology, Chinese Academy of Sciences, Chengdu 610041, China; jiangjp@cib.ac.cn
3 Shengnongjia National Park Administration, Huibei Provincial Key Laboratory on Conservation Biology of the Shennongjia Golden Snub-Nosed Monkey, Shennongjia 442421, China; snjgjgy@126.com
* Correspondence: owfaowfa@swu.edu.cn (T.Z.); sushengqi@swu.edu.cn (S.S.)

Simple Summary: Exploring species' potential suitable habitats is crucial for endangered species conservation, in particular under future global climate change conditions. The Wushan salamander (*Liua shihi*) is an endangered salamander in China, which is a national protected species (level II). Based on the distribution records of *L. shihi*, the main objective of this study was to predict the distribution of suitable habitats under current and future climate conditions for *L. shihi*. Our results showed that precipitation, cloud density, vegetation type, and ultraviolet radiation were the main environmental factors affecting the distribution of suitable habitat for *L. shihi*. At present, the suitable habitats are mainly distributed in the Daba Mountain area. Under the future climate conditions, the area of suitable habitats increased, which mainly occurred in central Guizhou Province. These findings provided important information for the conservation of *L. shihi*.

Abstract: Climate change has been considered to pose critical threats for wildlife. During the past decade, species distribution models were widely used to assess the effects of climate change on the distribution of species' suitable habitats. Among all the vertebrates, amphibians are most vulnerable to climate change. This is especially true for salamanders, which possess some specific traits such as cutaneous respiration and low vagility. The Wushan salamander (*Liua shihi*) is a threatened and protected salamander in China, with its wild population decreasing continuously. The main objective of this study was to predict the distribution of suitable habitat for *L. shihi* using the ENMeval parameter-optimized MaxEnt model under current and future climate conditions. Our results showed that precipitation, cloud density, vegetation type, and ultraviolet radiation were the main environmental factors affecting the distribution of *L. shihi*. Currently, the suitable habitats for *L. shihi* are mainly concentrated in the Daba Mountains, including northeastern Chongqing and western Hubei Provinces. Under the future climate conditions, the area of suitable habitats increased, which mainly occurred in central Guizhou Province. This study provided important information for the conservation of *L. shihi*. Future studies can incorporate more species distribution models to better understand the effects of climate change on the distribution of *L. shihi*.

Keywords: amphibian; maximum entropy; ENMeval; environmental factor; distribution pattern

Citation: Tao, J.; Hu, Y.; Jiang, J.; Yang, W.; Zhao, T.; Su, S. Prediction of Potential Suitable Distribution Areas for an Endangered Salamander in China. *Animals* **2024**, *14*, 1390. https://doi.org/10.3390/ani14091390

Academic Editor: Marco Mangiacotti

Received: 23 March 2024
Revised: 20 April 2024
Accepted: 29 April 2024
Published: 6 May 2024

1. Introduction

Understanding the effects of human-induced perturbations on biological diversity is one of the central concerns in modern ecology [1]. During the past hundreds of years, human activities have dramatically changed the environment on Earth, in particular the climate, which has strongly affected animals in recent decades [2]. Based on previous studies, over 27% of mammals, 13% of birds, 21% of reptiles, 41% of amphibians, and

37%of fishes are threatened due to climate change and other human disturbances [3]. For instance, although the Atlantic Forest network of protected areas (PAs) supports 18% of the amphibians in South America, the number of amphibian species in PAs is declining under changing climate conditions [4]. The reduction of precipitation will lead to a decrease in the reproductive success rate of shovel-nosed frog (*Hemisus marmoratus*) and natterjack toad (*Bufo calamita*), resulting in a decrease in their populations [5]. Climate change can also cause the alteration of amphibian phylogenetic composition and niche. For instance, most of the amphibians in PAs contracted their ranges and such responses are clade specific. Basal amphibian clades (e.g., Gymnophiona and Pipidae) were positively affected by climate change, whereas late-divergent clades (e.g., Cycloramphidae, Centrolenidae, Eleutherodactylidae, Microhylidae) were severely impacted [6]. From the wet season to the dry season, the vertical niche space of amphibians in Sierra Llorona has a clear downward trend in response to natural levels of climate variability [7]. In recent decades, increasing studies also indicated that climate change can lead to the shift of animals' geographical distribution. For instance, Nottingham et al. showed that the suitable habitats of Del Norte salamander (*Plethodon elongatus*) and Siskiyou Mountains salamander (*P. stormi*) will shift to the coast and out of the valley with a move north into the mountains under future climate change conditions in the Pacific Northwest of the United States [8]. Duan et al. demonstrated that amphibians in China would lose 20% of their original distribution ranges on average, and over 90% of species' suitable habitats will shift to the north when compared with their current distribution range. As a consequence, climate change can lead to significant changes in the spatial pattern of amphibian diversity in China [9].

Among all the vertebrates, amphibians are particularly sensitive to climate change as they cannot regulate their body temperature actively [10]. This is especially true for salamanders, which possess some specific traits such as cutaneous respiration and low vagility [11]. However, studies focused on the effects of climate change on salamanders are still limited (but see [12,13]). The Wushan salamander (*Liua shihi*) is a national protected (level II) salamander in China, which was classified as Near Threatened in the Red List of China. Although this species was listed as Least Concern (LC) in the International Union for Conservation of Nature (IUCN), the wild populations have declined continuously in recent years [14,15]. Based on the records, this species is widely distributed in montane streams of eastern Sichuan, Chongqing, western Hubei, and southern Shaanxi Provinces, with the elevation ranging from 900 to 2350 m [16], and it mainly feeds on aquatic insects and algae [17]. In recent years, the wild population of *L. shihi* has been decreasing continuously due to human-induced perturbations [18]. Therefore, it is urgent to understand the distribution of suitable habitats of this species, as well as how the suitable habitat will shift under future climate change conditions.

Species distribution models (SDMs) have been proved to be effective to predict the effects of climate change on species distribution patterns [19]. Based on species distribution points and environmental data, these models predict where species likely inhabit using approaches such as statistical and machine learning analyses [20]. Accordingly, SDMs are involved in several models such as Bioclim, random forest, maximum entropy, regression tree, and genetic algorithm [21,22]. Although none of the above models can be regarded as the best one, the maximum entropy model (MaxEnt) was considered to exhibit higher prediction accuracy, have a stronger ability to integrate multiple environmental variables, and provide more intuitive results [23,24]. Therefore, MaxEnt is increasingly used in ecological studies to investigate the responses of species distribution patterns to climate change. Using this model, Zhao et al. demonstrated that climate change can induce different effects on the evolutionarily significant units (ESUs) of Chinese giant salamander (*Andrias davidianus*) in China, with the northern ESU exhibiting more severe habitat loss [25]. Moreover, Zank et al. used MaxEnt to investigate the potential effects of climate change on 24 species of red-bellied toads (*Melanophryniscus*) in South America, and they found that 40% of the species may lose over 50% of their potential distribution area by 2080 [26]. However, most studies only used the default parameters provided by the MaxEnt model, despite the fact that MaxEnt is sensitive to sampling bias and prone to overfitting when using default parameters [23,27]. Therefore, it is essential to optimize the parameters of the MaxEnt before conducting the model analyses [27].

The main objective of the present study was to assess the effects of climate change on the distribution of suitable habitats for *L. shihi*. Specifically, we (1) investigated the distribution of suitable habitats for *L. shihi* under current climate conditions; (2) analyzed the key environmental factors affecting the distribution patterns of *L. shihi*; (3) revealed the shift of suitable habitats (i.e., the distribution patterns and the area) caused by climate change in the future. Based on previous studies (e.g., [28,29]), we predicted that the suitable habitats of *L. shihi* are mainly distributed in southwestern China at present. We also predicted that climate change will lead to the expansion of suitable habitats from the current distribution area to the southwest. In addition, the area of suitable habitats would decrease due to climate change.

2. Materials and Methods

2.1. Study Area

L. shihi is an endemic amphibian species in China. Although its distribution records were concentrated in the Daba Mountains, its potential suitable habitats could be widely distributed in China. Therefore, and in order to better protect this endangered species, we considered the whole of China as the study area.

2.2. Species Occurrence Data

The occurrence data of *L. shihi* in this study were obtained from published literature (Table S1), the Global Biodiversity Information Facility website (http://www.gbif.org, accessed on 28 May 2023) (Table S2), and our original field survey (Table S3). In total, 89 occurrence records of *L. shihi* were collected. To avoid spatial autocorrelation, redundant records within 5 × 5 km grids were excluded using SDMToolbox (version 2.4; [30]). Finally, a total of 53 occurrence records were obtained for further analyses (Figure 1).

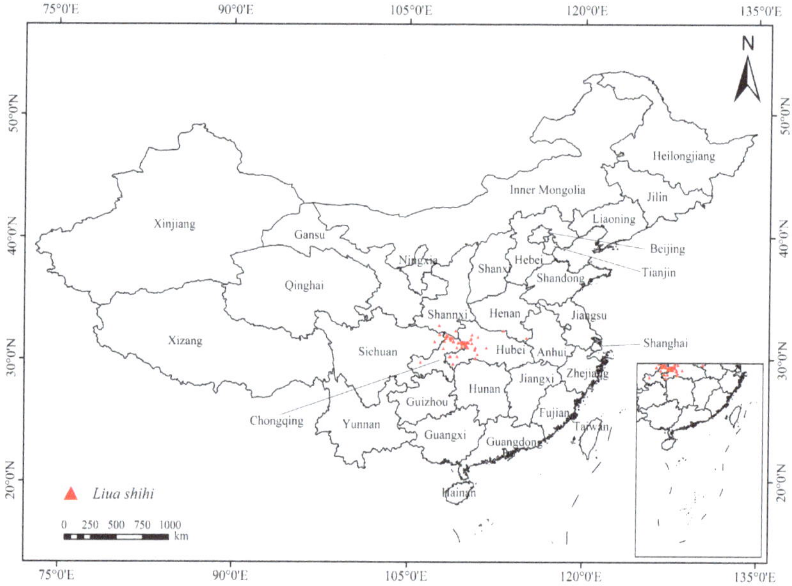

Figure 1. Distribution points of *L. shihi* (red triangles; after excluding autocorrelation).

2.3. Environmental Variables

Environmental variables were selected based on previous studies demonstrating that they may potentially affect the distribution of amphibians (e.g., [13,31,32]). These variables can be divided into five categories, including bioclimate, meteorology, vegetation, human disturbance, and topography. In total, we obtained 31 environmental variable raster layers (Table 1). Specifically, bioclimatic data were composed of 19 climate factors at a resolution of 2.5 min, which were derived from the WorldClim climate database (http://www.worldclim.org/, accessed on 26 May 2023) [33]. Meteorological factors were composed of ultraviolet-B (UV-B) radiation and cloud cover, which were derived from EarthEnv (https://www.earthenv.org/cloud, accessed on 3 June 2023) and Helmholtz Centre for Environmental Research (https://www.ufz.de/gluv, accessed on 3 June 2023), respectively. Vegetation data contained the percentages of tree coverage and vegetation types, which were from Global Map Data Archives (https://globalmaps.github.io/ptc.html, accessed on 3 June 2023) and Resources and Environmental Science Data Center (https://www.resdc.cn/, accessed on 3 June 2023), respectively. Human disturbance data were represented by the population density, which were downloaded from the Socioeconomic Data and Applications Centre (https://sedac.ciesin.columbia.edu/, accessed on 5 June 2023). Finally, topographic data included elevation, slope, and aspect at a resolution of 90 m, which were obtained from the Geospatial Data Cloud (https://www.gscloud.cn/, accessed on 5 June 2023). We unified their coordinate system as GCS_WGS_1984 and resampled them to obtain a consistent spatial resolution.

Table 1. Contribution and permutation importance of environmental variables in MaxEnt models.

Code	Envirnonmental Variable	Percentage Contribution (%)	Permutation Importance (%)
Bio14	Precipitation of driest month	29.7	6.9
Mseason	Cloud cover seasonal concentration	28.6	6.6
Veg	Vegetation type	15.8	0.5
UVB4	Mean UV-B of lowest month	5.6	13.2
Slope	Slope	5	5
Bio2	Mean diurnal range	3.5	0.6
Mspatial	Cloud cover spatial variability	3.5	1.6
Bio9	Mean temperature of driest quarter	3.1	14
Bio15	Precipitation seasonality	1.6	22
Bio4	Temperature seasonality	1.4	2.5
UVB3	Mean UV-B of highest month	1.2	25.6
Pdensity	Density of population	0.4	1.5
UVB1	Annual mean UV-B	0.4	0
Asp	Aspect	0	0.1
Plantcover	Density of trees on the ground	0	0

In order to reduce the influence of spatial correlation, environmental variables with high correlation but low contribution rate were removed before the model analyses [34]. Correlation analysis was performed using SPSS26.0 software. A Shapiro test was conducted using R software version 4.3.2 (https://www.r-project.org/, accessed on 7 November 2023) before the correlation analysis to identify the distribution of each variable [35]. Variables with a normal distribution were tested by Pearson correlations, and others were tested using Spearman correlations [36]. For the contribution rate, we performed a pre-simulation test in MaxEnt 3.4.4 with the distribution data of *L. shihi* and the 31 environmental variables. The contribution rate of the variables was tested using the jackknife test [37]. After that, variables with too high correlations ($|PCCs| \geq 0.8$; Figure 2) but a low contribution rate (<1%) were removed [38], and the rest of the variables were used for secondary simulation. Based on our results, 15 environmental variables were finally selected for constructing the final models, including five for bioclimate, five for meteorology, two for vegetation, one for human disturbance, and three for topography (Table 1).

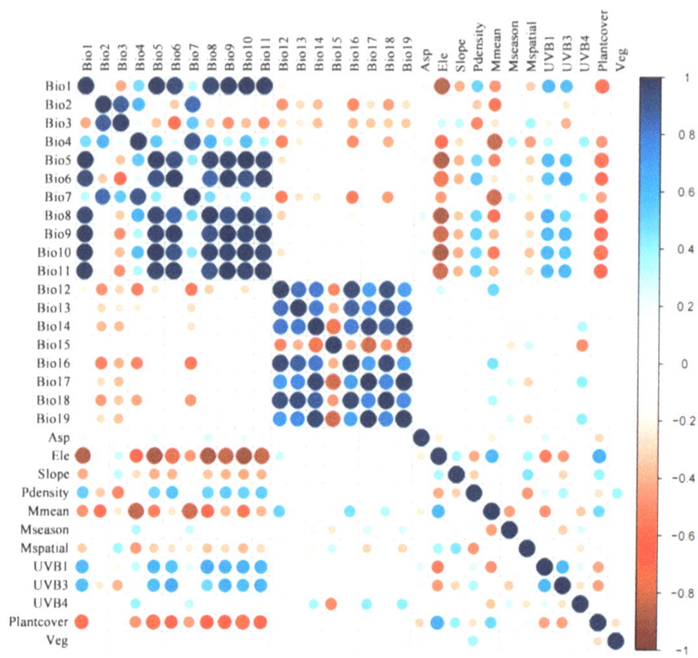

Figure 2. Correlation matrix between environmental variables. Bio1–19 are bioclimatic variables obtained from WorldClim website. Asp: aspect; Ele: elevation; Slope: slope; Pdensity: density of population; Mmean: mean annual cloud cover; Mseason: cloud cover seasonal concentration; Mspatial: cloud cover spatial variability; UVB1: annual mean UV-B; UVB3: mean UV-B of highest month; UVB4: mean UV-B of lowest month; Plantcover: density of trees on the ground; Veg: vegetation type. Positive correlations are displayed in blue and negative correlations in a red color. The color intensity and the size of the circle are proportional to the correlation coefficients.

The future climate data were obtained from the BCC-CSM2-MR climate system model [39]. These data contained two shared socioeconomic pathways (SSPs), SSP126 and SSP585, which are scenarios of global economic, demographic, and energy development in the future [40]. Specifically, SSP126 represents the combined effects of low vulnerability, mitigating stress, and radiative forcing. SSP585 represents the future socioeconomic path of high-emission, high-carbon (coal, oil, and natural gas) use [41]. In this study, two future climate scenarios (SSP126, SSP585) of three periods (2021–2040, 2041–2060, 2061–2080) were selected for projecting the future distribution area of *L. shihi*.

2.4. Parameter Optimization and Model Construction

There are five feature types in MaxEnt models, including linear (L), quadratic (Q), hinge (H), product (P), and threshold (T). For parameter adjustment, we computed the AICc values of the modeling parameters' regularization multiplier (RM) and feature combination (FC; the combination of the above five feature types) in R software using the ENMeval package [42]. In this study, we considered the range of RMs from 0.5 to 4.0 and selected six FC types (i.e., L, LQ, H, LQH, LQHP, and LQHPT). Then, we used the parameters corresponding to the minimum information criterion AICc value to construct the species distribution models [27].

The distribution data, environmental variables, and the optimized model parameters were input into MaxEnt3.4.4 software (New York, NY, USA, https://biodiversityinformatics.amnh.org/open_source/maxent/, accessed on 17 May 2023). The importance of environmental variables to the distribution of *L. shihi* was evaluated according to the relative contributions of environmental variables and the results of the jackknife test [40]. Twenty-five percent of the distribution data were randomly selected as the test set, while the rest were considered as the training set. The maximum number of background points was 10,000. A total of 10 runs were set for model construction, and the replicated run type was cross-validation.

We used the receiver operating characteristic (ROC) curve and the area under the ROC curve (AUC) to evaluate the accuracy of the model. The range of the AUC values was 0–1. A larger value indicates higher model accuracy, as well as higher credibility of the model. Models can be considered as having high prediction accuracy when the AUC value is greater than 0.8, and then the prediction results can be adopted [43]. An AUC value greater than 0.9 indicates that the prediction accuracy of the model is extremely high [23].

2.5. Parameter Optimization and Model Construction

We imported the average value of MaxEnt output results into ArcGIS 10.8 software and used a conversion tool to convert layers from asc format to raster data. The habitat suitability degree was divided into four levels, including high suitability area, moderate suitability area, low suitability area, and unsuitable area by natural breaks (Jenks) [44]. Finally, we calculated the area and proportion of suitable areas for each level. Moreover, we analyzed the change trend from current to future scenarios.

3. Results

3.1. Model Optimization and Accuracy Evaluation

For the current distribution models, the ΔAICc exhibited the lowest value when feature combination (FC) = LQHP and regulation multiplier (RM) = 2.5, indicating that the model was optimal with these parameters (Figure 3). This best model showed that the AUC value of the working curve of the subjects was 0.992 ± 0.004 (mean ± standard deviation), indicating the extremely high accuracy of the model prediction, thus the overfitting phenomenon could be effectively avoided (Figure 4).

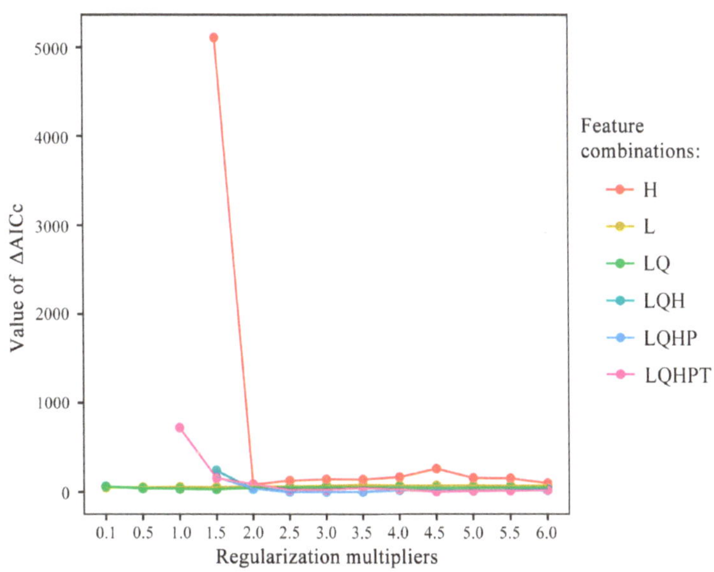

Figure 3. ΔAICc of the MaxEnt models under different regularization multipliers (RMs) and feature combinations (FCs).

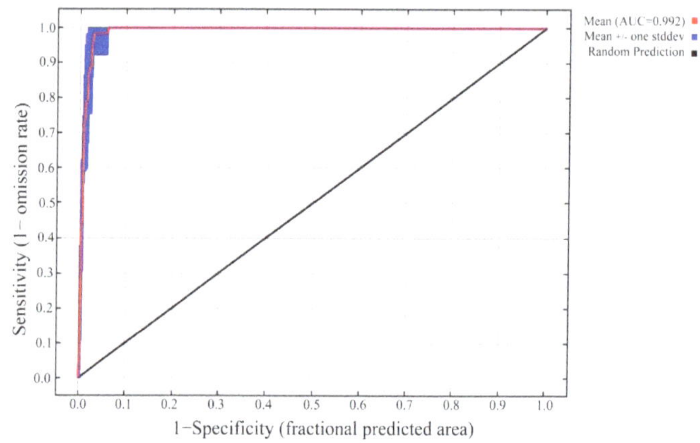

Figure 4. Receiver operating characteristic (ROC) curve and AUC value.

In terms of the future distribution models, the optimal parameters were FC = LQPH and RM = 2 for the SSP585 (2021–2040) scenario, while FC = LQ and RM = 0.5 for the rest of the scenarios. After applying the above parameters in MaxEnt to construct models, the results showed that the AUC values of the working curve of the subjects were all > 0.9.

3.2. *The Importance of Environmental Variables*

For the MaxEnt models constructed under the current climate scenario, the top five environmental variables accounted for 84.7% of the cumulative contribution, including precipitation of the driest month (Bio14, 29.7%), cloud cover seasonal concentration (Mseason, 28.6%), vegetation type (Veg, 15.8%), mean UV-B of the lowest month (UVB4, 5.6%), and slope (5%; Table 1). In terms of the permutation importance (the extent to which the model depends on the variable; [45]), the top five environmental variables were mean UV-B of the highest month (UVB3, 25.6%), precipitation seasonality (Bio15, 22%), mean temperature of the driest quarter (Bio9, 14%), mean UV-B of the lowest month (UVB4, 13.2%), and precipitation of the driest month (Bio14, 6.9%). For the jackknife test (Figure 5), the test gain value was 3.7 when considering all the environmental variables. When considering the variables individually, precipitation of the driest month (Bio14), vegetation type (Veg), precipitation seasonality (Bio15), cloud cover seasonal concentration (Mseason), and mean UV-B of the lowest month (UVB4) were the top five variables that exhibited the highest test gain values.

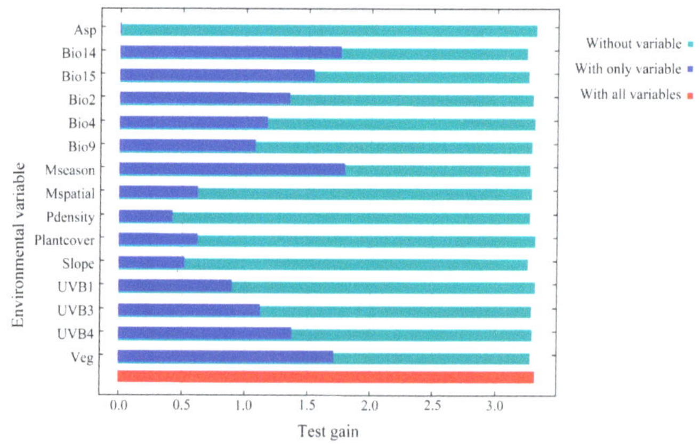

Figure 5. Jackknife of test gain for environmental variables in *L. shihi*.

3.3. Current Potential Suitable Habitats for *L. shihi*

Based on the results of MaxEnt models (Figure 6, Table 2), the potential suitable habitat for *L. shihi* was widely distributed in southwestern China, including Chongqing, Hubei, Sichuan, Shaanxi, Hunan, and Guizhou Provinces. In total, the suitable distribution area under current climate conditions for *L. shihi* was 45.61×10^4 km^2. Specifically, the high-suitability region was mainly concentrated in the Daba Mountains and Shennongjia National Park, which are located at the junction of Chongqing, Hubei, and Shaanxi Provinces. In addition, there were a small number of high-suitability regions scattered in central Sichuan Province. The size of the high-suitability area was 6.51×10^4 km^2, accounting for 14.3% of the total suitable habitat. The moderate-suitability region included the eastern part of Sichuan, southern Shaanxi, western Hubei, and eastern Chongqing Provinces, showing a ring shape, and the area was 9.77×10^4 km^2, accounting for 21.4% of the total suitable habitats. The low-suitability region was wrapped around the periphery of the moderate- and high-suitability areas, showing a strip shape. Moreover, the area was 29.31×10^4 km^2, accounting for 64.3% of the total suitable habitats.

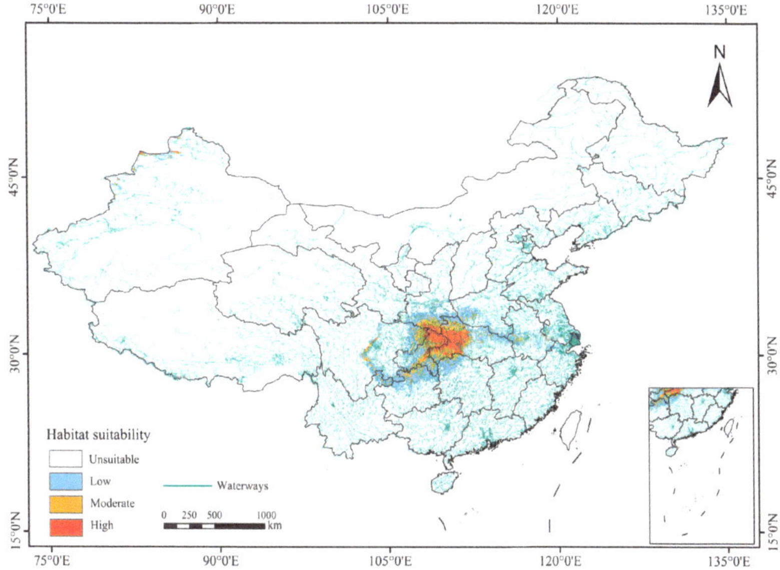

Figure 6. Potential suitable habitat for *L. shihi* under current climatic conditions.

Table 2. The area of suitable habitats in different periods (values in the brackets indicate the variations when compared with the area of the current period) ($\times 10^4$ km^2).

Grade	Current	2021–2040		2041–2060		2061–2080	
		SSP126	SSP585	SSP126	SSP585	SSP126	SSP585
Low	29.306	29.281 (−0.025)	26.800 (−2.506)	36.037 (6.731)	28.774 (−0.532)	30.802 (1.496)	35.229 (5.923)
Moderate	9.774	16.646 (6.872)	20.535 (10.761)	17.336 (7.562)	16.262 (6.488)	9.889 (0.115)	19.076 (9.302)
High	6.526	8.887 (2.361)	8.162 (1.636)	8.931 (−2.405)	8.457 (1.931)	8.137 (1.611)	9.307 (2.781)
Total	45.609	54.814 (9.205)	55.497 (9.888)	62.306 (16.697)	53.493 (7.884)	48.828 (3.219)	63.613 (18.004)

3.4. Future Distribution Patterns of the Suitable Habitats for *L. shihi*

In the 2021–2040 period under SSP126, the high-suitability area increased to 8.89×10^4 km^2, which was mainly contributed by the expansion in central Hubei Province and the junction of Chongqing and Guizhou Provinces. However, the high-suitability habitats in central Sichuan Province disappeared. The area of low-suitability habitats also decreased (244 km^2), associated with the loss in central Sichuan Province. In the 2041–2060 period under SSP126, the total area of suitable habitats increased to 62.31×10^4 km^2. Specifically, more high-suitability habitats occurred in Guizhou Province, despite the concentrated area in the Daba Mountains decreasing. The moderate-suitability habitats in central Sichuan Province disappeared, while there was no obvious change for low-suitability regions. In the 2061–2080 period under SSP126, a continuous decrease in high-suitability region was observed in the Daba Mountains, with the area being about 8.14×10^4 km^2. The moderate-suitability region in Guizhou Province was lost, while the low-suitability region can be only observed in central and south China (e.g., Henan, Hubei, and Guizhou Provinces; Figure 7, Table 2).

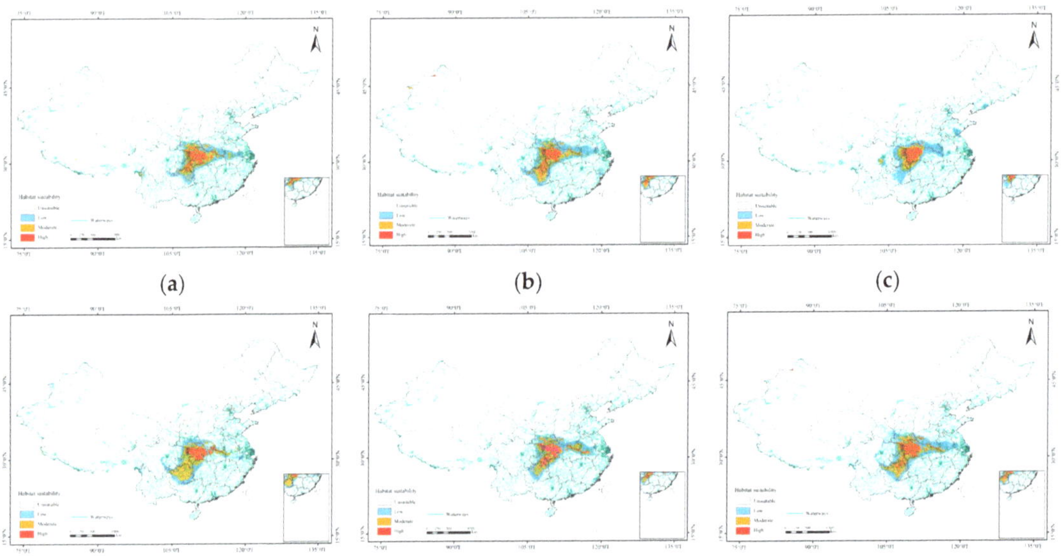

Figure 7. Potentially suitable climatic distribution of *L. shihi* under different climate change scenarios: (**a**) SSP126 from 2021–2040; (**b**) SSP126 from 2041–2060; (**c**) SSP126 from 2061–2082; (**d**) SSP585 from 2021–2040; (**e**) SSP585 from 2041–2060; (**f**) SSP585 from 2061–2080.

In the 2021–2040 period under SSP585, the high-suitability area increased to 8.16×10^4 km^2, which was mainly contributed by the expansion in the middle and north of Hubei Province. The moderate-suitability area expanded to the south, mainly located in the east of Chongqing Province and the north of Guizhou Province. The area of low-suitability habitats decreased (2.5×10^4 km^2), associated with the loss in the junction of Chongqing and Guizhou Provinces. In the 2041–2060 period under SSP585, the total area of suitable habitats increased to 53.50×10^4 km^2. Specifically, the high-suitability area expanded to Guizhou Province. The moderate-suitability area in the south of Henan Province expanded to 16.26×10^4 km^2, while the area of low-suitability habitats decreased (0.53×10^4 km^2). In the 2061–2080 period under SSP585, the highly suitable areas in the Daba Mountains were more concentrated. There was no obvious shift in the distribution pattern of the low-suitability area (Figure 7, Table 2).

4. Discussion

In the present study, we used optimized MaxEnt models to predict the distribution patterns of suitable habitats for *L. shihi* in China under current and future climate conditions. Based on the high AUC values, our models can be considered to have high accuracy in prediction [46]. Many previous studies only used the default parameters when conducting MaxEnt models (e.g., [47,48]). However, the default parameters will lead to over-fitting and high omission rates of the model. The "ENMeval" package developed by Muscarella et al. based on the R language has been widely used for optimizing the regularization multiplier (RM) and feature combination (FC) in the MaxEnt model to balance the complexity and avoid those defects [42]. Recently, increasing numbers of researchers have argued that MaxEnt models should be optimized before conducting predictions, as the default parameters may cause some bias [27,49]. Our results supported this claim as we found that the types of FC and the values of RM could change in different models. However, more theoretical work and field work are still needed to verify the effectiveness of parameter optimization in MaxEnt models.

As ectothermic animals, amphibians' growth and distribution are strongly affected by external environments, in particular the climate conditions [28,29]. This is especially true for salamanders, which are more sensitive to the change in climatic factors [13]. Among all the climatic variables, precipitation of the driest month was the most important one that determined the distribution of potential suitable habitats for *L. shihi*. Based on previous studies [17], the breeding period for this species is between March and April, associated with the dry season in the Daba Mountains. Therefore, sufficient precipitation can provide suitable spawning sites for *L. shihi* in montane streams, and permanent streams were critical for them to complete the life cycle [50]. High concentration of cloud cover, woody plant coverage, and low UV-B also were the main environmental variables that affect the distribution of suitable habitats for *L. shihi*. This is consistent with previous studies showing that ultraviolet light can cause oxidative stress, DNA damage, and egg death in salamanders [51,52]. Since an increase in ultraviolet rays may also lead to dramatic habitat reduction and connectivity fragmentation in other amphibian species that live in montane streams (e.g., spiny bellied frog: *Quasipaa boulengeri*; [31]), low UV-B could be an important factor driving the survival and distribution of aquatic amphibians. In the present study, it was found that high cloud density and forest coverage can effectively reduce the damage of ultraviolet rays to *L. shihi* [53], supporting the survival and distribution of this species. In addition, the influence of slope cannot be ignored, which was associated with the water flow rate and sunshine angle of the habitat, and salamanders usually preferred to select places with low water flow rate and sufficient light to grow and reproduce [54,55].

Our results showed that the high-suitability habitat for *L. shihi* was concentrated in the junction of Chongqing, Shaanxi, and Hebei Provinces, suggesting that this species may prefer some specific ecological conditions in this area [56]. Therefore, this region should be paid more attention for the protection of this species. For protected animals, a concentrated distribution pattern means they may be more easily threatened by climate change, and regional natural disasters and disease transmission will put the entire population at risk of extinction [57]. Interestingly, there were no distribution points recorded in some high-suitability regions (e.g., central Sichuan and central Chongqing Provinces), indicating that further field investigations can be carried out in these areas. In addition, a small number of existing distribution points were located in low-suitability or even non-suitable areas, suggesting that these populations should be paid more attention.

In the future, the total area of the suitable habitats for *L. shihi* will increase, although the main spatial distribution patterns did not change dramatically. This may be due to the unique climatic conditions (cool and humid all year round) in the Daba Mountains and Shennongjia National Park, which are climate transition regions between subtropical and northern warm temperate zones [58,59]. In two periods (2021–2040 and 2061–2080), the area of the suitable habitats under SSP126 was smaller than that of SSP585. This shows that the high-emission and high-carbon use scenario (SSP585) may cause an increase in the area of suitable habitat for *L. shihi*, which is similar to the finding of Wider et al. showing that the suitable range of the blue-spotted salamander (*Ambystoma laterale*) and the red-backed salamander (*P. cinereus*) increases with the increase in greenhouse gas concentration [60]. From the time point of view, the suitable area of *L. shihi* in the future is larger than the current results, and the high-suitability area under the SSP585 scenario will gradually increase with time. This increase may be the cumulative effect of climate change. This is contrary to previous studies showing that the area of suitable habitats of some other salamanders (e.g., leprous false brook salamander: *Pseudoeurycea leprosa*, streamside salamander: *A. barbouri*, and Cheat Mountain salamander: *P. nettingi*) will significantly decrease in the future [45,61]. We speculated that under this scenario, climate change has just reached the suitable conditions for *L. shihi* in some areas. It is worth noting that in the next three periods, the distribution range will be more concentrated. It indicates that the concentrated areas may have more important protection significance, as this region should be the refuge for *L. shihi* under future climate change.

5. Conclusions

In conclusion, the present study predicted the potential suitable habitats for *L. shihi* using a MaxEnt model with optimized parameters under current and future climate change scenarios for three time periods (SSP126 and SSP585). Our results indicated that precipitation of the driest month (Bio14), cloud cover seasonal concentration (Mseason), vegetation type (Veg), mean UV-B of the lowest month (UVB4), and slope are important environmental variables that have a great impact on the habitat suitability. The suitable habitats under the current situation are mainly distributed at the junction of Chongqing, Shaanxi, and Hubei Provinces. Under future climatic conditions, the total suitable area increased. The new suitable habitats were concentrated in the central part of Guizhou and Hubei Provinces. However, suitable habitats located in the central part of Sichuan and Chongqing Provinces were lost. The results of this study can help us better understand the distribution of *L. shihi* and can provide important information for determining the suitable areas of this species in China. Future studies can incorporate more species distribution models to better understand the effects of climate change on the distribution of suitable habitats for *L. shihi*.

Supplementary Materials: The following supporting information can be downloaded at: https://www.mdpi.com/article/10.3390/ani14091390/s1, Table S1: The distribution data of *Liua shihi* collected in published literatures; Table S2: The distribution data of *Liua shihi* collected on website; and Table S3: The distribution data of *Liua shihi* collected from field survey.

Author Contributions: Conceptualization, J.T. and T.Z.; data curation, J.T., Y.H., W.Y., and J.J.; formal analysis, J.T. and Y.H.; methodology, J.T. and Y.H.; resources, J.T. and Y.H.; software, J.T. and Y.H.; funding acquisition, J.J., T.Z. and S.S.; writing—original draft, J.T.; supervision, T.Z. and S.S.; validation, T.Z. and S.S.; visualization, W.Y.; writing—review and editing, J.T., T.Z. and S.S. All authors have read and agreed to the published version of the manuscript.

Funding: This research was funded by the Project of Background Resources Survey in Shennongjia National Park (SNJNP2022007), the Open Project Fund of Hubei Provincial Key Laboratory for Conservation Biology of Shennongjia Snub-nosed Monkeys (SNJGKL2022007), and the Chongqing Aquatic Science and Technology Innovation Alliance Research Program (CQFTIU2022-18).

Institutional Review Board Statement: Animal experiments were performed following the Guide for Care and Use of Laboratory Animals approved by the Committee of Laboratory Animal Experimentation of Southwest University (SWU2011032201, 22 March 2011).

Informed Consent Statement: Not applicable.

Data Availability Statement: The datasets presented in this study are available in Table S1.

Acknowledgments: We thank Chunlin Zhao and Da Kang for their help in constructing the MaxEnt models. We also thank Da Kang, Tianpeng Huang, and Jun Jiang for their help in collecting the distribution data of *L. shihi*.

Conflicts of Interest: The authors declare no conflicts of interest.

References

1. Marzin, A.; Archaimbault, V.; Belliard, J.; Chauvin, C.; Delmas, F.; Pont, D. Ecological assessment of running waters: Do macrophytes, macroinvertebrates, diatoms and fish show similar responses to human pressures? *Ecol. Indic.* **2012**, *23*, 56–65. [CrossRef]
2. Pacifici, M.; Foden, W.B.; Visconti, P.; Watson, J.E.M.; Butchart, S.H.M.; Kovacs, K.M.; Scheffers, B.R.; Hole, D.G.; Martin, T.G.; Akcakaya, H.R.; et al. Assessing species vulnerability to climate change. *Nat. Clim. Change* **2015**, *5*, 215–225. [CrossRef]
3. Luedtke, J.A.; Chanson, J.; Neam, K.; Hobin, L.; Maciel, A.O.; Catenazzi, A.; Borzee, A.; Hamidy, A.; Aowphol, A.; Jean, A.; et al. Ongoing declines for the world's amphibians in the face of emerging threats. *Nature* **2023**, *622*, 308. [CrossRef] [PubMed]
4. Lemes, P.; Melo, A.S.; Loyola, R.D. Climate change threatens protected areas of the Atlantic Forest. *Biodivers. Conserv.* **2014**, *23*, 357–368. [CrossRef]
5. Banks, B.; Beebee, T.; Cooke, A.S. Conservation of the natterjack toad Bufo-calamita in britain over the period 1970–1990 in relation to site protection and other factors. *Biol. Conserv.* **1994**, *67*, 111–118. [CrossRef]
6. Loyola, R.D.; Lemes, P.; Brum, F.T.; Provete, D.B.; Duarte, L.D.S. Clade-specific consequences of climate change to amphibians in Atlantic Forest protected areas. *Ecography* **2014**, *37*, 65–72. [CrossRef]
7. Basham, E.W.; Scheffers, B.R. Vertical stratification collapses under seasonal shifts in climate. *J. Biogeogr.* **2020**, *47*, 1888–1898. [CrossRef]
8. Nottingham, S.; Pelletier, T.A. The impact of climate change on western *Plethodon* salamanders' distribution. *Ecol. Evol.* **2021**, *11*, 9370–9384. [CrossRef] [PubMed]
9. Duan, R.; Kong, X.; Huang, M.; Varela, S.; Ji, X. The potential effects of climate change on amphibian distribution, range fragmentation and turnover in China. *Peerj* **2016**, *4*, e2185. [CrossRef]
10. Baskiera, S.; Gvozdik, L. Thermal independence of energy management in a tailed amphibian. *J. Vertebr. Biol.* **2020**, *69*, 20057.1-10. [CrossRef]
11. Barrett, K.; Guyer, C. Differential responses of amphibians and reptiles in riparian and stream habitats to land use disturbances in western Georgia, USA. *Biol. Conserv.* **2008**, *141*, 2290–2300. [CrossRef]
12. Ashrafzadeh, M.R.; Naghipour, A.A.; Haidarian, M.; Kusza, S.; Pilliod, D.S. Effects of climate change on habitat and connectivity for populations of a vulnerable, endemic salamander in Iran. *Glob. Ecol. Conserv.* **2019**, *19*, e00637. [CrossRef]
13. Zhang, P.; Dong, X.; Grenouillet, G.; Lek, S.; Zheng, Y.; Chang, J. Species range shifts in response to climate change and human pressure for the world's largest amphibian. *Sci. Total Environ.* **2020**, *735*, 139543. [CrossRef]
14. IUCN Red List of Threatened Species. Available online: https://www.iucnredlist.org/en (accessed on 21 June 2019).
15. Redlist of China's Biodiversity. Available online: https://www.mee.gov.cn/xxgk2018/xxgk/xxgk01/202305/W020230522536559098623.pdf (accessed on 16 July 2023).
16. McDonough, K. Amphibian Species of the World: An Online Reference (Version 6). *Ref. Rev.* **2014**, *28*, 32. [CrossRef]
17. Fei, L.; Ye, C.; Jiang, J. *Colorful Illustrations of Amphibians in China*, 1st ed.; Sichuan Publishing Group Sichuan Science and Technology Press: Chengdu, China, 2010; pp. 66–67. (In Chinese)
18. Ma, Q.; Wan, L.; Shi, S.; Wang, Z. Impact of Climate Change on the Distribution of Three Rare Salamanders (*Liua shihi*, *Pseudohynobius jinfo*, and *Tylototriton wenxianensis*) in Chongqing, China, and Their Conservation Implications. *Animals* **2024**, *14*, 672. [CrossRef] [PubMed]
19. Jimenez-Garcia, D.; Peterson, A.T. Climate change impact on endangered cloud forest tree species in Mexico. *Rev. Mex. Biodivers.* **2019**, *90*, e902781. [CrossRef]
20. Li, G.; Liu, C.; Liu, Y.; Yang, J.; Zhang, X.; Guo, K. Progress in theoretical research of species distribution models. *Acta Ecol. Sin.* **2013**, *33*, 4827–4835, (In Chinese with English Abstract).
21. Gama, M.; Crespo, D.; Dolbeth, M.; Anastacio, P. Predicting global habitat suitability for Corbicula fluminea using species distribution models: The importance of different environmental datasets. *Ecol. Model.* **2016**, *319*, 163–169. [CrossRef]
22. Padalia, I.; Srivastava, V.; Kushwaha, S.P.S. Modeling potential invasion range of alien invasive species, *Hyptis suaveolens* (L.) Poit. in India: Comparison of MaxEnt and GARP. *Ecol. Inform.* **2014**, *22*, 36–43. [CrossRef]
23. Phillips, S.J.; Anderson, R.P.; Schapire, R.E. Maximum entropy modeling of species geographic distributions. *Ecol. Model.* **2006**, *190*, 231–259. [CrossRef]
24. Phillips, S.J.; Dudik, M. Modeling of species distributions with Maxent:: New extensions and a comprehensive evaluation. *Ecography* **2008**, *31*, 161–175. [CrossRef]
25. Zhao, T.; Zhang, W.; Zhou, J.; Zhao, C.; Liu, X.; Liu, Z.; Shu, G.; Wang, S.; Li, C.; Xie, F.; et al. Niche divergence of evolutionarily significant units with implications for repopulation programs of the world's largest amphibians. *Sci. Total Environ.* **2020**, *738*, 140269. [CrossRef] [PubMed]
26. Zank, C.; Becker, F.G.; Abadie, M.; Baldo, D.; Maneyro, R.; Borges-Martins, M. Climate Change and the Distribution of Neotropical RedBellied Toads (*Melanophryniscus*, Anura, Amphibia): How to Prioritize Species and Populations? *PLoS ONE* **2014**, *9*, e94625. [CrossRef] [PubMed]
27. Zhu, G.; Qiao, H. Effect of the Maxent model's complexity on the prediction of species potential distributions. *Biodivers. Sci.* **2016**, *24*, 1189–1196, (In Chinese with English Abstract). [CrossRef]

28. Wu, J. The hazard and unsureness of reducing habitat ranges in response to climate warming for 91 amphibian species in China. *Acta Oecol* **2020**, *108*, 103640. [CrossRef]
29. Zhang, Z.; Mammola, S.; Liang, Z.; Capinha, C.; Wei, Q.; Wu, Y.; Zhou, J.; Wang, C. Future climate change will severely reduce habitat suitability of the Critically Endangered Chinese giant salamander. *Freshwater Biol.* **2020**, *65*, 971–980. [CrossRef]
30. Brown, J.L.; Bennett, J.R.; French, C.M. SDMtoolbox 2.0: The next generation Python-based GIS toolkit for landscape genetic, biogeographic and species distribution model analyses. *Peerj* **2017**, *5*, e4095. [CrossRef]
31. Yang, S.; Wang, X.; Hu, J. Mountain frog species losing out to climate change around the Sichuan Basin. *Sci. Total Environ.* **2022**, *806*, 150605. [CrossRef] [PubMed]
32. Jiang, D.; Zhao, X.; Lopez-Pujol, J.; Wang, Z.; Qu, Y.; Zhang, Y.; Zhang, T.; Li, D.; Jiang, K.; Wang, B.; et al. Effects of climate change and anthropogenic activity on ranges of vertebrate species endemic to the Qinghai-Tibet Plateau over 40 years. *Conserv. Biol.* **2023**, *37*, e14069. [CrossRef]
33. Hijmans, R.J.; Cameron, S.E.; Parra, J.L.; Jones, P.G.; Jarvis, A. Very high resolution interpolated climate surfaces for global land areas. *Int. J. Climatol.* **2005**, *25*, 1965–1978. [CrossRef]
34. Harisena, N.V.; Groen, T.A.; Toxopeus, A.G.; Naimi, B. When is variable importance estimation in species distribution modelling affected by spatial correlation? *Ecography* **2021**, *44*, 778–788. [CrossRef]
35. Aslam, M. Analysing Gray Cast Iron Data using a New Shapiro-Wilks test for Normality under Indeterminacy. *Int. J. Cast. Metal. Res.* **2021**, *34*, 1–5. [CrossRef]
36. Rovetta, A. Raiders of the Lost Correlation: A Guide on Using Pearson and Spearman Coefficients to Detect Hidden Correlations in Medical Sciences. *Cureus J. Med. Sci.* **2020**, *12*, e11794. [CrossRef] [PubMed]
37. Hosni, E.M.; Nasser, M.G.; Al-Ashaal, S.A.; Rady, M.H.; Kenawy, M.A. Modeling current and future global distribution of *Chrysomya bezziana* under changing climate. *Sci. Rep.* **2020**, *10*, 4947. [CrossRef] [PubMed]
38. Yang, X.; Kushwaha, S.P.S.; Saran, S.; Xu, J.; Roy, P.S. Maxent modeling for predicting the potential distribution of medicinal plant, *Justicia adhatoda* L. in Lesser Himalayan foothills. *Ecol. Eng.* **2013**, *51*, 83–87. [CrossRef]
39. Fick, S.E.; Hijmans, R.J. WorldClim 2: New 1-km spatial resolution climate surfaces for global land areas. *Int. J. Climatol.* **2017**, *37*, 4302–4315. [CrossRef]
40. Jiang, C.; Zhang, X.; Xie, W.; Wang, R.; Feng, C.; Ma, L.; Li, Q.; Yang, Q.; Wang, H. Predicting the potential distribution of the fall armyworm Spodoptera frugiperda (JE Smith) under climate change in China. *Glob. Ecol. Conserv.* **2022**, *33*, e01994. [CrossRef]
41. Riahi, K.; van Vuuren, D.P.; Kriegler, E.; Edmonds, J.; O'Neill, B.C.; Fujimori, S.; Bauer, N.; Calvin, K.; Dellink, R.; Fricko, O.; et al. The Shared Socioeconomic Pathways and their energy, land use, and greenhouse gas emissions implications: An overview. *Global Environ. Chang.* **2017**, *42*, 153–168. [CrossRef]
42. Muscarella, R.; Galante, P.J.; Soley-Guardia, M.; Boria, R.A.; Kass, J.M.; Uriarte, M.; Anderson, R.P. ENMeval: An R package for conducting spatially independent evaluations and estimating optimal model complexity for MAXENT ecological niche models. *Methods Ecol. Evol.* **2014**, *5*, 1198–1205. [CrossRef]
43. Fu, A.; Gao, E.; Tang, X.; Liu, Z.; Hu, F.; Zhan, Z.; Wang, J.; Luan, X. MaxEnt Modeling for Predicting the Potential Wintering Distribution of Eurasian Spoonbill (*Platalea leucorodia leucorodia*) under Climate Change in China. *Animals-Basel* **2023**, *13*, 856. [CrossRef]
44. Zhang, Y.; Hughes, A.C.; Zhao, Z.; Li, Z.; Qin, Y. Including climate change to predict the global suitable area of an invasive pest: Bactrocera correcta (Diptera: Tephritidae). *Glob. Ecol. Conserv.* **2022**, *34*, e02021. [CrossRef]
45. Sutton, W.B.; Barrett, K.; Moody, A.T.; Loftin, C.S.; DeMaynadier, P.G.; Nanjappa, P. Predicted Changes in Climatic Niche and Climate Refugia of Conservation Priority Salamander Species in the Northeastern United States. *Forests* **2015**, *6*, 1–26. [CrossRef]
46. Bartlow, A.W.; Giermakowski, J.T.; Painter, C.W.; Neville, P.; Schultz-Fellenz, E.S.; Crawford, B.M.; Lavadie-Bulnes, A.F.; Thompson, B.E.; Hathcock, C.D. Modeling the distribution of the endangered Jemez Mountains salamander (*Plethodon neomexicanus*) in relation to geology, topography, and climate. *Ecol. Evol.* **2022**, *12*, e9161. [CrossRef] [PubMed]
47. Li, R.; Xu, M.; Wong, M.H.G.; Qiu, S.; Li, X.; Ehrenfeld, D.; Li, D. Climate change threatens giant panda protection in the 21st century. *Biol. Conserv.* **2015**, *182*, 93–101. [CrossRef]
48. Stickley, S.F.; Fraterrigo, J.M. Microclimate species distribution models estimate lower levels of climate-related habitat loss for salamanders. *J. Nat. Conserv.* **2023**, *72*, 126333. [CrossRef]
49. Karuppaiah, V.; Maruthadurai, R.; Das, B.; Soumia, P.S.; Gadge, A.S.; Thangasamy, A.; Ramesh, S.V.; Shirsat, D.V.; Mahajan, V.; Krishna, H.; et al. Predicting the potential geographical distribution of onion thrips, *Thrips tabaci* in India based on climate change projections using MaxEnt. *Sci. Rep.* **2023**, *13*, 7934. [CrossRef] [PubMed]
50. Connette, G.M.; Osbourn, M.S.; Peterman, W.E. The Distribution of a Stream-breeding Salamander, *Desmognathus ocoee*, in Terrestrial Habitat Suggests the Ecological Importance of Low-order Streams. *Copeia* **2016**, *104*, 149–156. [CrossRef]
51. Lesser, M.P.; Turtle, S.L.; Farrell, J.H.; Walker, C.W. Exposure to ultraviolet radiation (290-400 nm) causes oxidative stress, DNA damage, and expression of *p53/p73* in laboratory experiments on embryos of the spotted salamander, *Ambystoma maculatum*. *Physiol. Biochem. Zool.* **2001**, *74*, 733–741. [CrossRef] [PubMed]
52. Blaustein, A.R.; Edmond, B.; Kiesecker, J.M.; Beatty, J.J.; Hokit, D.G. Ambient ultraviolet-radiation causes mortality in salamander eggs. *Ecol. Appl.* **1995**, *5*, 740–743. [CrossRef]
53. Xu, R.; Li, Y.; Teuling, A.J.; Zhao, L.; Spracklen, D.V.; Garcia-Carreras, L.; Meier, R.; Chen, L.; Zheng, Y.; Lin, H.; et al. Contrasting impacts of forests on cloud cover based on satellite observations. *Nat. Commun.* **2022**, *13*, 670. [CrossRef]

54. Huang, H.; Bu, R.; Xie, H.; Hou, S.; Wu, Z. Habitat selection by *Hynobius maoershanensis* during its breeding period. *Acta Ecol. Sin.* **2019**, *39*, 6443–6451, (In Chinese with English Abstract).
55. Zhang, Q.; Fang, R.; Deng, C.; Zhao, H.; Shen, M.H.; Wang, Q. Slope aspect effects on plant community characteristics and soil properties of alpine meadows on Eastern Qinghai-Tibetan plateau. *Ecol. Indic.* **2022**, *143*, 109400. [CrossRef]
56. Siemens, A.L.; Bogart, J.P.; Linton, J.E.; Norris, D.R. Predicting the occurrence of an endangered salamander in a highly urbanized landscape. *Endanger. Species Res.* **2023**, *52*, 81–95. [CrossRef]
57. Bakare, A.G.; Kour, G.; Akter, M.; Iji, P.A. Impact of climate change on sustainable livestock production and existence of wildlife and marine species in the South Pacific island countries: A review. *Int. J. Biometeorol.* **2020**, *64*, 1409–1421. [CrossRef] [PubMed]
58. Deng, Z.; Cao, J. Incorporating ecosystem services into functional zoning and adaptive management of natural protected areas as case study of the Shennongjia region in China. *Sci. Rep.* **2023**, *13*, 18870. [CrossRef] [PubMed]
59. Yao, Y.; Hu, Y.; Kou, Z.; Zhang, B. Spatial patterns of *Pinus tabulaeformis* and *Pinus massoniana* forests in Qinling-Daba Mountains and the boundary of subtropical and warm temperate zones. *J. Geogr. Sci.* **2020**, *30*, 1523–1533. [CrossRef]
60. Widmer, B.W.; Gehring, T.M.; Heumann, B.W.; Nicholson, K.E. Climate change and range restriction of common salamanders in eastern Canada and the United States. *J. Wildlife Manage* **2022**, *86*, e22235. [CrossRef]
61. Parra-Olea, G.; Martínez-Meyer, E.; de León, G. Forecasting climate change effects on salamander distribution in the highlands of central Mexico. *Biotropica* **2005**, *37*, 202–208. [CrossRef]

animals

MDPI

Article

Mitochondrial Variation of Bottlenose Dolphins (*Tursiops truncatus*) from the Canary Islands Suggests a Key Population for Conservation with High Connectivity within the North-East Atlantic Ocean

Daniel A. Gómez-Lobo [1,2], Agustín P. Monteoliva [2], Antonio Fernandez [3], Manuel Arbelo [3], Jesús de la Fuente [3], Mónica Pérez-Gil [4], Nuria Varo-Cruz [4], Antonella Servidio [4], Enrique Pérez-Gil [4], Yaisel J. Borrell [1] and Laura Miralles [1,2,*]

1 Department of Functional Biology, University of Oviedo, 33006 Oviedo, Spain; dagomezlobo@uc.cl (D.A.G.-L.); borrellyaisel@uniovi.es (Y.J.B.)
2 Department of Environmental Genetics, Ecohydros, 39600 Maliaño, Spain; apmonteoliva@ecohydros.com
3 Veterinary Histology and Pathology, Atlantic Center for Cetacean Research, Institute of Animal Health and Food Safety (IUSA), Veterinary School, University of Las Palmas de Gran Canaria, 35001 Las Palmas, Spain; antonio.fernandez@ulpgc.es (A.F.); manuel.arbelo@ulpgc.es (M.A.); jesus.delafuente@ulpgc.es (J.d.l.F.)
4 Cetaceans and Marine Research Institute of the Canary Islands (CEAMAR), 35509 Las Palmas, Spain; monica@ceamar.org (M.P.-G.); nuriavaro@hotmail.com (N.V.-C.); antonella@ceamar.org (A.S.); kikelanza@gmail.com (E.P.-G.)
* Correspondence: miralleslaura@uniovi.es

Simple Summary: The common bottlenose dolphin, *Tursiops truncatus*, is a worldwide cetacean species essential for marine ecosystems' health and balance. Understanding the genetic connectivity and structure of different populations is crucial for the correct management and conservation of a species, such as designing Special Areas of Conservation or Marine Protected Areas. In this study, we described the genetic composition of 49 bottlenose dolphins from the Canary Islands, which were previously unstudied, and compared them with individuals from the rest of the North-East Atlantic Ocean. The results showed that Canarian bottlenose dolphins have a remarkably diverse genetic composition, and this population is possibly part of a larger oceanic population in the North Atlantic. Therefore, the studied Special Areas of Conservation in the Canary Islands may correspond to a hotspot of genetic diversity and could be a strategic area for the conservation of the species.

Abstract: In recent decades, worldwide cetacean species have been protected, but they are still threatened. The bottlenose dolphin (*Tursiops truncatus*) is a vulnerable keystone species and a useful bioindicator of the health and balance of marine ecosystems in oceans all over the world. The genetic structure of the species is shaped by their niche specialization (along with other factors), leading to the classification of two ecotypes: coastal and pelagic. In this study, the genetic diversity, population structure, and ecotypes of bottlenose dolphins from the Canary Islands were assessed through the analysis of 49 new samples from biopsies and from stranded animals using the 636 bp portion of the mitochondrial control region and 343 individuals from databases (n = 392). The results reveal high genetic diversity in Canarian bottlenose dolphins (Hd = 0.969 and π = 0.0165) and the apparent lack of population genetic structure within this archipelago. High genetic structure (Fst, Φst) was found between the Canary Islands and coastal populations, while little to no structure was found with the pelagic populations. These results suggest that Canarian bottlenose dolphins are part of pelagic ecotype populations in the North Atlantic. The studied Special Areas of Conservation in the Canary Islands may correspond to a hotspot of genetic diversity of the species and could be a strategic area for the conservation of the oceanic ecotype of bottlenose dolphins.

Citation: Gómez-Lobo, D.A.; Monteoliva, A.P.; Fernandez, A.; Arbelo, M.; de la Fuente, J.; Pérez-Gil, M.; Varo-Cruz, N.; Servidio, A.; Pérez-Gil, E.; Borrell, Y.J.; et al. Mitochondrial Variation of Bottlenose Dolphins (*Tursiops truncatus*) from the Canary Islands Suggests a Key Population for Conservation with High Connectivity within the North-East Atlantic Ocean. *Animals* **2024**, *14*, 901. https://doi.org/10.3390/ani14060901

Academic Editor: Chunwang Li

Received: 1 February 2024
Revised: 8 March 2024
Accepted: 12 March 2024
Published: 14 March 2024

Keywords: *Tursiops truncatus*; Canary Islands; mitochondrial DNA; D-loop; special areas of conservation; conservation genetics; ecotypes

1. Introduction

The common bottlenose dolphin (*Tursiops truncatus*) is one of the most widely distributed cetacean species occurring in temperate and tropical waters worldwide [1]. As top predators, they are useful bioindicators of the health and status of marine ecosystems and play vital roles in maintaining the balance in such environments [2,3]. Thanks to their behavioral and ecological plasticity, bottlenose dolphins can inhabit a vast range of aquatic ecosystems, from deep oceanic waters to coastal estuarine ecosystems, even roaming into rivers [4,5]. This great ecological variability and the lack of apparent physical barriers to dispersal or gene flow in the marine environment make it challenging to define a population (stock) and its boundaries, which have important implications in both evolutionary and conservation biology.

The bottlenose dolphin is protected by the EU Habitats Directive 92/43/EEC. It is included in the Berne Convention as strictly protected fauna, and its coastal ecotype is present in the ACCOBAMS (Agreement on the Conservation of Cetaceans in the Black Sea, Mediterranean Sea and contiguous Atlantic area). In Spain, it is cataloged as Vulnerable (VU) in the National Catalog of Endangered Species in both peninsular waters (Order of 10 March 2000) and in those of the Canary Islands (Order of 9 June 1999). The bottlenose dolphin is also included as Vulnerable in the Red Book of Vertebrates, both in the waters of the European Union and in the Spanish Mediterranean. However, globally, the common bottlenose dolphin is cataloged as Least Concern (LC) in the IUCN Red List of Threatened Species [6], and different populations from distant geographical areas face different anthropogenic threats; therefore, such populations should be categorized and managed separately. For example, Mediterranean populations were classified as Vulnerable (VU) until 2021 [7], and currently, the Fiordland subpopulation in New Zealand is listed as Critically Endangered (CE) [8], raising special conservation concerns for small and resident coastal populations.

Coastal bottlenose dolphin populations are commonly found in shallow waters less than 40 m deep, while pelagic populations are observed in outer deeper oceanic waters (200 to 4000 m) [9,10], and several studies have noted differences in their distribution, diet, and skull morphology [11–15], leading to the idea of two different ecotypes. In addition, findings of significant genetic structure have reinforced this idea, with coastal populations presenting lower genetic diversity [16–23]. Site fidelity, along with resource specialization and different social and behavioral strategies, appears to be a strong driving force of genetic structuring in coastal resident bottlenose dolphins worldwide [12,15,16,19–21,23–26]. In the North Atlantic Ocean, pelagic populations show a highly diverse pattern with high levels of gene flow among extremely distant geographical regions, suggesting the existence of a single large panmictic oceanic population [16,18,20,22]. On the contrary, some coastal populations present fine-scale levels of genetic structure with low diversity [19,22,23,25,27], and even the recent extinction of a genetically isolated population (e.g., Humber estuary, UK) has occurred with no signs of repopulation so far [28]. Such contrasting patterns and the reduced population size of coastal bottlenose dolphins raise special concerns about the conservation of the species. Since coastal cetaceans could face more anthropogenic threats than oceanic ones [29], and their low effective population sizes might lead to a decrease in the adaptive potential to environmental changes [30,31], the identification of such threatened populations is crucial for the management of the species (e.g., designation of Special Areas of Conservation, SACs).

The Canary Islands is one of the major four archipelagos (Azores, Madeira, Canaries, and Cape Verde) within the Macaronesian region. This region is characterized by complex geomorphology, with several sea mountains, volcanic activity, and a rugged coastline [32].

Its bathymetry is typical of oceanic islands, rapidly reaching depths of 200 m near the coast (Figure 1). Many cetacean species inhabit and roam these oceanic waters, representing not only a hotspot of cetacean abundance and diversity [33] but also an important biological corridor for these large marine mammals due to their high dispersal capacities. Bottlenose dolphin populations observed in different SACs from the Canary Islands show high site fidelity patterns and are greater than populations of other archipelagos (e.g., Hawaii or Bahamas; [34]). Several individuals have been resighted off two or more islands and even in other archipelagos (Madeira), providing evidence of the long-distance movements (≈500 km) that these dolphins can undertake [34,35]. Nevertheless, to date, these populations remain unstudied in terms of genetic structure, connectivity with other regions, and ecotype assessment.

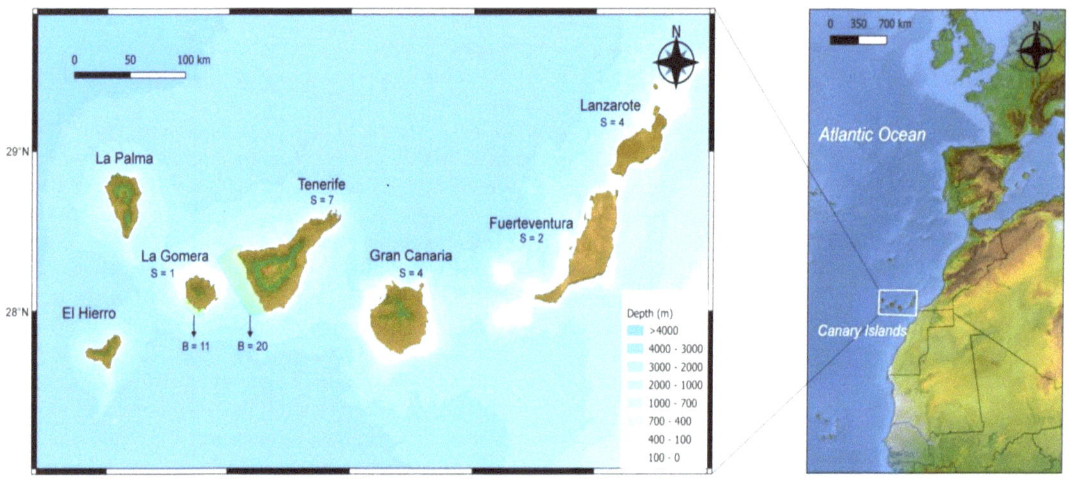

Figure 1. Map of the Canary Islands with sampling scheme of stranded (S) individuals and biopsy (B) samples. Areas in green highlight the Special Areas of Conservation (SACs) in La Gomera (ZEC-ES7020123) and Tenerife (ZEC-ES7020017) where biopsies were collected. Isobaths are plotted and denoted on a scale of blue.

In this context, the aims of this study were to determine the population structure within the Canary Islands and to assess the ecotype of bottlenose dolphins using molecular markers. Moreover, data relating to North-East Atlantic Ocean (NEAO) bottlenose dolphins were added to our analysis to study the phylogeographic relationships and possible connectivity with the other populations from the North-East Atlantic basin. Since no levels of genetic structure were found in other archipelagos (within and between Azores and Madeira) of the Macaronesian region [18], the high dispersal of Canarian bottlenose dolphins [34,35], and the broad connectivity of the pelagic ecotype in the North-East Atlantic [18,20], we hypothesized that none or negligible levels of genetic structure should be observed within the Canary Islands, clustering with the pelagic ecotype of the North-East Atlantic.

2. Materials and Methods

2.1. Sample Collection

A total of 49 bottlenose dolphin samples were collected from the Canary Islands from 2005 to 2022 (Table S1). Thirty-one biopsy samples were obtained from wild specimens in two different SACs (see studied area in Figure 1), the marine strips of Santiago-Valle Gran Rey in La Gomera island (ZEC-ES7020123) and Teno-Rasca in Tenerife Island (ZEC-ES7020017). The tissue size was 8 mm in diameter and length, and only adults were sampled. Photo identification was carried out simultaneously to spot individuals with site fidelity. Eighteen samples were obtained from stranded animals from five different locations (La Gomera, Tenerife, Gran Canaria, Fuerteventura, and Lanzarote). All stranding events were of single individuals (see Table S1), and the individuals were in the fresh decomposition stage (recently dead individuals), ensuring that death occurred near the coast. Tissue samples were either immediately preserved in ethanol or first frozen at −20 °C and later placed in ethanol for long-term storage.

2.2. Genetic Analysis: Population Structure and Diversity in the Canary Islands

DNA was extracted from the skin samples using DNeasy Blood and Tissue Kit (QIAGEN, Venlo, The Netherlands), following the manufacturer's instructions with modifications for small size samples (biopsies), such as longer lysis incubation (24 h), longer pre-elution incubation (5–10 min) and smaller elution volume (75 µL). All individuals were genetically sexed with the multiplexed SRY gene and ZFY/ZFX gene PCR [36]. A fragment of the mtDNA D-loop region was amplified using the primers described in the work of Dalebout et al. [37] following the protocol of Miralles et al. [38]. PCR sequencing in forward and reverse directions was carried out at Macrogen Inc. (Madrid, Spain) with a 3730XL DNA Analyzer (Applied Biosystems, Waltham, MA, USA). All of the obtained sequences were visualized, assembled, and checked for ambiguities in BioEdit, Version 7.0.5.3 [39]. The sequences were aligned and manually edited in BioEdit, producing a dataset of 49 sequences. Prior to molecular analyses, all of the sequences were corroborated using the Nucleotide BLAST tool (Basic Local Alignment Search tool, NCBI).

Genetic diversity and structure (*F*st, *Φ*st) indexes were assessed using ARLEQUIN, Version 3.5.1.2 [40]. *F*st may be an indicator of short-term or recent population processes, while *Φ*st may be an indicator of longer-term or older processes. Therefore, it is useful to calculate both types of indexes for any data set. Combining these statistics will enable more robust analyses of population structure than what is possible with only *F*st. Moreover, if they are different, it is possible that sample size and mutations have a larger influence on the results obtained. ARLEQUIN was also used to estimate the number of segregating sites (*S*), haplotypes (Nh), unique haplotypes (h), haplotype diversity (Hd), nucleotide diversity (π), and the average number of nucleotide differences between pairs of sequences (*k*). To determine if there were any deviations from the Wright–Fisher mutation-drift equilibrium due to population bottlenecks or expansions, Fu's Fs and Tajima's D indices were calculated in ARLEQUIN with their respective *p* values.

Phylogenetic relationships among the different haplotypes were inferred from a median-joining network constructed with PopArt, Version 1.7 [41,42], with the homoplasy parameter (ε) set to zero. To further visualize the possible genetic structure within the Canary Islands, non-metric Multidimensional Scaling (nMDS) analysis was conducted in PAST, Version 4.03 [43], using the mutation distribution of haplotypes and applying Tamura [44] for genetic distances and considering tolerable stress values <0.2 [45].

2.3. Genetic Analysis: Population Structure and Diversity in the NEAO

To study the phylogeographic relationships of the Canary Islands within the NEAO, the complete dataset of Louise et al. [20], except for the individuals of unknown origin, was downloaded from GenBank (n = 343). This dataset comprised four main groups containing several regions from the NEAO and Mediterranean Sea: the coastal south group (English Channel, Arcachon estuary, and South Galicia bottlenose dolphins), coastal north group (UK and Ireland resident or mobile coastal groups), pelagic Atlantic group (Azores archipelago and Bay of Biscay), and finally, the pelagic Mediterranean group (Gulf of Cadiz and Corsica) (see Louise et al. [20] for detailed description). The sequences were aligned using the Clustal W tool within *MEGA-X*, Version 10.0.5 [46], and trimmed to 636 pb to match our dataset. Since no polymorphism was present within the trimmed regions, no haplotype or information was lost, producing a final dataset of 392 sequences and defining 70 haplotypes, including the Canary Islands from this study. Genetic diversity and structure (*F*st, *Φ*st) indexes were calculated in ARLEQUIN in addition to an analysis of molecular variance (AMOVA) with 10,000 permutations. For Φst, the best-fit model of molecular evolution was determined using *MEGA-X*, which resulted in T92 +G + I [44], based on the Bayesian Information Criterion (BIC; [47]), with a gamma value of 0.46. Finally, a haplotype network was constructed using the median-joining algorithm in PopArt with ε set to zero.

3. Results

3.1. Population Structure and Genetic Diversity in the Canary Islands

In total, 49 mtDNA sequences of 636 bp were obtained, defining 28 haplotypes across the Canary Islands; 29 individuals corresponded to previously reported haplotypes, and 20 individuals presented 15 new unreported haplotypes (CAN1-CAN15, Table S1). New haplotypes were uploaded to GenBank under the accession numbers OQ656769-OQ656783.

Overall, mitochondrial haplotype and nucleotide diversities were high: Hd = 0.969 and π = 0.0165 (Table 1). Tenerife presented the highest number of haplotypes (Nh), unique haplotypes (h), and segregating sites (*S*), but it was also the location with the highest sample size (*n* = 27). Similar values of genetic variability in both largest samples in terms of π, Hd, and K were obtained despite the smaller sample size of La Gomera in comparison with Tenerife (Hd = 0.955 and 0.952, respectively) (Table 1). However, no significant population structure was found between these two localities (*F*st = 0.0008, Φst = 0.014; $p > 0.2$). In addition, no differences were found between biopsies and stranding samples (*F*st = 0.005, Φst = 0.049 $p > 0.05$), and all diversity indexes presented high similarity (Table S2, Supplementary Materials), discarding possible confounding effects between the two types of samples. Both Fu's Fs (Fs = −5.88, $p = 0.052$) and Tajima's D (D = 0.41, $p = 0.725$) were not significant.

Table 1. *Bottlenose dolphin* mitochondrial genetic diversity found in bottlenose dolphins from the Canary Islands, including sample size (*n*), segregating sites (*S*), number of haplotypes (Nh), number of unique haplotypes (h), haplotype diversity (Hd), nucleotide diversity (π), and average number of nucleotide differences (*k*). SD = standard deviation.

Populations	*n*	*S*	Nh	h	Hd (SD)	π (SD)	*k*
Tenerife	27	35	17	11	0.952 (0.025)	0.0166 (0.009)	10.571
La Gomera	12	31	9	2	0.955 (0.047)	0.0153 (0.008)	9.713
Lanzarote	4	24	4	3	1.000 (0.177)	0.0206 (0.014)	13.121
Gran Canaria	4	16	4	3	1.000 (0.177)	0.0149 (0.010)	9.491
Fuerteventura	2	9	2	1	1.000 (0.500)	0.0128 (0.014)	8.122
Total	49	43	28	/	0.969 (0.011)	0.0165 (0.009)	10.509

The median-joining network (Figure 2A) shows a highly diverse and reticulated pattern, with most individuals forming single haplotypes with multiple mutational steps. The two more distant haplotypes were separated by 47 bp. Only eight haplotypes were shared between individuals from different locations, where six were shared between Tenerife and La Gomera, one between Fuerteventura and La Gomera, and one between Gran Canaria and Lanzarote.

Non-metric Multidimensional Scaling analysis shows the lack of genetic structure within the Canary Islands since no clear clustering is observed, and all samples are scattered across the plot. The low stress value (0.09) indicates the validity of the analysis (Figure 3).

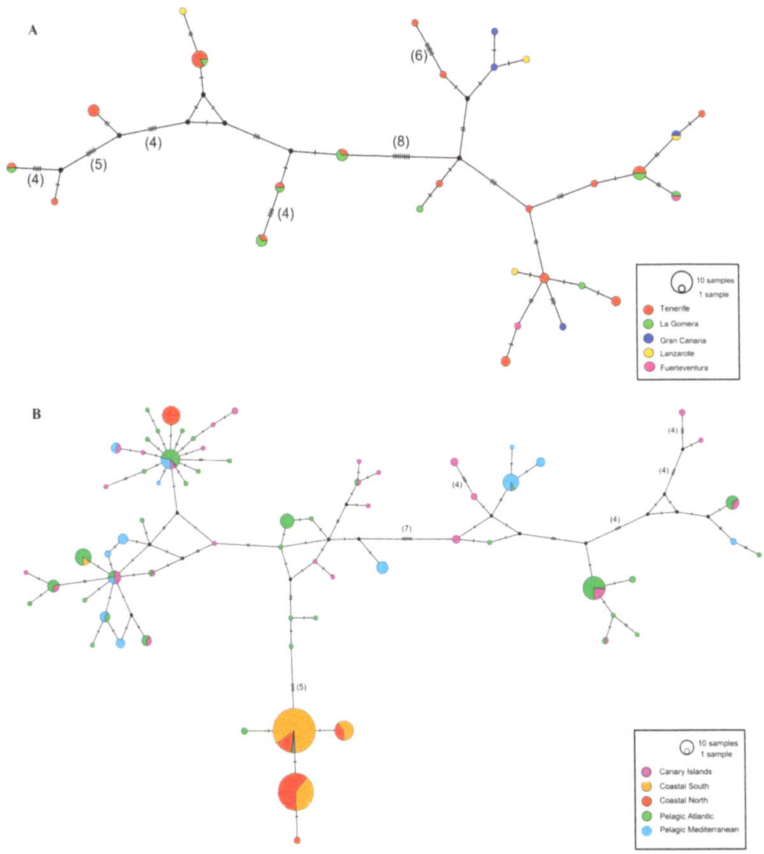

Figure 2. Median-joining network based on 636 bp of mtDNA haplotypes found in bottlenose dolphins from the Canary Islands (**A**), and from the North-East Atlantic and Mediterranean (**B**). Each circle represents a unique haplotype colored proportionally to the amount of individuals found in each location. Size of the circle is proportional to the haplotype frequencies. Black circles represent unsampled or extinct intermediate haplotypes. Hatch marks represent mutational steps. More than 3 mutational steps are denoted in parenthesis.

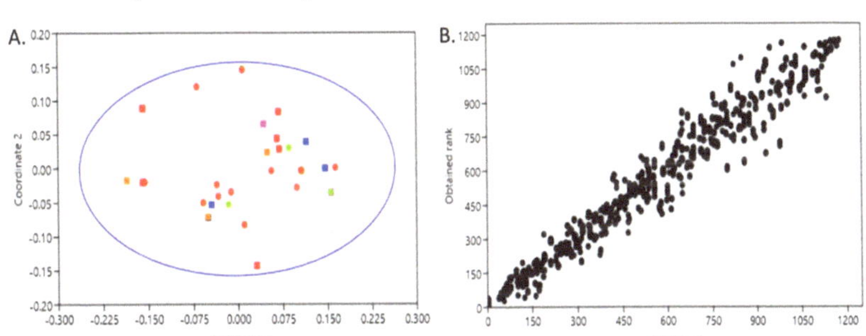

Figure 3. Non-metric Multidimensional Scaling of the five localities sampled in the Canary Islands. (**A**) Scatter plot showing necropsies (squares) and biopsies (circles) samples colored depending on the locality of collection. Red = Tenerife; Green = La Gomera; Blue = Gran Canaria; Orange = Lanzarote; Pink = Fuerteventura. Here, 95% ellipse is denoted with in a blue line. (**B**) Shepard plot. The stress value of the Shepard plot is 0.09.

3.2. Population Structure and Genetic Diversity in the Canary Islands

A dataset of 392 mtDNA sequences was obtained by combining this study (n = 49) and the work of Louis et al. [20] (n = 343), defining 70 haplotypes. With the inclusion of the highly diverse Canary population, the overall haplotypic diversity was augmented (Hd = 0.905) in relation to the values previously reported by Louis et al. [20] (Hd = 0.883). Out of the 70 haplotypes, 17 were private for the Canary Islands, being the second population with the most unique haplotypes after the Pelagic Atlantic samples (h = 25, Table 2). Despite having the smallest sample size, the Canary Islands presented the highest haplotypic diversity (Hd = 0.969) (Tables 1 and 2). Initially, an analysis of molecular variance (AMOVA) was tested with the Canary Islands as an independent group against pelagic and coastal populations, resulted in being not significant in the global structure (Φct = 0.377, p = 0.139; Fct = 0.075, p = 0.268), which is likely due to the large variability within the populations (Φst = 0.434, p < 0.0001; Fst = 0.219, p < 0.0001). A second test was performed, grouping the Canary Islands within the pelagic group, which also resulted in no significance (Φct = 0.436, p = 0.103; Fct = 0.136, p = 0.103). The pairwise comparisons of Fst and Φst obtained by Louis et al. [20] were replicated with the addition of the Canary Islands population, where the last was mainly differentiated from the coastal populations (Table 3). All of the comparisons among the Canary Islands and coastal populations were significant, with high Fst and Φst values (p < 0.001), while no structure was found when compared with the pelagic Atlantic populations (Fst and Φst values). However, one significant but low level of genetic structure (Fst = 0.057, p < 0.001) was found between the Canary Islands and the pelagic Mediterranean (but not the Φst value). It is the only comparison when Fst is significant but Φst is not.

Table 2. *Bottlenose dolphin* mitochondrial genetic diversity in the North-East Atlantic, including sample size (n), segregating sites (S), number of haplotypes (Nh), number of unique haplotypes (h), haplotype diversity (Hd), nucleotide diversity (π), and average number of nucleotide differences (k). Data from this work and from Louis et al. [20]. SD = standard deviation.

Populations	n	S	Nh	h	Hd (SD)	π (SD)	k
Canary Islands	49	43	28	17	0.969 (0.011)	0.0165 (0.009)	10.503
Coastal south	115	12	4	0	0.499 (0.044)	0.0014 (0.001)	0.889
Coastal north	76	13	5	2	0.667 (0.042)	0.0063 (0.003)	4.028
Pelagic Atlantic	101	41	38	25	0.929 (0.013)	0.0155 (0.007)	9.881
Pelagic Mediterranean	51	28	15	8	0.902 (0.022)	0.0137 (0.007)	8.680
Total	392	56	70	/	0.905 (0.009)	0.0140 (0.007)	8.894

Table 3. Population pairwise Fst (above diagonal) and Φst (below diagonal) values in terms of bottlenose dolphins from the Canary Islands.

Populations	Canary Islands	Coastal South	Coastal North	Pelagic Atlantic	Pelagic Mediterranean
Canary Islands	-	0.291 **	0.191 **	0.015	0.057 **
Coastal south	0.635 **	-	0.252 **	0.279 **	0.328 **
Coastal north	0.401 **	0.233 **	-	0.195 **	0.222 **
Pelagic Atlantic	0.004	0.541 **	0.349 **	-	0.071 **
Pelagic Mediterranean	0.040	0.671 **	0.446 **	0.056 **	-

** p < 0.01 after sequential Bonferroni correction.

A global haplotype network including all sequences from Louis et al. [20] was performed with the addition of the Canary Islands sequences (Figure 2B). All of the individuals clustered among the pelagic haplotypes in the upper part of the network, except for one stranded individual (CET0564), which showed the haplotype Ttrunc2, typical of the coastal bottlenose dolphins. It is worth mentioning that, in the upper-left side of the network, a coastal north haplotype (in red) from the UK and Ireland clustered with several pelagic haplotypes from the Atlantic (Azores and Bay of Biscay) and Mediterranean (Gulf of Cadiz and Corsica) as well as five Canarian haplotypes in a branch with a star-like pattern. In general, branches of the network were well defined in terms of several mutations between the closest haplotypes (e.g., four, five or seven positions).

4. Discussion

The genetic identification of natural populations is of crucial importance for the correct management and risk assessment of a species since small isolated populations are at increased risk of the effects of genetic drift and inbreeding [30], which can increase extinction probability. This is especially true in the case of bottlenose dolphins because coastal populations have been described to have low levels of genetic diversity, and even the extinction of an isolated population has been reported (e.g., Humber Estuary, UK) [28]. The results of this study would help to define key areas within the Macaronesian region for the management and long-term conservation of this relevant marine species protected in Europe under the EU Habitats Directive (92/43/CEE).

This study is the first to report the genetic structure of the population within the Canary Islands and to assess the ecotypes using molecular markers (i.e., mtDNA) and biopsies of free-ranging individuals. The overall mitochondrial haplotypic diversity found in this study (Hd = 0.969) is the highest reported in any previously studied bottlenose dolphin population in the North Atlantic [17,18,20,27]. Bottlenose dolphins from the Canary Islands were found to be remarkably diverse, with high genetic diversity indexes (Table 1). From a total of 49 samples, we found 28 haplotypes, meaning that more than half of the individuals sampled presented a different mtDNA sequence with multiple mutations between them (overall $k > 10$). Both Fu's Fs and Tajima's D were not significant, suggesting a relatively stable population size under mutation–drift equilibrium. No sign of genetic structuring among the islands was found in this work (Figure 2A). The haplotype network shows both patterns of high genetic diversity and the lack of a fine-scale structure, showing three major characteristics: (1) the presence of many haplotypes composed of single individuals, (2) multiple mutational steps among haplotypes, and (3) samples from different localities scattered across the network (Figure 2A). Additionally, non-metric Multidimensional Scaling analysis reinforced the evidence of a lack of genetic structure within the Canary Islands since no clear clustering is observed and all samples are scattered across the plot (stress value = 0.09; Figure 3). Although the lack of structure was expected, the small sample size of Lanzarote, Gran Canaria, and Fuerteventura, coupled with the absence of nuclear markers (microsatellites), might hinder the signals of a fine-scale genetic structure.

Previously, only Fernández et al. [27] reported genetic data of six stranded bottlenose dolphins from the Canary Islands. These authors found high nuclear and mitochondrial diversity. In our study, samples from the Canary Islands were grouped with the Azores, Basque Country, and Mainland Portugal, forming an offshore population in contrast with the genetically isolated population of Southern Galicia and the Sado estuary [27]. As reported in the Azores and Madeira archipelagos [18], along with photo identification data [34,35], our results support the hypothesis of the absence of a fine-scale genetic structure within the Canary Islands, with this population possibly grouping within the diverse large oceanic ecotypes.

The global haplotype network indicated that individuals from the Canary Islands are closely related to both pelagic Atlantic and pelagic Mediterranean populations by clustering within the upper pelagic mitochondrial lineage (Figure 2B). All of the individuals clustered among the pelagic haplotypes in the upper part of the network, except for one stranded individual (CET0564), showing haplotype Ttrunc2, which is typical of coastal populations. The Canary samples were scattered across the network, sharing ten and three haplotypes with pelagic Atlantic and pelagic Mediterranean populations, respectively, which could indicate current or historical gene flow, incomplete lineage sorting, or introgression [20]. In addition, despite having less than half of the sample size of the pelagic Atlantic population, the population of the Canary Islands possessed a remarkably high number of seventeen private haplotypes (i.e., haplotypes only found in that locality) in comparison to twenty-five (Table 2). The lack of genetic structure with pelagic populations, the deep bathymetry of the islands, and the high levels of haplotypic diversity support the hypothesis that bottlenose dolphins from the Canary Islands are part of a large oceanic population in the North-East Atlantic [18,20]. This connectivity among populations could be maintained by the high dispersal capacity of the species [35,48,49] and adaptations to deep oceanic environments [21]. However, one low but significant value in terms of genetic structure ($Fst = 0.057$, $p < 0.001$) was found between the Canary Islands and pelagic Mediterranean (but not the Φst value) (Table 3). It is known that the *F*st method is largely influenced by the presence of rare variants [50], while Φst statistics are not. Φst is derived from two different statistical distributions: the distribution of allele (haplotypes) frequencies among populations and the distribution of evolutionary distances among alleles [51]. When the significance of both markers differed, it is possible that samples size and/or mutation had a larger influence on the results obtained. After a population splits and until subpopulations have reached a stable equilibrium, *F*st is likely to increase first, indicating recent events. Only after new alleles have arisen and monophyletic clades of alleles have begun to arise in different subpopulations will Φst begin to increase substantially [51]. This way, it takes advantage of this additional information and provides greater insight into the patterns of relationships among the populations.

The results obtained in this work are in concordance with those obtained by Hildebrandt (unpub. data; [52]), in which Canarian bottlenose dolphins showed high diversity indexes, a lack of structure, and similarities with bottlenose dolphins from the North Atlantic Ocean. The Canary Islands are considered a hotspot of cetacean biodiversity [33], one of the most diverse places for cetaceans and the largest in Europe [53]. However, just three species dominated the sightings: bottlenose dolphins, pilot whales (*Globicephala macrorhynchus*) and spotted dolphins (*Stenella frontalis*) [53]. Comparing the results obtained here with these other two delphinid species, we observed the same lack of genetic structuring across the Canary Islands in spotted dolphins [54] but not in pilot whales [52]. On a broader scale, it has been described that spotted dolphins represent several distinct units in the Atlantic Ocean: Macaronesian group clustering, Canary Islands, Azores and Madeiran individuals [54].

Bottlenose dolphins are a highly endangered species due to coastal activities and fisheries. They are protected in Europe under the EU Habitats Directive (92/43/CEE), the Berne Convention and the ACCOBAMS, which requests the designation of SACs for their protection. Our results highlight the importance of the SACs in terms of managing and preserving bottlenose dolphins inhabiting the Canary Islands since this region seems to represent a hotspot of genetic diversity for a large pelagic population. The protection of these strategic areas could have positive impacts even in the outer parts of the marine reserve [23,55] thanks to the high connectivity of such pelagic ecotypes in the North-East Atlantic Ocean. This study provides baseline data for further investigations of the fine-scale genetic structure within the Canarian and Macaronesian region. Future studies that include nuclear markers (microsatellites) or genomics would provide higher-resolution information [56] on the connectivity among islands and detailed information for the future management of this protected species.

5. Conclusions

The analysis of 49 new samples, along with 343 individuals from databases, revealed a remarkable level of genetic diversity among Canarian bottlenose dolphins, as indicated by the highest reported mitochondrial haplotypic diversity in any North-East Atlantic bottlenose dolphin population. In line with our hypothesis, we found negligible levels of genetic structure within the Canary Islands, suggesting a cohesive population across the archipelago. The results align with the absence of fine-scale genetic structure reported in other oceanic archipelagos and support the hypothesis that Canarian bottlenose dolphins are part of a larger oceanic population in the North-East Atlantic.

Results from this research highlight the importance of Special Areas of Conservation (SACs) in the Canary Islands. The designation of SACs is crucial for preserving the genetic diversity of bottlenose dolphins, particularly considering their classification as a strategic area for the conservation of the oceanic ecotype. Additionally, we highlight the importance of incorporating nuclear markers (microsatellites) or SNPs to enhance the resolution of connectivity and provide detailed information for the ongoing conservation and management of this protected species.

Supplementary Materials: The following supporting information can be downloaded at: https://www.mdpi.com/article/10.3390/ani14060901/s1, Table S1: Detailed sampling data of bottlenose dolphins from the Canary Islands; Table S2: Comparison of genetic diversity indexes of Biopsies and Stranding samples.

Author Contributions: Conceptualization L.M., A.F., M.A. and J.d.l.F.; methodology D.A.G.-L., Y.J.B. and L.M.; software, D.A.G.-L.; formal analysis, D.A.G.-L. and L.M.; investigation, D.A.G.-L. and L.M.; sampling M.P.-G., N.V.-C., A.S., E.P.-G.; resources, A.P.M.; data curation, D.A.G.-L., Y.J.B. and L.M.; writing—original draft preparation, D.A.G.-L.; writing—review and editing, D.A.G.-L., Y.J.B. and L.M.; visualization, All authors; supervision, LM.; project administration, J.d.l.F.; funding acquisition, A.F. All authors have read and agreed to the published version of the manuscript.

Funding: This study is a result of the MARCET II Project (MAC2/4.6c/392), co-financed by the Interreg Madeira-Azores-Canarias (MAC) Territorial Cooperation Program 2014–2020, and coordinated by the University of Las Palmas de Gran Canaria. Dr. Miralles held a Torres Quevedo grant from the Ministry of Innovation and Science of Spain (2018–2022) referenced PTQ2018-010019.

Institutional Review Board Statement: This work got a positive evaluation from the Ethics Committee "Comité Ético de Experimentación Animal de la Universidad de Las Palmas de Gran Canaria (CEEA-ULPGC)" under the reference expedient: OEBA-ULPGC 34/2020R1. This work also got the governmental authorization from the "Ministerio para la Transición Ecológica y el Reto Demográfico" referenced: SGPM/BDM/AUTSPP/15/2020 and DGBBD/BDM/AUTSPP/19/2021.

Informed Consent Statement: Not applicable.

Data Availability Statement: Data are contained within the article and Supplementary Materials. New haplotypes were accessible in GenBank under the accession numbers OQ656769-OQ656783.

Acknowledgments: We thank Olga Filatova for her disposition and teaching of QGIS used for map building. Additionally, thanks to Tomás Fortuño for his advice with QGIS functions.

Conflicts of Interest: The authors declare no conflicts of interest.

References

1. Wells, R.S.; Scott, M.D. Common Bottlenose Dolphin: *Tursiops truncatus*. In *Encyclopedia of Marine Mammals*, 2nd ed.; Perrin, W.F., Würsig, B., Thewissen, J.G.M., Eds.; Academic Press: London, UK, 2009; pp. 249–255. ISBN 978-0-12-373553-9.
2. Wells, R.S.; Rhinehart, H.L.; Hansen, L.J.; Sweeney, J.C.; Townsend, F.I.; Stone, R.; Casper, D.R.; Scott, M.D.; Hohn, A.A.; Rowles, T.K. Bottlenose Dolphins as Marine Ecosystem Sentinels: Developing a Health Monitoring System. *EcoHealth* **2004**, *1*, 246–254. [CrossRef]
3. Torres, L.G.; Read, A.J.; Halpin, P. Fine-Scale Habitat Modeling of a Top Marine Predator: Do Prey Data Improve Predictive Capacity. *Ecol. Appl.* **2008**, *18*, 1702–1717. [CrossRef]
4. Ballance, L.T. Habitat Use Patterns and Ranges of the Bottlenose Dolphin in the Gulf of California, Mexico. *Mar. Mammal Sci.* **1992**, *8*, 262–274. [CrossRef]
5. Dos Santos, M.E.; Louro, S.; Couchinho, M.; Brito, C. Whistles of Bottlenose Dolphins (*Tursiops truncatus*) in the Sado Estuary, Portugal: Characteristics, Production Rates, and Long-Term Contour Stability. *Aquat. Mamm.* **2005**, *31*, 453–462. [CrossRef]
6. Wells, R.; Natoli, A.; Braulik, G. Tursiops truncatus (Errata Version Published in 2019). In *The IUCN Red List of Threatened Species*; IUCN: Cambridge, UK, 2019. [CrossRef]
7. Natoli, A.; Genov, T.; Kerem, D.; Gonzalvo, J.; Lauriano, G.; Holcer, D.; Labach, H.; Marsili, L.; Mazzariol, S.; Moura, A.E.; et al. Tursiops truncatus (Mediterranean Subpopulation) (Errata Version Published in 2022). In *The IUCN Red List of Threatened Species*; IUCN: Cambridge, UK, 2021.

8. Currey, R.J.C.; Dawson, S.M.; Slooten, E. An Approach for Regional Threat Assessment under IUCN Red List Criteria That Is Robust to Uncertainty: The Fiordland Bottlenose Dolphins Are Critically Endangered. *Biol. Conserv.* **2009**, *142*, 1570–1579. [CrossRef]
9. Certain, G.; Ridoux, V.; van Canneyt, O.; Bretagnolle, V. Delphinid Spatial Distribution and Abundance Estimates over the Shelf of the Bay of Biscay. *ICES J. Mar. Sci.* **2008**, *65*, 656–666. [CrossRef]
10. Hammond, P.S.; Macleod, K.; Berggren, P.; Borchers, D.L.; Burt, L.; Cañadas, A.; Desportes, G.; Donovan, G.P.; Gilles, A.; Gillespie, D.; et al. Cetacean Abundance and Distribution in European Atlantic Shelf Waters to Inform Conservation and Management. *Biol. Conserv.* **2013**, *164*, 107–122. [CrossRef]
11. Walker, W.A. *Geographical Variation in Morphology and Biology of Bottlenose Dolphins (Tursiops) in the Eastern North Pacific*; NOAA Administrative Report LJ-81-0003c.; National Marine Fisheries Service, Southwest Fisheries Center: La Jolla, CA, USA, 1981.
12. Mead, J.G.; Potter, C.W. Recognizing Two Populations off the Bottlenose Dolphin (*Tursiops truncatus*) of the Atlantic Coast of North America—Morphologic and Ecologic Considerations. In *International Marine Biology Research Institute: IBI Reports*; International Marine Biology Research Institute: Kamogawa, Japan, 1995; pp. 31–44.
13. Bearzi, M.; Saylan, C.A.; Hwang, A. Ecology and Comparison of Coastal and Offshore Bottlenose Dolphins (*Tursiops truncatus*) in California. *Mar. Freshw. Res.* **2009**, *60*, 584–593. [CrossRef]
14. Perrin, W.F.; Thieleking, J.L.; Walker, W.A.; Archer, F.I.; Robertson, K.M. Common Bottlenose Dolphins (*Tursiops truncatus*) in California Waters: Cranial Differentiation of Coastal and Offshore Ecotypes. *Mar. Mammal Sci.* **2011**, *27*, 769–792. [CrossRef]
15. Oudejans, M.G.; Visser, F.; Englund, A.; Rogan, E.; Ingram, S.N. Evidence for Distinct Coastal and Offshore Communities of Bottlenose Dolphins in the North East Atlantic. *PLoS ONE* **2015**, *10*, e0122668. [CrossRef]
16. Hoelzel, A.R.; Potter, C.W.; Best, P.B. Genetic Differentiation between Parapatric 'Nearshore' and 'Offshore' Populations of the Bottlenose Dolphin. *Proc. R. Soc. Lond. B Biol. Sci.* **1998**, *265*, 1177–1183. [CrossRef]
17. Natoli, A.; Birkun, A.; Aguilar, A.; Lopez, A.; Hoelzel, A.R. Habitat Structure and the Dispersal of Male and Female Bottlenose Dolphins (*Tursiops truncatus*). *Proc. R. Soc. B Biol. Sci.* **2005**, *272*, 1217–1226. [CrossRef]
18. Quérouil, S.; Silva, M.A.; Freitas, L.; Prieto, R.; Magalhães, S.; Dinis, A.; Alves, F.; Matos, J.A.; Mendonça, D.; Hammond, P.S.; et al. High Gene Flow in Oceanic Bottlenose Dolphins (*Tursiops truncatus*) of the North Atlantic. *Conserv. Genet.* **2007**, *8*, 1405–1419. [CrossRef]
19. Fruet, P.F.; Secchi, E.R.; Daura-Jorge, F.; Vermeulen, E.; Flores, P.A.C.; Simões-Lopes, P.C.; Genoves, R.C.; Laporta, P.; Di Tullio, J.C.; Freitas, T.R.O.; et al. Remarkably Low Genetic Diversity and Strong Population Structure in Common Bottlenose Dolphins (*Tursiops truncatus*) from Coastal Waters of the Southwestern Atlantic Ocean. *Conserv. Genet.* **2014**, *15*, 879–895. [CrossRef]
20. Louis, M.; Viricel, A.; Lucas, T.; Peltier, H.; Alfonsi, E.; Berrow, S.; Brownlow, A.; Covelo, P.; Dabin, W.; Deaville, R.; et al. Habitat-Driven Population Structure of Bottlenose Dolphins, *Tursiops truncatus*, in the North-East Atlantic. *Mol. Ecol.* **2014**, *23*, 857–874. [CrossRef]
21. Louis, M.; Fontaine, M.C.; Spitz, J.; Schlund, E.; Dabin, W.; Deaville, R.; Caurant, F.; Cherel, Y.; Guinet, C.; Simon-Bouhet, B. Ecological Opportunities and Specializations Shaped Genetic Divergence in a Highly Mobile Marine Top Predator. *Proc. R. Soc. B Biol. Sci.* **2014**, *281*, 20141558. [CrossRef] [PubMed]
22. Gaspari, S.; Scheinin, A.; Holcer, D.; Fortuna, C.; Natali, C.; Genov, T.; Frantzis, A.; Chelazzi, G.; Moura, A.E. Drivers of Population Structure of the Bottlenose Dolphin (*Tursiops truncatus*) in the Eastern Mediterranean Sea. *Evol. Biol.* **2015**, *42*, 177–190. [CrossRef]
23. Nykänen, M.; Louis, M.; Dillane, E.; Alfonsi, E.; Berrow, S.; O'Brien, J.; Brownlow, A.; Covelo, P.; Dabin, W.; Deaville, R.; et al. Fine-Scale Population Structure and Connectivity of Bottlenose Dolphins, *Tursiops truncatus*, in European Waters and Implications for Conservation. *Aquat. Conserv. Mar. Freshw. Ecosyst.* **2019**, *29*, 197–211. [CrossRef]
24. Wiszniewski, J.; Allen, S.J.; Möller, L.M. Social Cohesion in a Hierarchically Structured Embayment Population of Indo-Pacific Bottlenose Dolphins. *Anim. Behav.* **2009**, *77*, 1449–1457. [CrossRef]
25. Mirimin, L.; Miller, R.; Dillane, E.; Berrow, S.D.; Ingram, S.; Cross, T.F.; Rogan, E. Fine-Scale Population Genetic Structuring of Bottlenose Dolphins in Irish Coastal Waters. *Anim. Conserv.* **2011**, *14*, 342–353. [CrossRef]
26. Allen, S.J.; Bryant, K.A.; Kraus, R.H.S.; Loneragan, N.R.; Kopps, A.M.; Brown, A.M.; Gerber, L.; Krützen, M. Genetic Isolation between Coastal and Fishery-impacted, Offshore Bottlenose Dolphin (*Tursiops* spp.) Populations. *Mol. Ecol.* **2016**, *25*, 2735–2753. [CrossRef] [PubMed]
27. Fernández, R.; Santos, M.B.; Pierce, G.J.; Llavona, Á.; López, A.; Silva, M.A.; Ferreira, M.; Carrillo, M.; Cermeño, P.; Lens, S.; et al. Fine-Scale Genetic Structure of Bottlenose Dolphins, *Tursiops truncatus*, in Atlantic Coastal Waters of the Iberian Peninsula. *Hydrobiologia* **2011**, *670*, 111–125. [CrossRef]
28. Nichols, C.; Herman, J.; Gaggiotti, O.E.; Dobney, K.M.; Parsons, K.; Hoelzel, A.R. Genetic Isolation of a Now Extinct Population of Bottlenose Dolphins (*Tursiops truncatus*). *Proc. R. Soc. B Biol. Sci.* **2007**, *274*, 1611–1616. [CrossRef]
29. Crespo, E.A.; Notarbartolo di Sciara, G.; Reeves, R.R.; Smith, B.D. *Dolphins, Whales and Porpoises: 2002–2010 Conservation Action Plan for the World's Cetaceans*; IUCN: Gland, UK, 2003; ISBN 978-2-8317-0656-6.
30. Frankham, R.; Ballou, J.D.; Briscoe, D.A. Introduction to Conservation Genetics. Available online: https://www.cambridge.org/highereducation/books/introduction-to-conservation-genetics/696B4E558C93F7FBF9C33D6358EA7425 (accessed on 19 December 2023).

31. Hare, M.P.; Nunney, L.; Schwartz, M.K.; Ruzzante, D.E.; Burford, M.; Waples, R.S.; Ruegg, K.; Palstra, F. Understanding and Estimating Effective Population Size for Practical Application in Marine Species Management. *Conserv. Biol.* **2011**, *25*, 438–449. [CrossRef]
32. Troll, V.R.; Carracedo, J.C. Chapter 1—The Canary Islands: An Introduction. In *The Geology of the Canary Islands*; Troll, V.R., Carracedo, J.C., Eds.; Elsevier: Amsterdam, The Netherlands, 2016; pp. 1–41. ISBN 978-0-12-809663-5.
33. Correia, A.M.; Gil, Á.; Valente, R.F.; Rosso, M.; Sousa-Pinto, I.; Pierce, G.J. Distribution of Cetacean Species at a Large Scale—Connecting Continents with the Macaronesian Archipelagos in the Eastern North Atlantic. *Divers. Distrib.* **2020**, *26*, 1234–1247. [CrossRef]
34. Tobeña, M.; Escánez, A.; Rodríguez, Y.; López, C.; Ritter, F.; Aguilar, N. Inter-Island Movements of Common Bottlenose Dolphins *Tursiops truncatus* among the Canary Islands: Online Catalogues and Implications for Conservation and Management. *Afr. J. Mar. Sci.* **2014**, *36*, 137–141. [CrossRef]
35. Dinis, A.; Molina, C.; Tobeña, M.; Sambolino, A.; Hartman, K.; Fernandez, M.; Magalhães, S.; dos Santos, R.P.; Ritter, F.; Martín, V.; et al. Large-Scale Movements of Common Bottlenose Dolphins in the Atlantic: Dolphins with an International Courtyard. *PeerJ* **2021**, *9*, e11069. [CrossRef]
36. Rosel, P.E. PCR-based sex determination in Odontocete cetaceans. *Conserv. Gen.* **2003**, *4*, 647–649. [CrossRef]
37. Dalebout, M.L.; Robertson, K.M.; Frantzis, A.; Engelhaupt, D.; Mignucci-Giannoni, A.A.; Rosario-Delestre, R.J.; Baker, C.S. Worldwide Structure of mtDNA Diversity among Cuvier's Beaked Whales (*Ziphius cavirostris*): Implications for Threatened Populations. *Mol. Ecol.* **2005**, *14*, 3353–3371. [CrossRef]
38. Miralles, L.; Lens, S.; Rodríguez-Folgar, A.; Carrillo, M.; Martín, V.; Mikkelsen, B.; Garcia-Vazquez, E. Interspecific Introgression in Cetaceans: DNA Markers Reveal Post-F1 Status of a Pilot Whale. *PLoS ONE* **2013**, *8*, e69511. [CrossRef]
39. Hall, T.A. BioEdit: A User-Friendly Biological Sequence Alignment Editor and Analysis Program for Windows 95/98/NT. In *Nucleic Acids Symposium Series*; Oxford University Press: Oxford, UK, 1999; Volume 41, pp. 95–98.
40. Excoffier, L.; Laval, G.; Schneider, S. Arlequin (Version 3.0): An Integrated Software Package for Population Genetics Data Analysis. *Evol. Bioinform.* **2005**, *1*. [CrossRef]
41. Bandelt, H.J.; Forster, P.; Röhl, A. Median-Joining Networks for Inferring Intraspecific Phylogenies. *Mol. Biol. Evol.* **1999**, *16*, 37–48. [CrossRef] [PubMed]
42. Leigh, J.W.; Bryant, D. Popart: Full-Feature Software for Haplotype Network Construction. *Methods Ecol. Evol.* **2015**, *6*, 1110–1116. [CrossRef]
43. Hammer, O.; Harper, D.; Ryan, P. PAST: Paleontological Statistics Software Package for Education and Data Analysis. *Palaeontol. Electron.* **2001**, *4*, 1–9.
44. Tamura, K. Estimation of the Number of Nucleotide Substitutions When There Are Strong Transition-Transversion and G+C-Content Biases. *Mol. Biol. Evol.* **1992**, *9*, 678–687. [CrossRef] [PubMed]
45. Oksanen, J.; Blanchet, F.G.; Kindt, R.; Legendre, P.; Minchin, P.; O'Hara, R.; Simpson, G.; Solymos, P.; Stevens, M.; Wagner, H. *Vegan: Community Ecology Package. R Package Version. 2.0-10*; CRAN: Vienna, Austria, 2013.
46. Kumar, S.; Stecher, G.; Li, M.; Knyaz, C.; Tamura, K. MEGA X: Molecular Evolutionary Genetics Analysis across Computing Platforms. *Mol. Biol. Evol.* **2018**, *35*, 1547–1549. [CrossRef]
47. Guindon, S.; Gascuel, O. A Simple, Fast, and Accurate Algorithm to Estimate Large Phylogenies by Maximum Likelihood. *Syst. Biol.* **2003**, *52*, 696–704. [CrossRef] [PubMed]
48. Wells, R.S.; Rhinehart, H.L.; Cunningham, P.; Whaley, J.; Baran, M.; Koberna, C.; Costa, D.P. Long Distance Offshore Movements of Bottlenose Dolphins. *Mar. Mammal Sci.* **1999**, *15*, 1098–1114. [CrossRef]
49. Genov, T.; Železnik, J.; Bruno, C.; Ascheri, D.; Fontanesi, E.; Blasi, M.F. The Longest Recorded Movement of an Inshore Common Bottlenose Dolphin (*Tursiops truncatus*). *Mamm. Biol.* **2022**, *102*, 1469–1481. [CrossRef]
50. Bhatia, G.; Patterson, N.; Sankararaman, S.; Price, A.L. Estimating and Interpreting FST: The Impact of Rare Variants. *Genome Res.* **2013**, *23*, 1514–1521. [CrossRef]
51. Holsinger, K.E.; Weir, B.S. Genetics in Geographically Structured Populations: Defining, Estimating and Interpreting FST. *Nat. Rev. Genet.* **2009**, *10*, 639–650. [CrossRef] [PubMed]
52. Hildebrandt, S. Estructura Genética de las Poblaciones de Cetáceos del Archipiélago Canario: Secuenciación de la Región Control y los Genes COI y NADH5 del ADN Mitocondrial [Genetic Structure of the Cetacean Populations from the Canarian Archipelago: Sequencing of the Control Region and the COI and NADH5 Genes of the Mitochondrial DNA]. Ph.D. Dissertation, Universidad de Las Palmas de Gran Canaria, Las Palmas, Spain, 2002.
53. Herrera, I.; Carrillo, M.; Cosme de Esteban, M.; Haroun, R. Distribution of cetaceans in the canary islands (Northeast Atlantic ocean): Implications for the natura 2000 network and future conservation measures. *Front. Mar. Sci.* **2021**, *8*, 669790. [CrossRef]
54. Do Amaral, K.B.; Barragán-Barrera, D.C.; Mesa-Gutiérrez, R.A.; Farías-Curtidor, N.; Caballero Gaitán, S.J.; Méndez-Fernandez, P.; Oliveira Santos, M.C.; Rinaldi, C.; Rinaldi, R.; Siciliano, S.; et al. Seascape genetics of the Atlantic Spotted Dolphin (*Stenella frontalis*) based on mitochondrial DNA. *J. Hered.* **2021**, *112*, 646–662. [CrossRef]

55. Hooker, S.K.; Cañadas, A.; Hyrenbach, K.D.; Corrigan, C.; Polovina, J.J.; Reeves, R.R. Making Protected Area Networks Effective for Marine Top Predators. *Endanger. Species Res.* **2011**, *13*, 203–218. [CrossRef]
56. Allendorf, F.W.; Hohenlohe, P.A.; Luikart, G. Genomics and the Future of Conservation Genetics. *Nat. Rev. Genet.* **2010**, *11*, 697–709. [CrossRef]

animals

MDPI

Article

Habitat Suitability, Distribution Modelling and GAP Analysis of Przewalski's Gazelle Conservation

Dongni Liang [1,2] and Chunwang Li [1,2,*]

1 Key Laboratory of Animal Ecology and Conservation Biology, Institute of Zoology, Chinese Academy of Sciences, Beijing 100101, China; liangdongni20@mails.ucas.ac.cn
2 College of Life Sciences, University of Chinese Academy of Sciences, Beijing 100049, China
* Correspondence: licw@ioz.ac.cn

Simple Summary: The population of Przewalski's gazelle (*Procapra przewalskii*) has increased over the past decades, but it is still threatened by a variety of environmental factors and human disturbance. Most of the suitable habitats for Przewalski's gazelle are limited to the vicinity of Qinghai Lake. Moreover, most of the suitable habitat for Przewalski's gazelle is not included in the scope of the reserve. Thus, conservation translocation may be an effective way of protecting Przewalski's gazelle.

Abstract: Although the population of Przewalski's gazelle (*Procapra przewalskii*) has increased, this species is still threatened by a variety of risk factors, such as habitat loss and fragmentation, grassland fencing, grazing conflict, the segmentation of different populations, and declines in population genetic diversity. In order to determine the potential suitable habitat of Przewalski's gazelle and find a new suitable location for its conservation translocation, we used the MaxEnt model to predict the suitable habitats in Qinghai Province, Gansu Province, and the Ordos Plateau in Inner Mongolia and other regions with historical distribution records. On the basis of the MaxEnt model's prediction of the potential suitable habitat of Przewalski's gazelle, we used GAP analysis to determine the existing protection gaps and provide a new reference for the future protection of Przewalski's gazelle. We found that altitude, temperature, vegetation type, and distance from roads were the main environmental factors affecting the geographical distribution of Przewalski's gazelle. Most of the suitable habitat of Przewalski's gazelle is confined around Qinghai Lake. GAP analysis revealed that most of the suitable habitats of Przewalski's gazelle are not included in the established reserves, and Qinghai Lake National Nature Reserve only covers a small area around Qinghai Lake. The whole reserve only accounts for 7.11% of the area of the suitable habitat for Przewalski's gazelle and 15.79% of the area of the highly suitable habitat for Przewalski's gazelle. We suggest that conservation translocation for Przewalski's gazelle should be put on the agenda. It is necessary to consider reintroducing these gazelles into their potential suitable habitats as a feasible way of establishing new populations and saving this species.

Keywords: habitat suitability; MaxEnt; GAP analysis; reintroduction; conservation translocation

Citation: Liang, D.; Li, C. Habitat Suitability, Distribution Modelling and GAP Analysis of Przewalski's Gazelle Conservation. *Animals* **2024**, *14*, 149. https://doi.org/10.3390/ani14010149

Academic Editors: Mathew Crowther and Greg S. Baxter

Received: 24 September 2023
Revised: 14 December 2023
Accepted: 18 December 2023
Published: 2 January 2024

1. Introduction

A suitable habitat is crucial for the survival and reproduction of wild animals. At present, some of the most significant threats that wildlife face are the destruction and fragmentation of their habitats, the expansion of human activities and social and economic development, and infrastructure construction, such as roads and railways; these processes have had certain impacts on the activities and habitats of wildlife [1].

Przewalski's gazelle (*Procapra przewalskii*) is one of the most threatened species. It is an endangered ungulate endemic to the Qinghai–Tibet Plateau and the flagship species in the Qinghai Lake Basin. Przewalski's gazelle was once widely distributed in China in western Inner Mongolia, Qinghai, Gansu, and Ningxia [2,3]. But due to human population growth, economic growth, the development of animal husbandry, and the large-scale exploitation and utilization of grassland, the ecological environment of the gazelle distribution area has undergone great changes. The population size and distribution area of Przewalski's gazelle have been shrinking noticeably [4,5].

Recently, the population of Przewalski's gazelle exhibited a general increase due to protective measures, rising from approximately 200 individuals in 1994 to over 2700 in 2021 [6]. However, this species' distribution area has not been significantly expanded and remains confined to the vicinity of Qinghai Lake [7]. Despite some progress, threats to Przewalski's gazelle persist. Global warming is likely to induce changes in the annual average temperature and precipitation, vegetation coverage, vegetation types, and river flow rates in the Qinghai Lake area [8,9]. This may ultimately lead to a decline in habitat quality [8]. In addition, human activities are also increasingly affecting the survival of Przewalski's gazelle. Examples include increased tourism and overgrazing and the presence of grassland fences that divide the ownership of a grassland [10]. As a result, their habitat is continuously being compressed and fragmented [10,11].

It is therefore crucial to plan for the protection area of Przewalski's gazelle and consider establishing new populations by reintroducing them to their historical distribution range. Prior to this, an assessment of the species' suitable habitat is necessary, and Species Distribution Models (SDMs), particularly the Maximum Entropy Model (MaxEnt), are essential tools for studying species distribution and suitability. The MaxEnt model can calculate the distribution probability and possible distribution of species in a predicted area when the entropy is maximum [12,13]. Since the release of the MaxEnt 3.4.4 software product [14], due to its good performance and many advantages, its application in predicting suitable habitats for numerous species has steadily increased [15–17]. In addition, conducting further research on conservation gaps for Przewalski's gazelle is also crucial for this species' protection. GAP analysis, a geographical approach to conserving biological diversity, involves identifying factors such as vegetation types and species that are underrepresented or absent within a protected area system, with the aim of defining and addressing these gaps [18]. This approach has been widely applied in conservation projects across numerous countries and regions [19–22].

Hu and Jiang (2011) [9] analyzed the nationwide habitat suitability of Przewalski's gazelle. In addition to their study, by conducting field surveys and using the latest data, we focus on the prediction and analysis of the historical distribution range and conservation translocation of Przewalski's gazelle. We combined MaxEnt and GAP analysis for the first time to investigate and predict the suitable habitat for Przewalski's gazelle in its historical distribution areas, namely, Qinghai, Gansu, Ningxia, and west Inner Mongolia, to determine the suitable habitat and protection vacancy for the survival and reproduction of Przewalski's gazelle and provide a further basis and reference for the protection planning, management, and conservation translocation of Przewalski's gazelle.

2. Materials and Methods

2.1. Study Area

According to research records, Nikolay M. Przhevalsky was the first to collect a specimen of Przewalski's gazelle in the Ordos Plateau of China in 1875, and Przewalski's gazelle was once distributed in the Inner Mongolia, Ningxia, Gansu, and Qinghai regions of China [2,23]. From 1995 to 1997, several field investigations were carried out in the Ordos Plateau and its surrounding areas, as well as other areas corresponding to the historical distribution of Przewalski's gazelle. However, no evidence of living gazelles in Inner Mongolia, China, was found [24]. Therefore, Qinghai, Ningxia, Gansu, and the disputed Ordos Plateau area of Inner Mongolia were included in our research. The terrain and landforms in the study area include mountains, deserts, lakes, and farmlands.

Regarding the current distribution range of Przewalski's gazelle around Qinghai Lake (36°90′–37°56′ N, 97°50′–101°60′ E) (Figure 1), the average annual temperature is

0.3 °C−1.1 °C, and the annual precipitation is 350–450 mm [7]. The vegetation types include alpine shrub-steppe, alpine meadow, and desert shrub-grassland [7].

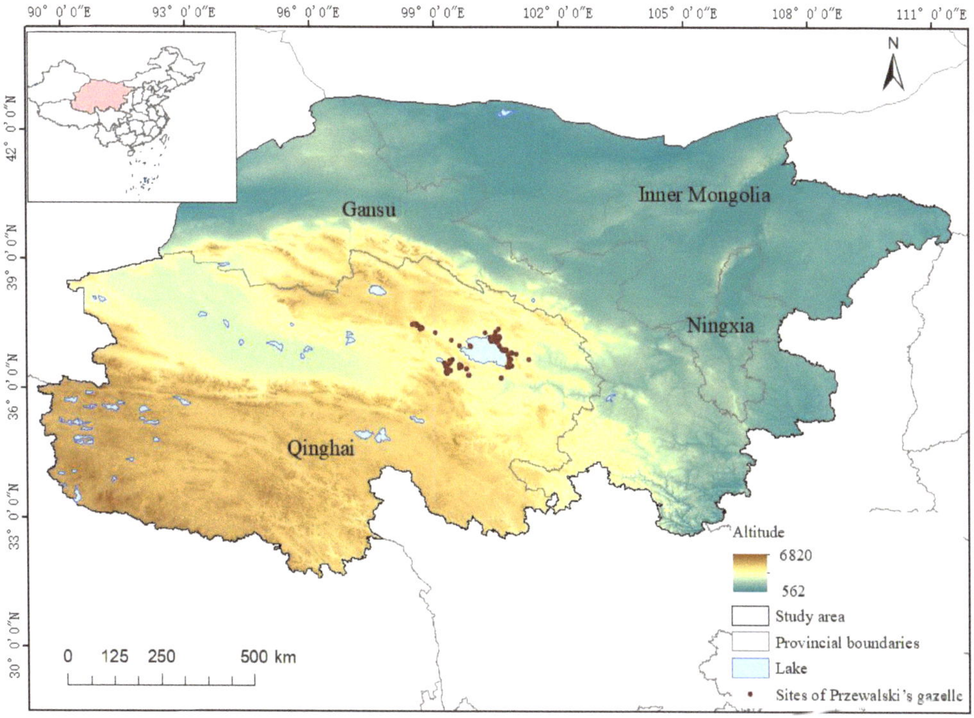

Figure 1. The current distribution sites of Przewalski's gazelles.

2.2. Data Collection

We compiled Przewalski's gazelle presence records from a field survey conducted in 2019 by our research group, data from publications [7], and an online database, the GBIF (the Global Biodiversity Information Facility, http://www.gbif.org, accessed on 21 March 2022) (Table 1). In total, we collected 194 sites of Przewalski's gazelle. After conducting filtering and screening using ENM Tools (v1.4), duplicate or similar sites were deleted. Finally, 136 sites (all from a field survey conducted in 2019 by our research group) were retained and used for modeling in CSV format according to the requirements of MaxEnt.

Table 1. Sources of Przewalski's gazelle presence records.

The Number of Records of Przewalski's Gazelle's Presence	Source	Years
13	GBIF	2016–2021
1	[7]	2018
180	This field survey	2019

The Geographic variables originated from the Geospatial Data Cloud Platform of Computer Network Information Center of Chinese Academy of Sciences (http://www.gscloud.cn, accessed on 3 April 2022). From this website, we obtained data on altitude, rivers, and other information regarding the study area, and we calculated the distance from the roads and rivers using ArcGIS 10.8.

We downloaded 19 climate data (Bio1-Bio19) from the Worldclim database (https://www.worldclim.org/data/worldclim21.html, accessed on 5 April 2022) for the contemporary time period (1970–2000), with a spatial resolution of 30″ and about 1 km [25].

From the Resource and Environmental Science Data Center of Institute of Geographical Sciences and Resources, Chinese Academy of Sciences (https://www.resdc.cn/Default.aspx, accessed on 15 April 2022), we obtained the spatial distribution data of China's monthly 1 km Normalized Difference Vegetation Index (NDVI) in 2020, which represents the vegetation coverage [26], and downloaded the spatial distribution data of China's 1:1,000,000 vegetation types on this website, which include grasslands, meadows, deserts, swamps, tundra, evergreen shrubs, deciduous shrubs, evergreen coniferous forests, deciduous coniferous forests, and mixed coniferous forests.

We also downloaded the road layer from the Geospatial Data Cloud Platform of Computer Network Information Center of Chinese Academy of Sciences (http://www.gscloud.cn, accessed on 19 April 2022) and used the Euclidean distance calculation in the toolbox of ArcGIS10.8 to measure distance from roads. From the Wildlife Conservation (WCS) and Center for International Earth Science Information Network (CIESIN) (https://sedac.ciesin.columbia.edu/, accessed on 22 April 2022) websites, we obtained the human footprint index. The population density data for 2020 were downloaded from LanScan website (https://landscan.ornl.gov/, accessed on 23 April 2022).

2.3. Environment Variable Filtering

In this survey, there were 28 environmental factors used in MaxEnt v3.4.4, including 19 climatic factors (Bio1–Bio19), 4 geographical factors (Slope, Aspect, Altitude, Dis_river), 2 vegetation factors (NDVI, Veg) and 3 anthropogenic factors (Footprint, Dis_road, Pop) (Table 2).

Table 2. Summary of environmental variables.

Types	Variables	Description	Units
Climatic factors	Bio1	Annual Mean Temperature	°C
	Bio2	Mean Diurnal Range (Mean of monthly (max temp-min temp))	°C
	Bio3	Isothermality (Bio2/Bio7) (×100)	%
	Bio4	Temperature Seasonality (standard deviation × 100)	%
	Bio5	Max Temperature of Warmest Month	°C
	Bio6	Min Temperature of Coldest Month	°C
	Bio7	Temperature Annual Range (BIO5-BIO6)	°C
	Bio8	Mean Temperature of Wettest Quarter	°C
	Bio9	Mean Temperature of Driest Quarter	°C
	Bio10	Mean Temperature of Warmest Quarter	°C
	Bio11	Mean Temperature of Coldest Quarter	°C
	Bio12	Annual Precipitation	mm
	Bio13	Precipitation in Wettest Month	mm
	Bio14	Precipitation in Driest Month	mm
	Bio15	Precipitation Seasonality (Coefficient of Variation)	%
	Bio16	Precipitation in Wettest Quarter	mm
	Bio17	Precipitation in Driest Quarter	mm
	Bio18	Precipitation in Warmest Quarter	mm
	Bio19	Precipitation in Coldest Quarter	mm
Vegetation factor	NDVI	Normalized Difference Vegetation Index	-
	Veg	Vegetation Type	
Geographical factors	Altitude	Altitude	m
	Dis_river	Distance from river	m
	Slope	Slope Degree	°
	Aspect	Slope Aspect	-
Anthropogenic factor	Dis_road	Distance from Road	m
	Footprint	Human Footprint Index	-
	Pop	Population density	people/km^2

In order to reduce the effects of interference and over fitting of multicollinearity among multiple factors on model analysis, all climate variables were pretested, and the contribution rates of all climate variables were analyzed using MaxEnt (Figure 2). In the preliminary analysis, to retain as many variables as possible, we chose 0.1% as the criterion for excluding the contribution rate [27]. That is, after comparing and ranking the contribution rates of various climate variables, the climate variables with contribution rates of less than 0.1% were eliminated. Then, we used ENM Tools software (v1.4) to test the correlation between environmental factors. If the absolute value of the correlation coefficient (r) between the two factors (|r|) was >0.8 [16,28], we removed one of the strongly correlated variables (Table 3). Finally, combined with the ranking of contribution rates and the results of correlation analysis, 7 climate variables were finally selected for the operation of the model, namely, Daily temperature range (Bio2), Seasonal change rate of temperature (Bio4), Lowest temperature in the coldest month (Bio6), Average temperature in the hottest season (Bio10), Precipitation in the driest month (Bio14), Seasonal change of precipitation (Bio15), and Precipitation in the hottest season (Bio18).

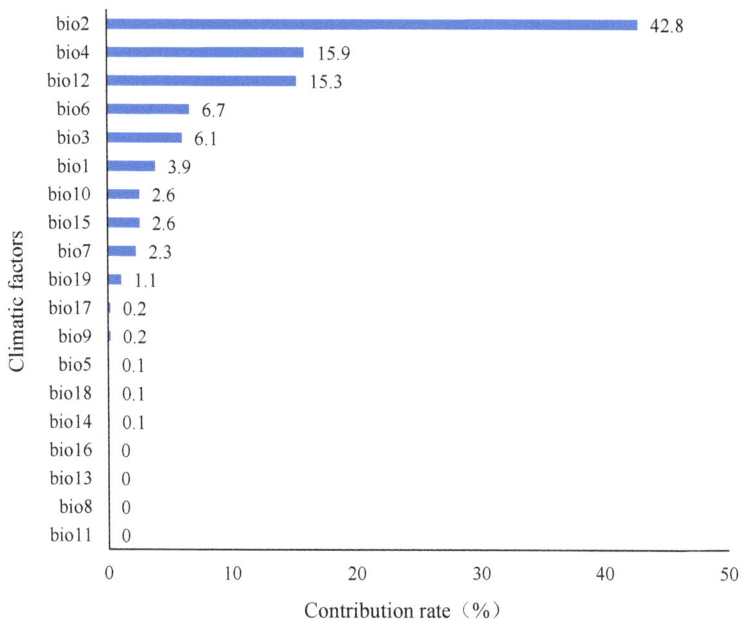

Figure 2. Ranking of contribution rates of climate variables.

As the environmental variable data were grid data that have different resolutions, we resampled variables in arcgis10.8 to 1 km resolution. The projection coordinates were set to WGS1984 UTM Zone 47 N and then converted to the ASCII format required by MaxEnt.

2.4. Model Parameter Optimization

Multiple studies have highlighted the importance of considering parameter optimization when utilizing MaxEnt for model analysis, as default parameters may not yield optimal results for different datasets [29,30]. The RM (regularization multiplier) and FC (feature combination multiplier) parameters in MaxEnt can be adjusted to optimize model analysis. There are five selectable features: Linear (L), Quadratic (Q), Hinge (H), Product (P), and Threshold (T). The default parameters in MaxEnt are RM = 1 and FC = LQHP. To evaluate model complexity, using ENMeval, a package in R, we calculated the values of AICc (Akaike Information Criterion with correction) as a measure [31]. Smaller AICc values indicate lower model complexity and greater excellence [32]. In this study, we utilized the "block" method in ENMeval and set RM to 0.5–4 (increasing by 0.5 each time), paired with five feature combinations (FC), namely, L, LQ, LQH, LQHP, and LQHPT. Finally, when the optimal model parameters were RM = 3 and FC = LQHPT, the AICC value was the lowest.

Table 3. Correlation coefficients (r) of climatic factors.

	bio19	bio1	bio2	bio3	bio4	bio5	bio6	bio7	bio8	bio9	bio10	bio11	bio12	bio13	bio14	bio15	bio16	bio17	bio18
bio19	0	0.59	0.71	0.12	0.45	0.29	0.69	0.57	0.26	0.66	0.40	0.66	0.84	0.71	0.97	0.60	0.73	0.98	0.63
bio1	0	0.00	0.59	0.15	0.30	0.81	0.91	0.42	0.84	0.88	0.89	0.91	0.64	0.57	0.62	0.53	0.58	0.62	0.55
bio2	0	0	0	0.29	0.47	0.23	0.74	0.67	0.33	0.57	0.40	0.67	0.85	0.80	0.76	0.46	0.81	0.75	0.78
bio3	0	0	0	0	0.67	0.51	0.10	0.49	0.48	0.21	0.49	0.16	0.05	0.08	0.18	0.09	0.04	0.16	0.02
bio4	0	0	0	0	0	0.30	0.66	0.97	0.22	0.64	0.17	0.66	0.55	0.46	0.43	0.30	0.51	0.44	0.50
bio5	0	0	0	0	0	0	0.50	0.18	0.96	0.50	0.98	0.51	0.24	0.21	0.33	0.36	0.19	0.32	0.16
bio6	0	0	0	0	0	0	0	0.76	0.56	0.95	0.63	0.99	0.77	0.68	0.72	0.56	0.71	0.72	0.67
bio7	0	0	0	0	0	0	0	0	0.09	0.70	0.02	0.75	0.70	0.61	0.57	0.37	0.66	0.57	0.64
bio8	0	0	0	0	0	0	0	0	0	0.52	0.97	0.56	0.32	0.33	0.33	0.29	0.31	0.32	0.30
bio9	0	0	0	0	0	0	0	0	0	0	0.60	0.97	0.66	0.55	0.65	0.58	0.58	0.65	0.53
bio10	0	0	0	0	0	0	0	0	0	0	0	0.63	0.40	0.37	0.44	0.40	0.36	0.44	0.33
bio11	0	0	0	0	0	0	0	0	0	0	0	0	0.74	0.65	0.68	0.54	0.68	0.68	0.64
bio12	0	0	0	0	0	0	0	0	0	0	0	0	0	0.95	0.86	0.43	0.97	0.86	0.94
bio13	0	0	0	0	0	0	0	0	0	0	0	0	0	0	0.73	0.21	0.99	0.73	0.98
bio14	0	0	0	0	0	0	0	0	0	0	0	0	0	0	0	0.60	0.75	0.99	0.66
bio15	0	0	0	0	0	0	0	0	0	0	0	0	0	0	0	0	0.27	0.61	0.20
bio16	0	0	0	0	0	0	0	0	0	0	0	0	0	0	0	0	0	0.75	0.98
bio17	0	0	0	0	0	0	0	0	0	0	0	0	0	0	0	0	0	0	0.66
bio18	0	0	0	0	0	0	0	0	0	0	0	0	0	0	0	0	0	0	0

Note: The different colors show different degrees of correlation. corresponds to $r \leq 0.10$, corresponds to $0.10 < r \leq 0.20$, corresponds to $0.20 < r \leq 0.30$, corresponds to $0.30 < r \leq 0.40$, corresponds to $0.40 < r \leq 0.50$, corresponds to $0.50 < r \leq 0.60$, corresponds to $0.60 < r \leq 0.70$, corresponds to $0.70 < r \leq 0.80$, corresponds to $0.80 < r \leq 0.90$, and corresponds to $r > 0.90$.

2.5. Model Operation

The distribution point data for Przewalski's gazelle and the 16 environmental variables screened were imported into MaxEnt, and the jackknife operation was used to test the contribution rates of various environmental factors [33]. The response curve function was also selected to determine the relationship between distribution probability and environmental factors. We randomly selected 25% of the sample distribution points as the model test data and 75% as the model training data and set the model so that it would repeat 10 times, with "Cloglog" as the output method.

2.6. Result Threshold Division

The distribution prediction data output by MaxEnt were imported into ArcGIS, converted to raster format, and reclassified. To convert data from the continuous suitability index maps to binary habitat and no-habitat maps, a probability threshold is needed to determine potential changes in habitat for species. There are six types of threshold results for the MaxEnt model: (1) minimum training presence threshold; (2) 10-percent training presence threshold; (3) equal training sensitivity and specificity threshold; (4) maximum training sensitivity plus specificity threshold; (5) balance training omission predicted area and threshold value threshold; and (6) equal entropy of threshold and original distributions threshold.

Some studies have proved that "Maximum training sensitivity plus specificity threshold" is the optimal threshold division standard with high accuracy [34]. Therefore, we used this threshold as the division threshold for suitable and non-suitable areas. In this study, our model analysis revealed that this threshold was 0.1985. Then, we used the reclassification tool in ArcGIS10.8 to divide the suitable areas of Przewalski's gazelle into three grades, namely, non-suitable areas (0–0.1985), suitable areas (0.1985–0.6), and highly suitable areas (0.6–0.99) [35], and calculate and analyze the ecological suitable area changes for Przewalski's gazelle.

2.7. Model Accuracy Evaluation

The accuracy and effectiveness of the prediction results yielded by the model were evaluated using the area under the curve (AUC) value under the receiver operating characteristic curve (ROC). The abscissa of the ROC test curve is 1-specificity, and the ordinate is 1-omission rate. The AUC value is the area enclosed by the ROC curve and abscissa, and its size can represent the accuracy of model prediction results. The larger the AUC value, the more the distribution of species deviates from the random distribution, and the better the prediction effect of the model. The evaluation result of AUC value is not affected by the threshold value, so the evaluation result is more reliable. If the AUC value is above 0.9, this means the accuracy of the model is high and the model is performing well.

2.8. GAP Analysis

Based on the obtained habitat suitability distribution map and map of Przewalski's gazelle reserve, the potential suitable habitat outside the reserve was determined to be the area of protection vacancy.

The data on national and provincial nature reserves were obtained from the Resource and Environmental Science Data Center of the Institute of Geographical Sciences and Resources, Chinese Academy of Sciences (https://www.resdc.cn/Default.aspx, accessed on 6 May 2022).

The data sets consisting of species distribution point, species distribution prediction results, and nature reserve layer used in the GAP analysis process were input into ArcGIS10.8. We utilized the overlay analysis function of ArcGIS to overlay the distribution map of Przewalski's gazelle obtained from the MaxEnt analysis with the existing protected area layer, aiming to identify the gaps in the protection of Przewalski's gazelle. These gaps refer to areas that are either unprotected or have weaker protection measures. The potential suitable habitat distribution map generated by MaxEnt was then subjected to Gap analysis with the Qinghai Lake protected area layer. The potential suitable habitats located outside the protected area boundary were considered the protection gap areas.

3. Results

3.1. MaxEnt Result Accuracy Analysis

The ROC analysis results showed that the average AUC value of the training set was 0.989, indicating that the potential ecologically suitable area of Przewalski's gazelle predicted by this model had high reliability (Figure 3).

3.2. Main Environmental Variables Affecting the Distribution of Przewalski's Gazelle

The importance of various environmental variables for the geographical distribution of Przewalski's gazelle was assessed using the knife-cutting method (Figure 4). The results showed that altitude, vegetation type, daily temperature range (Bio2), seasonal change rate for temperature (Bio4), lowest temperature in the coldest month (Bio6), and average temperature in the hottest season (Bio10) have a great impact on the geographical distribution of Przewalski's gazelle. It was shown that altitude, temperature, vegetation type, and road distance are the main environmental factors affecting the geographical distribution of Przewalski's gazelle. According to the response curve of the environmental variables (Figure 5), it can be gleaned that the range of suitable altitude for the survival of Przewalski's gazelle is about 3000~3400 m. Additionally, the Mean Daily Range (Bio2) for their habitat is between 9 and 12 °C, with the Min Temperature of the Coldest Month (Bio6) ranging between −23 and −20 °C. In terms of slope, the ideal range for this gazelle's habitat is 2~5°.

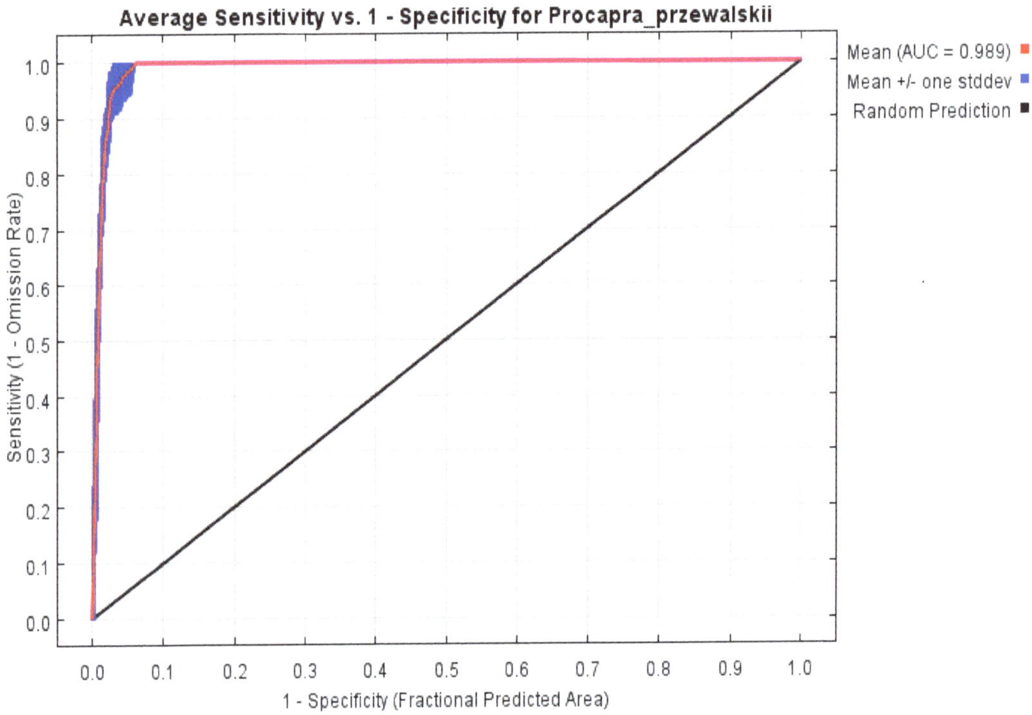

Figure 3. ROC curve of prediction results.

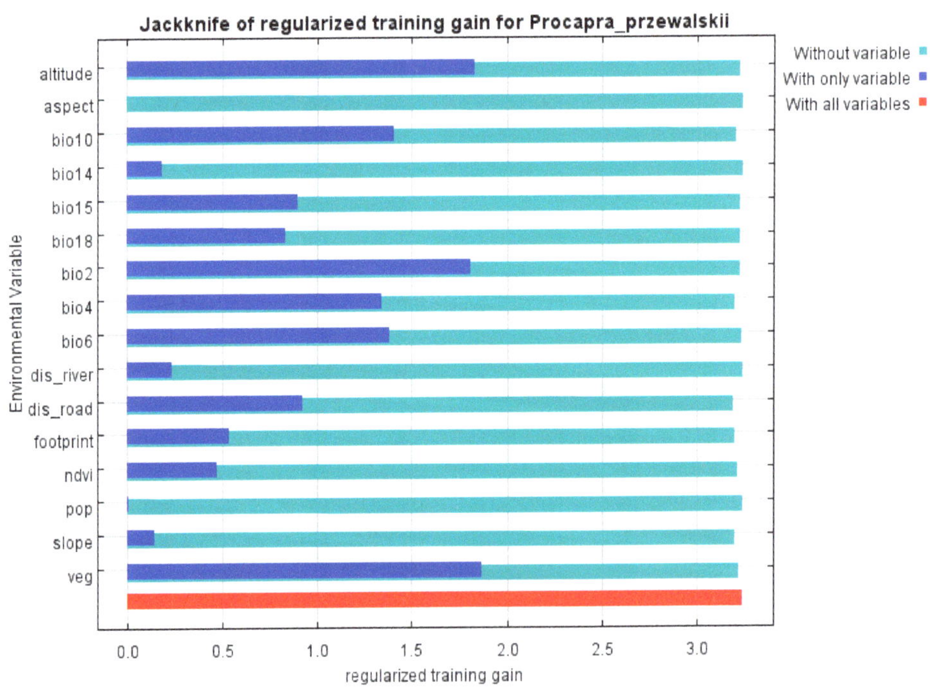

Figure 4. Importance of various environment variables determined using the knife-cutting method.

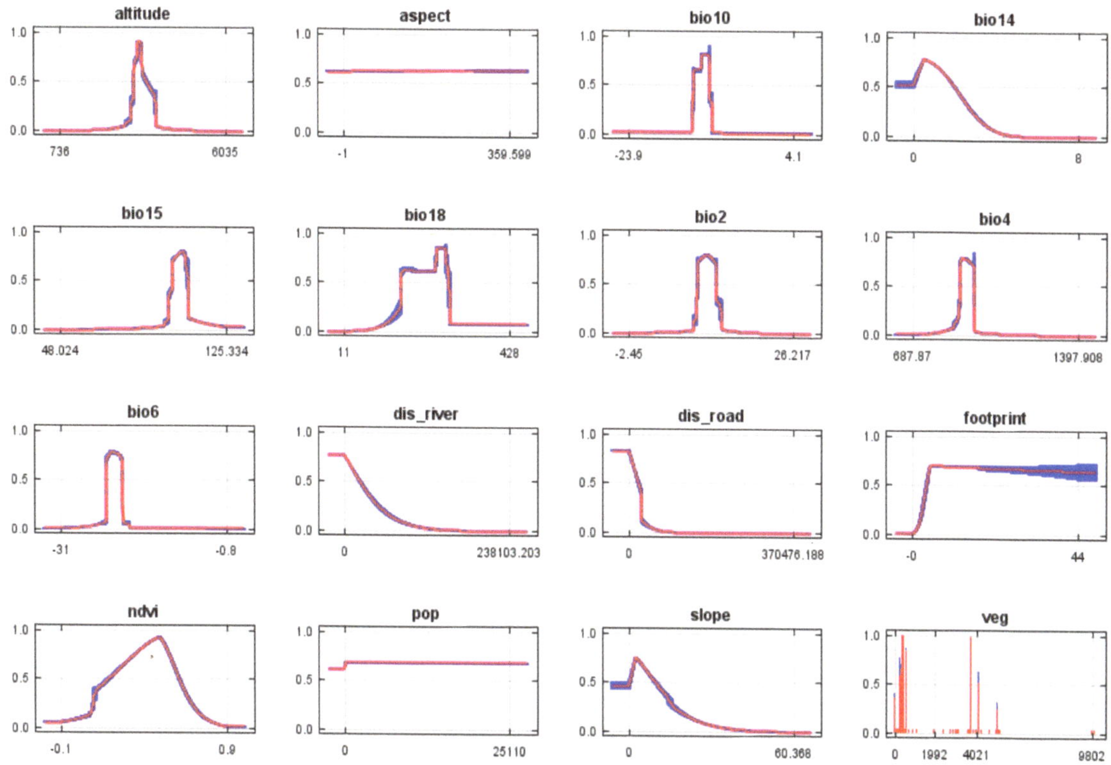

Figure 5. Response curves of environmental variables.

3.3. *Potential Suitable Distribution Area of Przewalski's Gazelle*

After we imported the distribution prediction data into ArcGIS10.8 and converted them into raster format, we divided the suitable areas of Przewalski's gazelle into three levels: non-suitable areas, suitable areas, and highly suitable areas. By calculating the grid area, the highly suitable area of Przewalski's gazelle was determined to be 11,441.45 km^2, accounting for 0.61% of the total area of the study area. Furthermore, the suitable area and non-suitable area were 21,506.85 km^2 and 1,839,440.22 km^2, respectively, accounting for 1.15% and 98.24% of the total area of the study area. This finding indicates that the highly suitable area of Przewalski's gazelle is narrow, and most of this area is distributed around Qinghai Lake; another suitable habitat is at the Xidahe Reservoir and its surrounding area (Figures 6 and 7).

3.4. *GAP Analysis Results*

According to the overlay analysis of the layers of highly suitable area, suitable area and Qinghai Lake National Nature Reserve in ArcGIS10.8, the scope of Qinghai Lake National Nature Reserve is small at present, which is only distributed around Qinghai Lake, and most of the suitable habitats of Przewalski's gazelle are not included in the field of the reserve. We extracted the overlapped part by using the mask and calculated that the protected area accounts for 1529.04 km^2 of the suitable area, which is 7.11% of the suitable area; The reserve occupies 1806.59 km^2 of the highly suitable area, accounting for 15.79% of the highly suitable area (Figure 7).

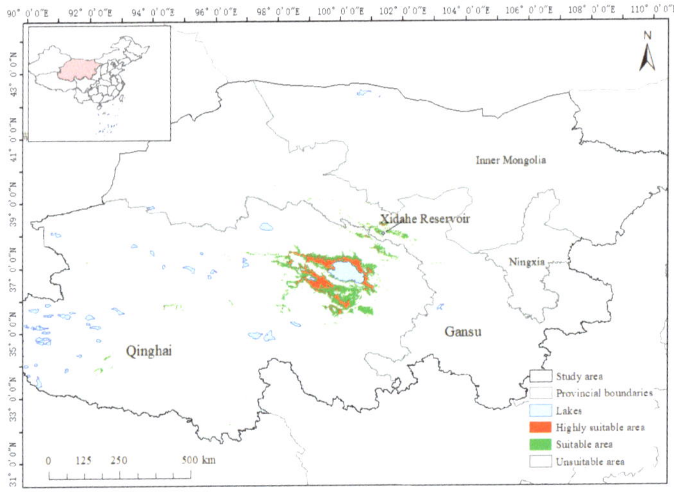

Figure 6. Distribution of potential suitable habitats of Przewalski's gazelle.

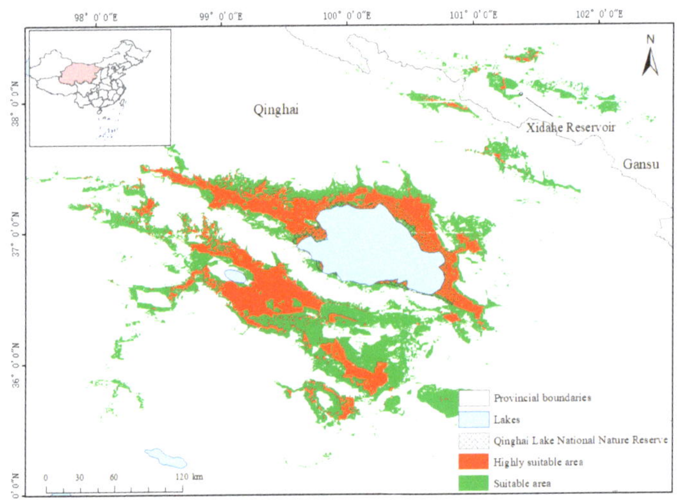

Figure 7. GAP analysis of potential suitable habitat of Przewalski's gazelle and Qinghai Lake National Nature Reserve.

4. Discussion

Via MaxEnt modeling, this study identifies altitude, temperature, vegetation type, and distance from roads as the primary environmental factors influencing the geographic distribution of Przewalski's gazelle. First, altitude and temperature are basic factors affecting species distribution, and this finding is similar to other research results [36]. Additionally, vegetation type is one of the important factors affecting the distribution of Przewalski's gazelle. Previous researchers have studied the reasons behind the failure of releasing Mohor gazelle (*Gazella dama mhorr*) and found that the suitable habitat for its survival consists of areas with low grass levels rather than areas with high grass levels and a dense canopy [37]. This finding underscores the critical role of vegetation type in the habitat selection of gazelles, presenting significant implications for future relocation efforts. Moreover, distance from roads was identified as another major factor influencing the distribution of Przewalski's gazelle. Studies have consistently shown that roads can have detrimental effects on species distribution [38]. In the case of Przewalski's gazelle, roads can impede population connectivity, resulting in habitat fragmentation, decreased quality, and increased mortality for this species [11,39]. Different roads, such as railways, expressways, provincial roads, and township roads, will also have different impacts [40]. As there are Qinghai–Tibet railway and expressways, as well as provincial roads and township roads, in the areas where Przewalski's gazelle lives, we suggest that further research on how these different roads will affect their distribution is required.

Aside from temperature, climate, and vegetation, slope and aspect were also key environmental factors that concerned and were mentioned by other researchers [27,41], but we found that Przewalski's gazelle did not seem to have a strong dependence on slope and aspect, which may be due to the small changes in grassland slope and the small impact of slope on grassland vegetation. Meanwhile, some studies have found that different species have different preferences for different slopes in different states (resting or moving) [42]. Besides that, the distribution range of some species also varies with the seasons [43]. This study focused exclusively on the distribution of the gazelle without considering seasonal variations. It is possible that the preferred slope and direction for Przewalski's gazelle may differ with the changing seasons.

Furthermore, our GAP analysis demonstrated that the Qinghai Lake National Nature Reserve only encompasses a limited range around Qinghai Lake, leaving most of the suitable habitats for Przewalski's gazelle unprotected. This highlights the limited coverage area of the reserve and suggests the existence of significant protection gaps. In other words, there is still a substantial area of protection gaps with respect to the protection of Przewalski's gazelle. Despite the rapid increase in the number of nature reserves, certain endangered species still lack effective protection within these designated areas [44]. GAP analysis allows us to identify conservation gaps in a timely manner, making it a valuable tool in conservation ecology. In this study, GAP analysis was conducted based on the suitable habitat of Przewalski's gazelle. However, further improvements are needed, such as analyzing the local species diversity. To further strengthen the in situ protection of Przewalski's gazelle in the surrounding areas of Qinghai Lake, it is necessary to scientifically plan the coverage and layout of the Qinghai Lake National Nature Reserve; reasonably divide the protection core area, buffer area, and experimental area; balance the relationship between the tourism industry and the ecological protection of Qinghai Lake; and achieve both the protection of the local ecological environment and biodiversity. At the same time, this will also provide a mutual benefit (win–win results) for the local economy and tourism industry.

In recent years, translocation has been considered an effective conservation strategy for protecting endangered animals [45,46]. Given the ongoing efforts in conservation, conservation translocation is a promising approach to preserving Przewalski's gazelle. According to the results yielded by the MaxEnt model concerning the distribution area of the Przewalski's gazelle, we found that the majority of highly suitable habitats were concentrated around Qinghai Lake. It was found that there are minimal suitable habitats for this species in the Ordos region of west Inner Mongolia, China, although researchers have collected specimens of Przewalski's gazelle in this area [47,48]. Besides the surrounding areas of Qinghai Lake, only a few areas, such as the Xidahe Reservoir and its surrounding area in Gansu, were found to be suitable habitats for Przewalski's gazelle under the current climatic conditions. This makes Xidahe Reservoir and its surrounding areas crucial for future protection and migration efforts centered on Przewalski's gazelle. If this site is selected for conservation translocation, it is vital to conduct follow-up investigations (e.g., examining carrying capacity and residents' attitudes toward Przewalski's gazelle) in these areas. In addition, this involves conducting cross-provincial assessments and discussions between multiple government departments and experts. For relocation protection, manual means may be needed for relocation, and multiple professional assessments are also required in this regard.

Due to limited time, resources, and capabilities, there are many shortcomings of this study. The limitations of this study include its lack of consideration of livestock density and grazing frequency. Additionally, this study is based on model analysis, which has its own limitations. Further on-site investigations and comprehensive scientific evaluations are needed to address these limitations.

5. Conclusions

Our data showed that most of the suitable habitat for Przewalski's gazelle is not within the scope of the established reserve. Translocation should be considered a feasible way of establishing new populations and saving Przewalski's gazelle. It is necessary to consider reintroducing these gazelles into an area from which they have disappeared to establish several new populations. Furthermore, although there are still many difficulties in the development of Qinghai Lake National Park [49], the conservation translocation of Przewalski's gazelle should be considered in China's national park system.

Author Contributions: C.L. conceived the idea for this research and designed the study; D.L. and C.L. investigated and collected the data; D.L. analyzed the data; D.L. and C.L. wrote the manuscript. All authors have read and agreed to the published version of the manuscript.

Funding: This research was financially supported by the Science & Technology Fundamental Resources Investigation Program of the Ministry of Science and Technology of China (Grant No. 2023FY100305), the Third Comprehensive Scientific Expedition of Xinjiang Uyghur Autonomous Region of the Ministry of Science and Technology of China (Grant No. 2022xjkk0802), the National Key Research and Development Program of China-Science & Technology Cooperation Project of Chinese and Russian Government (2023YFE0111300), and the Biodiversity Survey and Assessment Program of the Ministry of Ecology and Environment of China (Grant No. 2019HB2096001006).

Institutional Review Board Statement: Not applicable.

Informed Consent Statement: Not applicable.

Data Availability Statement: The data presented in this study are available on request from the corresponding author. The data are not publicly available due to government policy restrictions.

Acknowledgments: We sincerely thank the staff of the Qinghai Lake National Nature Reserve, the staff of the Forestry and Grassland Bureau of Delingha City of Qinghai, and the staff of the Forestry Bureau of Shandan County of Gansu for their help in our field survey and data collection. We also thank Xiaoge Ping, Qi Zhang, Songkai Guo, Yonglin Wu, and Yaxin Liu for their help with our study.

Conflicts of Interest: The authors declare no conflicts of interest.

References

1. Wilson, M.C.; Chen, X.; Corlett, R.T.; Didham, R.K.; Ding, P.; Holt, R.D.; Holyoak, M.; Hu, G.; Hughes, A.C.; Jiang, L.; et al. Habitat fragmentation and biodiversity conservation: Key findings and future challenges. *Landsc. Ecol.* **2016**, *31*, 219–227. [CrossRef]
2. Allen, G.M. The Mammals of China and Mongolia. Part 2. In *Journal of Mammalogy*; American Museum of Natural History: New York, NY, USA, 1940; pp. 1216–1222.
3. Zhao, K. Ungulates from Inner Mongolia. *J. Inn. Mong. Univ.* **1960**, *1*, 53–60.
4. Jiang, Z.; Feng, Z.; Wang, Z.; Chen, L.; Cai, P.; Li, Y. Historical and current distribution of Przewalski's gazelle. *Acta Theriol. Sin.* **1995**, *15*, 241–245.
5. Jiang, Z.; Li, D.; Wang, Z.; Zhu, S.; Wei, W. Population structure of Przewalski's gazelle in Qinghai Lake area. *Acta Zool. Sin.* **2001**, *47*, 158–162.
6. Liang, D.; Liu, Y.; Ping, X.; Jiang, Z.; Li, C. Ensuring recovery for the Przewalski's gazelle. *Science* **2021**, *374*, 163. [CrossRef] [PubMed]
7. Ping, X.; Li, C.; Li, C.; Tang, S.; Fang, H.; Cui, S.; Chen, J.; Wang, E.; He, Y.; Cai, P.; et al. The distribution, population and conservation status of Przewalski's gazelle (*Procapra przewalskii*). *Biodiversity* **2018**, *26*, 177–184.
8. Li, Z.X.J. The Impact of Climate Change on the Breeding Population and Habitat Quality of the Spotted Goose in Qinghai Lake. Master's Thesis, Beijing Forestry University, Beijing, China, 2020.
9. Hu, J.; Jiang, Z. Climate change hastens the conservation urgency of an endangered ungulate. *PLoS ONE* **2011**, *6*, e22873. [CrossRef]
10. You, Z.; Jiang, Z.; Li, C.; Mallon, D. Impacts of grassland fence on the behavior and distribution range of the critically endangered Przewalski's gazelle on the Qinghai-Tibetan Plateau. *Chin. Sci. Bull.* **2013**, *58*, 2262–2268. [CrossRef]
11. Li, C.; Jiang, Z.; Feng, Z.; Yang, X.; Yang, J.; Chen, L. Effects of highway traffic on diurnal activity of the critically endangered Przewalski's gazelle. *Wildl. Res.* **2009**, *36*, 379–385. [CrossRef]

12. Elith, J.; Phillips, S.J.; Hastie, T.; Dudík, M.; Chee, Y.E.; Yates, C.J. A statistical explanation of MaxEnt for ecologists. *Divers. Distrib.* **2011**, *17*, 43–57. [CrossRef]
13. Li, W.; Xu, Z.; Shi, M.; Chen, J. Prediction of potential geographical distribution patterns of *Salix tetrasperma* Roxb. in Asia under different climate scenario. *Acta Ecol. Sin.* **2019**, *39*, 3224–3234.
14. Phillips, S.J.; Dudík, M.; Schapire, R.E. A maximum entropy approach to species distribution modeling. In *Machine Learning, Proceedings of the Twenty-first International Conference (ICML 2004), Banff, AB, Canada, 4–8 July 2004*; Elsevier: Amsterdam, The Netherlands, 2004; pp. 83–88.
15. Anand, V.; Oinam, B.; Singh, I.H. Predicting the current and future potential spatial distribution of endangered *Rucervus eldii eldii* (Sangai) using MaxEnt model. *Environ. Monit. Assess.* **2021**, *193*, 147. [CrossRef]
16. Pant, G.; Maraseni, T.; Apan, A.; Allen, B.J. Predicted declines in suitable habitat for greater one-horned rhinoceros (*Rhinoceros unicornis*) under future climate and land use change scenarios. *Ecol. Evol.* **2021**, *11*, 18288–18304. [CrossRef]
17. Gull, E.; Fareen, A.; Mahmood, T.; Bodlah, I.; Rashid, A.; Khalid, A.; Mahmood, S. Modeling potential distribution of newly recorded ant, *Brachyponera nigrita* using MaxEnt under climate change in Pothwar region, Pakistan. *PLoS ONE* **2022**, *17*, e0262451.
18. Scott, J.M.; Davis, F.W.; Csuti, B.A.; Noss, R.F.; Groves, C.R.; Anderson, H.N.; Caicco, S.L.; Edwards, T.C.; Ulliman, J.J.; Wright, R.G. Gap analysis: A geographic approach to protection of biological diversity. *Wildl. Monogr.* **1993**, *123*, 3–41.
19. Jennings, M.D. Gap analysis: Concepts, methods, and recent results. *Landsc. Ecol.* **2000**, *15*, 5–20. [CrossRef]
20. Stoms, D.M. Gap management status is and regional indicators of threats to diversity. *Landsc. Ecol.* **2000**, *15*, 21–33. [CrossRef]
21. Maiorano, L.; Falcucci, A.; Boitani, L. Gap analysis of terrestrial vertebrates in Italy: Priorities for conservation planning in a human dominated landscape. *Biol. Conserv.* **2006**, *133*, 455–473. [CrossRef]
22. van Breugel, P.; Kindt, R.; Barnekow Lillesø, J.P.; van Breugel, M. Environmental Gap analysis to prioritize conservation efforts in eastern Africa. *PLoS ONE* **2015**, *10*, e0121444. [CrossRef]
23. Feng, L.; Duan, W.; Gao, W.; Run, S. *Atlas of Rare and Endangered Animals in Mongolia*; China Agricultural Science and Technology Press: Beijing, China, 1991; pp. 182–185.
24. Bi, J.; Feng, L. Investigation on the historical distribution and current status of Przewalski's gazelle in Inner Mongolia. In *Chinese Animal Science Research, Zoological Society of China Edition; Proceedings of the 14th Member Congress of the Zoological Society of China and the 65th Annual Conference of the Zoological Society of China*; China Forestry Publishing House: Beijing, China, 1999; pp. 311–314.
25. Fick, S.E.; Hijmans, R.J. WorldClim 2: New 1-km spatial resolution climate surfaces for global land areas. *Int. J. Climatol.* **2017**, *37*, 4302–4315. [CrossRef]
26. Xu, X. *Spatial Distribution Dataset of Monthly Normalized Difference Vegetation Index (NDVI) in China*; Data Registration and Publishing System of Resource and Environmental Science Data Center; Chinese Academy of Sciences: Beijing, China, 2018.
27. Li, H.; Niu, C.; Yuan, X.; Huo, A.; Shang, W.; Liu, J.; Han, P.; Zhang, Y.; Jian, N. Predicting the Potential Distribution Area of *Moschus moschiferus* Using the MaxEnt Model in the Huanglong Mountains, Shaanxi Province. *J. Ecol. Rural Environ.* **2021**, *37*, 1616–1622.
28. Abolmaali, M.R.S.; Tarkesh, M.; Bashari, H. MaxEnt modeling for predicting suitable habitats and identifying the effects of climate change on a threatened species, *Daphne mucronata*, in central Iran. *Ecol. Inform.* **2018**, *43*, 116–123. [CrossRef]
29. Zhu, G.; Qiao, H. Effect of the MaxEnt model's complexity on the prediction of species potential distributions. *Biodivers. Sci.* **2016**, *24*, 1189–1196. [CrossRef]
30. Kong, W.; Li, X.; Zou, H. Optimizing MaxEnt model in the prediction of species distribution. *Chin. J. Appl. Ecol.* **2019**, *30*, 2116–2128.
31. Warren, D.L.; Seifert, S.N. Ecological niche modeling in MaxEnt: The importance of model complexity and the performance of model selection criteria. *Ecol. Appl.* **2011**, *21*, 335–342. [CrossRef] [PubMed]
32. Muscarella, R.; Galante, P.J.; Soley-Guardia, M.; Boria, R.A.; Kass, J.M.; Uriarte, M.; Anderson, R.P. ENMeval: An R package for conducting spatially independent evaluations and estimating optimal model complexity for MaxEnt ecological niche models. *Methods Ecol. Evol.* **2014**, *5*, 1198–1205. [CrossRef]
33. Zhang, T.; Huang, Z.; Peng, Y.; Wang, Y.; Wang, P.; Wang, S.; Cui, G. Prediction of potential suitable areas of *Actinidia chinensis* based on MaxEnt model. *Acta Ecol. Sin.* **2020**, *40*, 4921–4928.
34. Jiménez-Valverde, A.; Lobo, J.M. Threshold criteria for conversion of probability of species presence to either—Or presence-absence. *Acta Oecol.* **2007**, *31*, 361–369. [CrossRef]
35. Kaushalendra, K.J.; Radhika, J. Study of Vulture Habitat Suitability and Impact of Climate Change in Central India Using MaxEnt. *J. Resour. Ecol.* **2021**, *12*, 30–42.
36. Helenbrook, W.D.; Valdez, J.W. Species distribution and conservation assessment of the black-headed night monkey (*Aotus nigriceps*): A species of Least Concern that faces widespread anthropogenic threats. *Primates* **2021**, *62*, 817–825. [CrossRef]
37. Moreno, E.; Sane, A.; Benzal, J.; Ibáñez, B.; Sanz-Zuasti, J.; Espeso, G. Changes in habitat structure may explain decrease in reintroduced mohor gazelle population in the Guembeul Fauna Reserve, Senegal. *Animals* **2012**, *2*, 347–360. [CrossRef] [PubMed]
38. Makki, T.; Fakheran, S.; Moradi, H.; Iravani, M.; Senn, J. Landscape-scale impacts of transportation infrastructure on spatial dynamics of two vulnerable ungulate species in Ghamishloo Wildlife Refuge, Iran. *Ecol. Indic.* **2013**, *31*, 6–14. [CrossRef]
39. Wright, P.; Coomber, F.; Bellamy, C.; Perkins, S.E.; Mathews, F. Predicting hedgehog mortality risks on British roads using habitat suitability modelling. *PeerJ* **2020**, *7*, e8154. [CrossRef] [PubMed]

40. Feng, T.; Wu, X.; Zhang, H. Distribution of ungulate in different grades of highway and influential factors concerned in Qinling Mountains. *Shaanxi For. Sci. Technol.* **2019**, *47*, 1–6.
41. Khadka, K.K. Habitat Ecology, Trophic Interactions, and Distribution of Endangered Himalayan Musk Deer in the Nepal Himalaya. Graduate Thesis, University of Arkansas, Fayetteville, AR, USA, 2017.
42. Liu, Y.; Niu, Y.; Zhou, S.; Zhang, Z.; Liang, Z.; Yang, J.; Ju, D. Winter moving, bedding habitat selection and evaluation of roe deer in southern Laoyeling, Heilongjiang Province. *Acta Ecol. Sin.* **2021**, *41*, 6913–6923.
43. Shams-Esfandabad, B.; Ahmadi, A.; Yusefi, T. Seasonal changes in distribution of suitable habitats for Persian goitered gazelle (*Gazella subgutturosa*) in Isfahan province. *J. Wildl. Biodivers.* **2019**, *3*, 58–65.
44. Wei, F.; Ping, X.; Hu, Y.; Nie, Y.; Zeng, Y.; Huang, G. Main Achievements, Challenges, and Recommendations of Biodiversity Conservation in China. *Bull. Chin. Acad. Sci.* **2021**, *36*, 375–383.
45. Hellmann, J.J. Translocation as a conservation strategy. In *Encyclopedia of Biodiversity*, 2nd ed.; Levin, S.A., Ed.; Academic Press: Waltham, MA, USA, 2013; pp. 236–240.
46. Lloyd, N.A.; Keating, L.M.; Friesen, A.J.; Cole, D.M.; McPherson, J.M.; Akçakaya, H.R.; Moehrenschlager, A. Prioritizing species conservation programs based on IUCN green status and estimates of cost-sharing potential. *Conserv. Biol.* **2023**, *37*, e14051. [CrossRef]
47. Ellerman, J.R.; Morrison-Scott, T.C.S. *Checklist of Palaearctic and Indian Mammals-New Name 1758–1946*; The British Museum: London, UK, 1951.
48. Corbet, G.B. *The Mammals of the Palaearctic Regioni a Taxonomic Review*; Cornell University Press: London, UK, 1978.
49. Chen, P.; He, D. Research on key and difficult problems in the establishment of Qinghai Lake National Park. *J. Natl. Acad. For. Grassl. Adm.* **2023**, *22*, 29–39.

animals

MDPI

Article

Conservation Genetics of the Loggerhead Sea Turtle, *Caretta caretta*, from the Central Mediterranean: An Insight into the Species' Reproductive Behaviour in Maltese Waters

Adriana Vella * and Noel Vella

Conservation Biology Research Group, Department of Biology, Faculty of Science, University of Malta, MSD 2080 Msida, Malta
* Correspondence: adriana.vella@um.edu.mt

Simple Summary: This work represents the first scientific study using genetic markers to better understand the reproductive behaviour of the loggerhead sea turtle from the Central Mediterranean through dead hatchlings and undeveloped dead embryos collected from recorded nesting sites between 2020 and 2022. Data generated from the genetically analysed specimens were used for parentage analyses. Results of this research show that some turtles laid more than one nest within the same nesting season while we also detected an instance of multiple paternity. These findings contribute to a better understanding of the biology of the species in Maltese waters, which is a requirement for effective conservation management.

Abstract: Loggerhead sea turtle, *Caretta caretta* (Linnaeus, 1758), nestlings were investigated through specimens found dead either after hatching or unhatched (n = 120) from eight nests around the Maltese islands (Central Mediterranean). Molecular genetics was used to conduct maternity and paternity tests of the collected specimens utilizing expanded mitochondrial DNA sequences from the control region (858 bp) and 25 microsatellite loci (12 dinucleotide loci and 13 tetranucleotide loci). Mitochondrial data produced two haplotypes, CC-A2.1 and CC-A3.1, with the most common haplotype being present in seven nests. Microsatellite data revealed the identity of six different females that were involved in the deposition of the eggs in the eight turtle nests analysed. This confirms that two females laid multiple nests. Additionally, microsatellite data allowed for the determination of multiple paternity, with one clutch being sired by two fathers. These results are useful for monitoring the genetic diversity of loggerhead sea turtle nestlings and of the turtle mothers and fathers contributing to future turtle offspring, which rely on Maltese sandy beaches for their successful start to life. Effective conservation management benefits from merging scientific knowledge with effective measures at potential nesting sites to avoid losses of nestlings caused by human negligence.

Keywords: maternity test; paternity test; mtDNA control region; microsatellites; philopatry; polyandry; multiple paternity; renesting

Citation: Vella, A.; Vella, N. Conservation Genetics of the Loggerhead Sea Turtle, *Caretta caretta*, from the Central Mediterranean: An Insight into the Species' Reproductive Behaviour in Maltese Waters. *Animals* **2024**, *14*, 137. https://doi.org/10.3390/ani14010137

Academic Editor: Chunwang Li

Received: 24 November 2023
Revised: 26 December 2023
Accepted: 28 December 2023
Published: 30 December 2023

1. Introduction

The Mediterranean Sea hosts the green sea turtle *Chelonia mydas* (Linnaeus, 1758) and the loggerhead sea turtle *Caretta caretta* (Linnaeus, 1758) [1–3], both of which are known to nest in the region [2,4]. *Chelonia mydas* nesting sites are mostly restricted to the Eastern Mediterranean [2,5] while the major nesting sites for *C. caretta* were historically also associated with the Eastern Mediterranean; however, in the last few decades, the geographical centre for Mediterranean *C. caretta* nesting sites shifted towards the Central Mediterranean due to increasing records in the Western Mediterranean [2,6,7]. The Mediterranean Sea is also frequented by leatherback turtles, *Dermochelys coriacea* (Vandelli, 1761), migrating occasionally from the Atlantic Ocean [3,8,9]; however, there are no nest records for this species in the region.

According to the IUCN, the loggerhead sea turtle is listed as vulnerable at a global level [10] and as least concern at the Mediterranean subpopulation level [11]. It has also been enlisted in a number of international frameworks, including the Convention for the Protection of the Mediterranean Sea against Pollution (Barcelona Convention—Annex II) [12], the Convention on Migratory Species (CMS—Appendix I) [13], the Convention on the Conservation of European Wildlife and Natural Habitats (Bern Convention—Appendix II) [14], the Convention on International Trade in Endangered Species (CITES—Appendix I) [15], and the Protocol of the Barcelona Convention concerning Specially Protected Areas and Biological Diversity in the Mediterranean (SPA/BD—Annex II) [16]. Moreover, the International Commission for the Conservation of Atlantic Tunas (ICCAT) and the General Fisheries Commission for the Mediterranean (GFCM) provide recommendations to fishery managers in relation to the incidental bycatch of sea turtles in the Mediterranean Sea [17,18] while the Marine Strategy Framework Directive within the Biodiversity Descriptor allows for good environmental status assessments of sea turtles [19].

Globally, the loggerhead sea turtle is split into a number of regional management units (RMUs), which have been identified through mitochondrial DNA (mtDNA) and nuclear DNA (nDNA) [3,20]. This structure is mostly based on nesting populations due to female philopatry, exhibiting nesting site fidelity. Nonetheless, population structures are more complex due to the male-mediated gene flow and overlapping populations, especially during migration. Mitochondrial DNA control region (mtDNA CR) data for the Mediterranean Sea indicate that *Caretta caretta* populations have colonized the Mediterranean Sea during the Pleistocene [21], have survived the glacial periods in warm refugia within the south-eastern parts of the Mediterranean Sea [21], and have undergone multiple colonization events [20]. Additionally, the mtDNA CR also exhibits enough variation and structure at a small scale that the RMU of the Mediterranean Sea can be split into smaller management units [2,21,22], which may be necessary for the management of rookeries.

The species is prone to various anthropogenic pressures. The effects of climate change, including rising sea levels, changes in tides, waves, precipitation patterns, and changing temperatures, impose threats to the development of sea turtle embryos [23–26] while rising sand temperatures skew the primary sex ratios towards females, leading to the potential feminization of several sea turtle populations while lowering the hatching success and hatchling fitness, including those of *C. caretta* [20,27,28]. These climatic changes together with other human-induced threats, such as coastal development, disturbances, and vessel traffic, also affect the behaviours of adults, including their approach to land [24,29]. Therefore, there is an increasing requirement for more scientific data collection on the reproductive biology of sea turtles to design conservation management strategies that improve the resilience and survival of these species.

Therefore human activities impact both the nestlings and free-swimming sea turtles through fast-changing and deteriorating marine environments with greater pollution, disturbances, and injuries caused by boat propellers and fishing gear [1,2,30–34]. The Mediterranean Sea is highly exposed to such anthropogenic activities [35], with most of them increasing in frequency during late spring and summer, which coincides with breeding migration and the mating period of *C. caretta*, which peaks between April and May [2]. Therefore, the cumulative effect of these threats would result in an increased mortality rate and smaller population sizes, reducing reproductive fitness while disturbing the natural behaviours of this species. Consequently, anthropogenic effects on the nesting sites, nests, nesting behaviours, mating behaviours, and migratory patterns influence the reproductive potential of this species.

The objectives of this first scientific research paper on loggerhead sea turtle nestlings from the Maltese islands are to (1) understand the genetic relationship between different nests and whether the females return to nest within the same season (renesting) and site fidelity and (2) estimate the frequency of multiple paternity. Here, mtDNA sequences and nuclear DNA genotyping were used as tools to allow for the computational sibship and parentage investigation [36,37] of all the recorded nests between 2020 and 2022.

2. Materials and Methods

2.1. Study Area and Background Information

This study focuses on *C. caretta* nests from the Maltese archipelago. This archipelago is situated in the Central Mediterranean, approximately 95 km south of Sicily and 285 km from the Tunisian coast (Figure 1), and is characterized by two main islands, Malta and Gozo. In recent years there have been thirteen reported nests, one each in 2012, 2016, and 2018; six in 2020; one each in 2021 and 2022; and an additional two in 2023 (Table 1).

We looked into the remains from nests laid between 2018 and 2022. Given that the 2018 nest did not contain any dead individuals then, the genetic analyses focused on eight nests laid between summer 2020 and 2022, representing four beaches around the Maltese islands (Figure 1). For six of the nests analysed, the local Environment and Resource Authority (ERA, *pers. comm.*) was alerted about the nests when the respective mothers laid their eggs and the nests were monitored until the juveniles hatched. In one of these cases, the eggs were dug up and transferred to an incubator, given that the integrity of the nest was assessed as being compromised by adverse environmental conditions (ERA, *pers. comm.*). In two instances, the nest was unknown prior to the discovery of the hatchlings.

Locally, the species is strictly protected under *Flora, Fauna, and Natural Habitats protection regulations* [38] and a number of marine Natura 2000 sites have been designed to further protect the species, including MT0000113; MT0000115; and MT0000116 [39]. Consequently, any dead hatched and dead unhatched specimens were collected from the respective nests by the local authority ERA; then, they were handed on for tissue sampling and scientific investigation by AV in accordance with handling and research ERA permits.

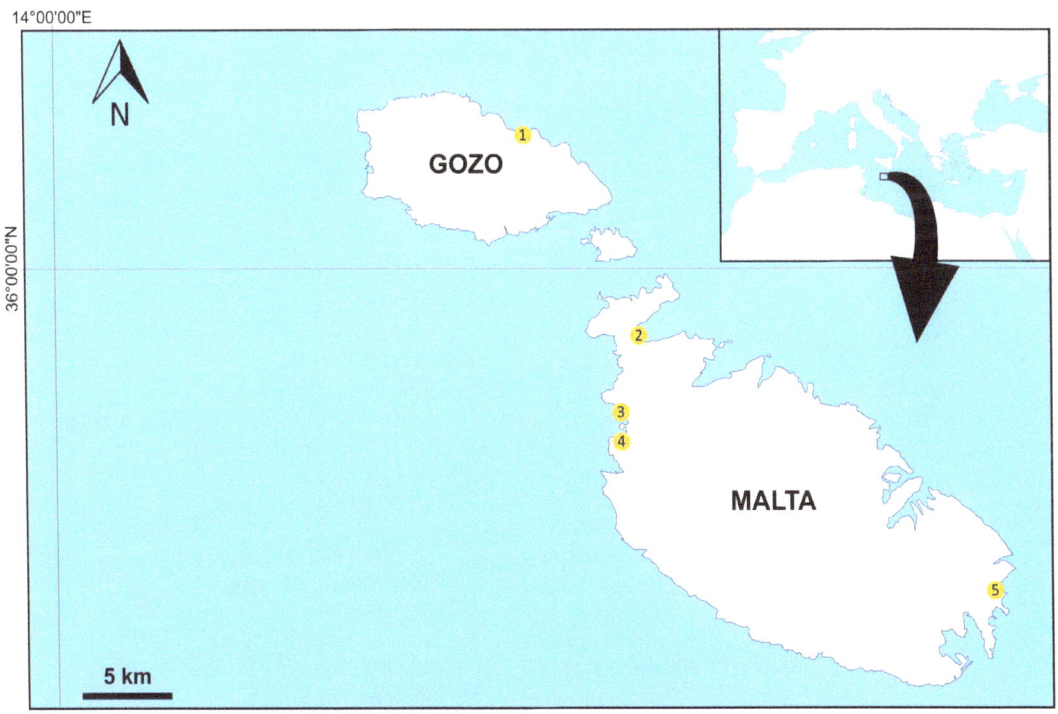

Figure 1. Map indicating the five nesting beaches recorded on the Maltese islands between 2012 and 2023. 1. Ramla Bay; 2. Għadira Bay; 3. Golden Bay; 4. Ġnejna Bay; 5. Fajtata Bay.

Table 1. A list of recently recorded nesting sites in the Maltese islands. The table includes the nests' codes and the number of dead individuals used from each respective nest during the current study.

Nesting Site	Nesting Year	Nest Reference Code in This Study	Specimens Analysed
Ġnejna Bay	2012	-	-
Golden Bay	2016	-	-
Ġnejna Bay	2018	-	-
Ramla Bay	2020	CRA	1
Ramla Bay	2020	CRB	7
Golden Bay	2020	CMA	5
Fajtata Bay	2020	CFA	23
Għadira Bay	2020	CGA	44
Għadira Bay	2020	CGB	7
Ramla Bay	2021	CRC	7
Ramla Bay	2022	CRD	26
Ramla Bay	2023	-	-
Gnejna Bay	2023	-	-

2.2. *Sample Collection, DNA Extraction, PCR Amplification, and Sequencing*

Tissue samples from dead unhatched individuals or dead hatchlings were excised and stored in 100% ethanol. The total genomic DNA was then extracted from tissue samples using the GF-1 Tissue DNA Extraction Kit (Vivantis Technologies, Shah Alam, Malaysia) following the manufacturer's manual. The concentration of the purified DNA was estimated using Qubit (ThermoFisher Scientific, Waltham, MA, USA).

Given that all samples within a clutch come from the same mother, for mtDNA, analyses of two specimens per nest were randomly chosen, except for the nest CRA, where only one individual was available and, thus, the same individual was analysed twice. For the selected specimens, the mtDNA CR was amplified using LCM15382 and H950 [40] following the work of Shamblin et al. [41]. The PCR products were then purified and sequenced with their respective forward and reverse primers via the ABI3730XL sequencer (Applied Biosystems, Waltham, MA, USA).

Twelve dinucleotide microsatellite loci [42–46] and thirteen tetranucleotide microsatellite loci [47,48] were selected for more detailed maternal and paternal analyses of each nest. Each sampled specimen was analysed for the 25 microsatellites (Table 1) that were all tagged by M13 tails; fluorescently labelled using 6-FAM, VIC, NED, or PET; and amplified following published temperature profiles [46–48] (Supplementary Table S1). PCR products were size-scored through Applied Biosystems ABI3730XL, using Liz600 as the fluorescent size standard. During these analyses, 12 specimens, representing 10% of the sample size, were randomly chosen and were run twice for all microsatellites to estimate the error rate.

2.3. *Data Analyses*

Mitochondrial DNA sequences were manually trimmed and the complementary sequences of each individual were assembled using Geneious R10 [49]. The sequences of each individual and those within each nest were checked for consistency. The genetic sequence obtained for each nest was compared to other publicly available sequences through BLASTn [50,51] to identify the mtDNA lineage of the locally nesting turtles to those found in other regions of the Mediterranean.

Microsatellite allele sizes were scored with Geneious R10 [49] and binned using FlexiBin v2 [52]. For each microsatellite locus at each nest, the number of alleles detected and the observed heterozygosity (H_o) were estimated through Arlequin v3.5 [53]. Genotypes were checked for scoring errors due to stuttering, large allele dropouts, and null alleles using Micro-Checker 2.2.3 [54]. Analysis of paternity was initially checked visually by evaluating the multi-locus genotypes and the number of alleles per locus at each nest. Then, the data were analysed through the software COLONY v2.0.6.8 [37] to computationally assign sibship and parentage among individuals using likelihood methods through multi-locus genotype data with a less than 1% error rate.

3. Results

3.1. The Nests

After decades of no records of *C. caretta* nests in the Maltese islands [1], there was an unsuccessful nesting event in 2012 and a successful one in 2016 where 83.5% of eggs counted by the local authority ERA had hatched. These were followed by another successful nesting event in 2018, where all individuals hatched except one inviable egg.

The genetic results of the 120 *C. caretta* specimens presented here represent the first valuable output derived from using dead specimen samples from eight Maltese turtle nest clutches laid in the summers between 2020 and 2022. In 2020, six nests from four different beaches were identified. In four of these instances, the nest was identified when the mother laid the eggs and, thus, the nests were protected throughout the natural incubation period; meanwhile, the other two nests were not protected throughout their incubation period. The latter two were represented by the nest CRB at Ramla Bay, which is a sizable and popular sandy beach (coastline: ~400 m), and the nest CFA at Fajtata Bay, which is a small sandy beach (coastline: ~23 m) highly frequented by bathers. In 2021 and 2022, there was one recorded nesting event per year and, in 2023, two recorded nesting events (Table 1).

3.2. Genetic Data

In this study, a total of 858 bp of the mtDNA CR sequence was analysed. As expected, within the same clutch, all specimens exhibited the same haplotype and, overall, the clutches were represented by two haplotypes that differed from each other by 1 bp. The two haplotypes identified in this study were CC-A2.1, which represented all the studied nests, except the nest CRD, which was represented by CC-A3.1. Sequences were deposited in GenBank under accession numbers PP056536 – PP056543.

In this study, 25 microsatellite loci were analysed (Table 2), with more than 76% of the individuals producing positive scores for each locus. Analyses through Micro-Checker [54] indicated that there is no evidence for scoring errors due to stuttering, no large allele dropout, and no indication of null alleles while the genotypes of the replicates were 100% identical. Consequently, all loci were used for subsequent analyses. The number of alleles per locus varied from four alleles (Cc-2, Cc-10, Cc-17, Cc-28 and CcP5C11) to thirteen alleles in locus (CcP7D04 and Cc8E07). The mean number of alleles per locus was 7.6 $\pm$ 2.9 SD (dinucleotide loci 5.6 $\pm$ 1.6 SD; tetranucleotide loci 9.4 $\pm$ 2.7 SD). The overall heterozygosity ranged between 0.479 (Cc-2) and 0.983 (CcP7F06), with the mean heterozygosity per locus being 0.758 $\pm$ 0.144 SD (dinucleotide loci 0.701 $\pm$ 0.135 SD; tetranucleotide loci 0.810 $\pm$ 0.136 SD).

Table 2. The genetic data per nest, including the mtDNA haplotypes, the sample sizes per locus (n), the number of alleles identified per locus (N_a), and the observed heterozygosity per locus (H_o).

Nest Code	CFA	CGA	CGB	CMA	CRA	CRB	CRC	CRD	Overall
mtDNA	CC-A2.1	CC-A2.1	CC-A2.1	CC-A2.1	CC-A2.1	CC-A2.1	CC-A2.1	CC-A3.1	
cc141									
n	23	44	7	5	1	7	7	25	119
N_a	2	3	3	2	2	3	3	4	6
H_o	0.478	0.682	0.571	0.600	1.000	0.714	0.714	0.680	0.638
cc7									
n	23	44	7	5	1	7	7	25	119
N_a	4	4	4	3	2	2	4	4	9
H_o	1.000	1.000	1.000	1.000	1.000	0.714	1.000	0.720	0.924
Ccar176									
n	23	44	7	5	1	7	7	25	119
N_a	4	2	2	3	2	2	2	2	6
H_o	1.000	0.455	0.714	1.000	1.000	0.857	0.857	0.280	0.613
cc117									
n	22	43	7	5	1	7	7	26	118
N_a	3	2	2	3	2	2	3	3	7
H_o	0.955	0.535	0.286	1.000	1.000	0.571	0.857	0.577	0.653
Cc-2									
n	23	44	7	5	1	7	7	24	118
N_a	2	2	2	1	1	2	2	4	4
H_o	0.652	0.409	0.143	0.000	0.000	0.286	0.429	0.750	0.483
Cc-8									
n	23	44	7	5	1	7	7	26	120
N_a	2	3	3	2	1	2	2	2	5
H_o	0.565	0.773	0.857	0.400	0.000	0.714	0.571	0.500	0.641
Cc-10									
n	23	42	7	5	1	7	6	25	116
N_a	3	4	4	3	2	3	3	3	4
H_o	1.000	0.929	1.000	1.000	1.000	0.571	0.667	0.880	0.905
Cc-17									
n	23	42	7	4	1	7	6	25	115
N_a	1	2	2	3	2	2	2	3	4
H_o	0.000	1.000	1.000	1.000	1.000	0.571	0.333	0.400	0.609
Cc-22									
n	23	44	7	5	1	7	6	26	119
N_a	4	2	2	2	1	4	3	3	7
H_o	0.957	0.636	0.714	0.600	0.000	1.000	0.667	0.692	0.731
Cc-25									
n	23	42	7	5	1	7	7	26	118
N_a	2	2	2	3	2	3	2	4	6
H_o	0.609	0.452	0.571	0.800	1.000	1.000	0.286	1.000	0.653
Cc-28									
n	23	44	7	5	1	7	7	26	120
N_a	3	2	2	3	2	2	2	3	4
H_o	0.870	0.545	0.714	0.800	1.000	0.571	0.286	0.846	0.683
Cc-30									
n	22	43	7	5	1	7	7	26	118
N_a	2	4	4	4	2	2	4	2	5
H_o	1.000	1.000	1.000	1.000	1.000	0.143	1.000	0.692	0.881

Table 2. *Cont.*

Nest Code	CFA	CGA	CGB	CMA	CRA	CRB	CRC	CRD	Overall
Cc1G02									
n	23	44	7	4	1	7	7	26	119
N_a	3	3	3	3	2	4	3	7	12
H_o	0.696	0.727	0.857	0.750	1.000	1.000	1.000	1.000	0.824
Cc1G03									
n	23	44	7	5	1	7	7	26	120
N_a	3	3	3	4	2	4	3	4	10
H_o	0.696	0.705	0.714	1.000	1.000	1.000	0.429	0.884	0.758
Cc5H07									
n	23	41	7	4	1	7	5	26	114
N_a	4	5	4	3	1	3	4	5	11
H_o	0.957	0.976	1.000	0.500	0.000	1.000	1.000	1.000	0.956
Cc7E11									
n	22	42	7	5	1	7	6	26	116
N_a	4	2	2	3	2	4	3	6	9
H_o	1.000	0.452	0.429	1.000	1.000	0.857	1.000	0.923	0.741
Cc2H12									
n	23	43	7	5	1	7	6	25	117
N_a	3	4	3	2	1	3	3	4	9
H_o	0.957	0.627	0.857	0.400	0.000	0.857	0.333	0.680	0.701
Cc7B07									
n	23	44	7	5	1	7	7	26	120
N_a	3	3	3	3	2	4	4	6	10
H_o	0.957	1.000	1.000	1.000	1.000	1.000	0.714	0.962	0.967
Cc7G11									
n	23	42	7	5	1	7	5	24	114
N_a	3	2	2	4	2	3	4	5	8
H_o	1.000	0.452	0.571	1.000	1.000	0.857	0.800	0.708	0.692
Cc8E07									
n	22	44	7	5	1	7	7	26	119
N_a	4	3	3	3	2	3	2	6	13
H_o	1.000	1.000	1.000	0.800	1.000	0.571	0.571	1.000	0.941
CcP1F09									
n	23	39	7	4	1	7	6	24	111
N_a	4	4	4	1	1	3	3	4	9
H_o	0.652	0.949	1.000	0.000	0.000	0.857	0.500	0.958	0.820
CcP5C11									
n	23	44	7	3	1	7	7	26	118
N_a	2	2	2	3	2	3	3	2	4
H_o	0.609	0.500	0.429	0.667	1.000	1.000	0.571	0.500	0.559
CcP7D04									
n	23	43	7	4	1	7	6	25	116
N_a	4	5	4	3	2	4	4	5	13
H_o	0.913	0.953	1.000	0.500	1.000	1.000	0.667	0.960	0.922
CcP7F06									
n	23	42	7	5	1	7	6	25	116
N_a	3	4	4	2	2	4	3	5	9
H_o	1.000	1.000	1.000	1.000	1.000	1.000	1.000	0.920	0.983
CcP7H10									
n	23	43	7	5	1	7	7	26	119
N_a	2	3	3	2	2	2	3	4	5
H_o	0.522	0.721	1.000	0.600	1.000	0.571	0.857	0.577	0.664

3.3. Sibship and Parentage Analyses

Analyses of mtDNA haplotypes allowed for the conclusion that the female that laid the eggs in the nest CRD is different from those involved in the other nests, an observation that was further confirmed through nuclear data analyses. The analyses of microsatellites through COLONY [37] indicated that the eight studied nests had originated from six mothers and seven fathers.

We found two instances where the female returned to lay the second nest within the same nesting season. The nest clutches CGA and CGB that were both laid in Għadira Bay, on 30 July 2020 and 10 August 2020, respectively, belonged to the same parents (Mother 2 and Father 2; Table 3). Therefore 11 days after laying the first clutch of 79 eggs, the mother returned to the same beach to lay a second clutch of 86 eggs. Likewise, during the same nesting season, a second female laid two nests (Mother 3 and Father 3; Table 3); the first nest having 102 eggs was laid at Ramla Bay (CRA) on 29 May 2020 and the second nest containing 92 eggs was laid at Golden Bay (CMA) on 5 July 2020. This means that Mother 3 laid two nests within 37 days, around 22 km apart.

Table 3. A table showing the mother and father per nest.

Nest Code	CFA	CGA	CGB	CMA	CRA	CRB	CRC	CRD
Mother	Mother 1	Mother 2	Mother 2	Mother 3	Mother 3	Mother 4	Mother 5	Mother 6
Father	Father 1	Father 2	Father 2	Father 3	Father 3	Father 4	Father 5	Father 6 Father 7

Additionally, evident from the occurrence of more than four alleles per locus for the nestlings from the nest CRD and confirmed through COLONY [37] was the presence of multiple sires for the clutch. From the twenty-six analysed specimens, we noted that eighteen specimens (69.2%) belonged to Father 6 while eight specimens (30.8%) were fertilized by a secondary male (Father 7). We did not find evidence of polyandry in the other nests; although, some of them were represented by a few specimens, making the detection of polyandry more difficult.

4. Discussion

This is the first parentage study of *C. caretta* nests from Malta documenting renesting events and multiple paternity. Genetic data from the eight analysed nests indicated that the most common mtDNA CR haplotype is CC-A2.1, which is commonly proposed as the ancestral lineage for the Mediterranean Sea and was possibly introduced from the Atlantic Ocean by colonizing females in the last post-glacial period [20,21,41,55]. Currently, CC-A2.1 is the most commonly encountered haplotype, present in more than 60% of the free-living Mediterranean loggerhead sea turtles [21,56–60]. This haplotype dominates most Mediterranean nests and has been detected in all nesting areas studied [22,41,59,61,62]. Haplotype CC-A3.1 is the second most commonly encountered haplotype in the Mediterranean Sea and accounts for around 20% of the free-living individuals studied [21,56–58]. This haplotype has been recorded in nests from the eastern and southern-central Mediterranean areas, namely, Turkey, Cyprus, Lebanon, Greece and Crete, Cyprus, Libya, and Tunisia. CC-A3.1 was found to occur in high frequencies in two nesting sites, Dalyan and Dalaman in Turkey, where, in the latter, it was more common than CC-A2.1 [41]. The other haplotypes found in the Mediterranean occur at much lower frequencies in both free-living turtles and in analysed nests [41,59,60].

Nuclear data analyses showed that there were two instances where the female returned twice to a nesting beach to lay separate clutches of eggs within the same nesting season. This phenomenon is known to be common in sea turtles [63,64]. In *C. caretta*, mean renesting intervals have been reported to range between 12.7 days and 19.9 days [63]. While one female turtle returned to lay another clutch 11 days later, on the same beach, the other female turtle took much longer than reported in some other studies, laying her second clutch after 37 days and on a different beach. A comparable actual record of a similar extended renesting period of 34 days has been reported in Turkey [65]. Such delayed renesting events and the choice of a totally different nesting area may be the result of human presence and disturbance, which affect the females' decision when opting for a nesting beach [66]. In this instance, we cannot exclude the possibility that Mother 3 laid other unrecorded nests in the interval between CMA and CRA, given that female *C. caretta* can lay more than two clutches in a season [67]. We were unable to detect cases of a female returning to nest in a different nesting season twice, up to now, since female *C. caretta* have a nesting interval of two to three years and the nests we analysed covered a shorter period of time [2,24].

In one of the nests, we were also able to detect multiple paternity resulting from polyandry. Polyandry is a common mating behaviour in sea turtles [68–71], including *C. caretta* [44,72–77]. While there are benefits to this behaviour [78], including fertilization assurance and genetic diversity benefits [79–81], it remains debatable whether this is a consequence of high male–female encounters. Within this scenario, low frequencies of polyandry may indicate lower chances of mating encounters, even though the sex ratios for adults in the Mediterranean are balanced [82]. In the current study, we encountered multiple paternity in 12.5% of the nesting events studied. This percentage is much lower than that noted in other Mediterranean countries, where, in Greece, more than one father was detected in nineteen out of twenty nests, with two clutches representing the contribution of at least five males [77]. A similar study in Turkey [76] revealed that multiple paternity was present in 18 out of 25 nests analysed. In the latter study, whenever multiple paternity was recorded, the majority of the offspring had a primary sire, which, on average, contributed to 62.7% of the clutch, followed by a secondary sire which, on average, contributed to 30.9%, and a tertiary sire contributing to 7.2% [76]. In our case, the primary sire contributed to 69.2% of the clutch sample analysed while the secondary sire contributed to the remaining 38.8%. Unlike some other species of turtles [83,84], *C. caretta* does not store sperm across breeding seasons [85]. Therefore the occurrence of multiple paternity is the result of multiple mating encounters and sperm storage for each breeding season, as loggerhead sea turtles rarely mate between nesting events within the season [67]. This is consistent with the observations noted in this study where the instances of renesting were represented by the same parentage. Table 3 shows that, in most instances, the mothers studied here exhibited monandry, except for the nest CRD. Consequently, the diversity of fathers noted was almost equal to that of the mothers contributing to the turtle clutches in Malta and Gozo. The genetic results of this work reveal the first details of the reproductive behaviour of *C. caretta* around the Maltese islands.

For decades, there have been no records of turtle nests from the Maltese islands [1]; although, one cannot exclude the possibility of unrecorded nests, even in some of the most human-frequented beaches, as noted in 2020. Since the first recent record in 2012, these islands experienced an increase in recorded turtle nests, with the identification of five nesting beaches, four on the island of Malta and one on the island of Gozo (Table 1; Figure 1). While increasing awareness and reporting by citizens may partially explain the increasing records of nesting events, the long absence of nesting events followed by almost yearly events between 2018 and 2023 clearly indicates an increasing trend in nesting activity around this archipelago. A trend similar to that noted in the Western Mediterranean, where, during the last decade, *C. caretta* has been expanding its nesting range and increasing nesting events in the region [6,7,86], possibly in response to global warming [7].

Impacts of human activities on nesting beaches may hinder female turtles from laying eggs or even interrupt their egg-laying activity, leading to their return later on to the same beach or to seek a quieter beach. The Maltese nests studied here were all found on beaches that are highly frequented by locals and tourists, especially during the summer months, coinciding with the peak of *C. caretta*'s nesting season between May and August [2]. Consequently, as noted in several other regions of the Mediterranean Sea, the recorded nests are highly exposed to anthropogenic impacts [7], possibly because beaches that are not frequented by humans are more difficult to monitor. Anthropogenic influence on turtle behaviour became clearer in the summer of 2020, when Malta, similar to elsewhere in the Mediterranean Sea [7], experienced a spike in nesting activities, a phenomenon that may be associated with the COVID-19 reduction in mass tourism, lowering pressures on *C. caretta* distribution and access to breeding habitats [87]. Scientific evidence shows that this species is capable of noticing changes and adjusting accordingly, moving away from the shore in the presence of disturbances [87] and reducing the number of nesting attempts in the increasing presence of artificial light [66]. While females tend to show a high variability in nest site selection [88], the presence of humans may interfere with the female turtles' behaviour, limiting their choice. Research shows that the hatching success decreased towards the waterline; thus, nest site selection is crucial [88].

Undetected and unprotected nests are highly prone to being trampled on by beach users. One such instance was the nest CFA, in this study, which was only found after the turtles hatched, got trapped under the sand, and were found by chance. This nest was characterized by a high percentage of dead corpses of hatched individuals that found difficulty in making it through the highly compacted sand above, possibly due to sunbathers.

Knowledge of reproductive behaviour is crucial to better understand the species' biology in an area and to be in a better position to design and manage Natura 2000 sites [39].

5. Conclusions

This work provides the first detailed insights into the *C. caretta* reproductive and nesting behaviour around Maltese islands through the use of genetic markers. The use of these genetic tools to understand flagship and vulnerable species in the Central Mediterranean offers an opportunity to better assess the importance of this area vis-à-vis the various behavioural aspects related to the biology of species in the region [89–91]. Sustained turtle research on the number of turtle egg clutches laid, nesting sites, hatching success rate, and developmental ecology are required, side by side with field research and conservation genetics research, for a complete conservation status assessment.

Supplementary Materials: The following supporting information can be downloaded at: https://www.mdpi.com/article/10.3390/ani14010137/s1. Table S1: List of primer, sequences (forward and reverse primers, including M13 sequence), fluorescent dye used and references.

Author Contributions: Conceptualization, A.V.; methodology, A.V. and N.V.; software, N.V.; validation, A.V. and N.V.; formal analysis, A.V. and N.V.; investigation, A.V. and N.V.; resources, A.V.; data curation, A.V. and N.V.; writing—original draft preparation, A.V. and N.V.; writing—review and editing, A.V. and N.V.; visualization, A.V. and N.V.; project administration, A.V.; funding acquisition, A.V. All authors have read and agreed to the published version of the manuscript.

Funding: This work was financially supported by the BioCon_Innovate Research Excellence Grant from the University of Malta, [grant No. I18LU06-01] awarded to A.V.

Institutional Review Board Statement: For this study on protected species, approval was sought and obtained from the local Environment and Resource Authority, Malta (ERA). The dead specimens analysed were picked from their natural habitat by ERA and handed over to A.V. for research purposes in accordance with ERA permits **NP0386/18**, **NP0503/19**, **EP1028/21**, and **EP1470/22**, covering the years 2018 to 2023.

Informed Consent Statement: Not applicable.

Data Availability Statement: Mitochondrial DNA data related to the analyses conducted during this study are available on GenBank under accession numbers PP056536–PP056543.

Acknowledgments: The authors would like to thank the local authorities, namely, the ERA, for making these dead specimens available for academic scientific research.

Conflicts of Interest: The authors declare no conflicts of interest.

References

1. Gramentz, D. Marine Turtles in the Central Mediterranean Sea. *Centro* **1989**, *1*, 41–56.
2. Casale, P.; Broderick, A.C.; Camiñas, J.A.; Cardona, L.; Carreras, C.; Demetropoulos, A.; Fuller, W.J.; Godley, B.J.; Hochscheid, S.; Kaska, Y.; et al. Mediterranean Sea turtles: Current knowledge and priorities for conservation and research. *Endanger. Species Res.* **2018**, *36*, 229–267. [CrossRef]
3. Wallace, B.P.; DiMatteo, A.D.; Hurley, B.J.; Finkbeiner, E.M.; Bolten, A.B.; Chaloupka, M.Y.; Hutchinson, B.J.; Abreu-Grobois, F.A.; Amorocho, D.; Bjorndal, K.A.; et al. Regional Management Units for marine turtles: A novel framework for prioritizing conservation and research across multiple scales. *PLoS ONE* **2010**, *5*, e15465. [CrossRef] [PubMed]
4. Deidun, A.; Schembri, P.J. A report of nesting on a Maltese beach by the loggerhead turtle *Caretta caretta* (Linnaeus, 1758) (Reptilia: Cheloniidae). *Cent. Mediterr. Nat.* **2005**, *4*, 137–138.
5. Sönmez, B.; Karaman, S.; Turkozan, O. Effect of predicted sea level rise scenarios on green turtle (*Chelonia mydas*) nesting. *J. Exp. Mar. Bio. Ecol.* **2021**, *541*, 151572. [CrossRef]
6. Mancino, C.; Canestrelli, D.; Maiorano, L. Going west: Range expansion for loggerhead sea turtles in the Mediterranean Sea under climate change. *Glob. Ecol. Conserv.* **2022**, *38*, e02264. [CrossRef]
7. Hochscheid, S.; Maffucci, F.; Abella, E.; Bradai, M.N.; Camedda, A.; Carreras, C.; Claro, F.; de Lucia, G.A.; Jribi, I.; Mancusi, C.; et al. Nesting range expansion of loggerhead turtles in the Mediterranean: Phenology, spatial distribution, and conservation implications. *Glob. Ecol. Conserv.* **2022**, *38*, e02194. [CrossRef]
8. Vella, N.; Vella, A. The first genetic analyses of the Leatherback turtle, *Dermochelys coriacea* from a stranding in the Central Mediterranean. *Rapp. Comm. Int. Mer Méditerr.* **2016**, *41*, 500.
9. Karaa, S.; Jribi, I.; Bouain, A.; Girondot, M.; Bradai, M.N. On the occurrence of leatherback turtles *Dermochelys coriacea* (Vandelli, 1761), in Tunisian waters (Central Mediterranean Sea) (Testudines: Dermochelydae). *Herpatozoa* **2013**, *26*, 65–75.
10. Casale, P.; Tucker, A.D. *Caretta caretta* (Amended Version of 2015 Assessment). The IUCN Red List of Threatened Species 2017. Available online: https://www.iucnredlist.org/species/3897/119333622 (accessed on 8 November 2023).
11. Casale, P. *Caretta caretta* (Mediterranean Subpopulation). The IUCN Red List of Threatened Species 2015. Available online: https://www.iucnredlist.org/species/83644804/83646294 (accessed on 8 November 2023).
12. United Nations Convention for the Protection of the Mediterranean Sea against Pollution. Available online: https://www.unep.org/unepmap/who-we-are/barcelona-convention-and-protocols (accessed on 8 November 2023).
13. CMS Appendices I and II of the Convention on the Conservation of Migratory Species of Wild Animals. Available online: https://www.cms.int/sites/default/files/basic_page_documents/appendices_cop13_e_0.pdf (accessed on 8 November 2023).
14. Convention on the Conservation of European Wildlife and Natural Habitats (Bern Convention) Annex III. Available online: https://rm.coe.int/168097eb57 (accessed on 8 November 2023).
15. CITES Convention on International Trade in Endangered Species of Wild Fauna and Flora. Appendices I, II and III. Available online: https://cites.org/sites/default/files/eng/app/2023/E-Appendices-2023-05-21.pdf (accessed on 8 November 2023).
16. UNEP/MAP-SPA/RAC SPA-BD Protocol Annex II: List of Endangered or Threatened Species. Available online: https://www.rac-spa.org/sites/default/files/annex/annex_2_en_20182.pdf (accessed on 8 November 2023).
17. ICCAT Recommendation by ICCAT Amending Recommendation 10-09 on the By-Catch of Sea Turtles in ICCAT Fisheries. 2011. Available online: https://www.iccat.int/Documents/Recs/compendiopdf-e/2013-11-e.pdf (accessed on 8 November 2023).
18. GFCM Recommendation GFCM/35/2011/4 on the Incidental Bycatch of Sea Turtles in Fisheries in the GFCM Area of Application. 2011. Available online: https://faolex.fao.org/docs/pdf/mul201486.pdf (accessed on 8 November 2023).
19. Girard, F.; Girard, A.; Monsinjon, J.; Arcangeli, A.; Belda, E.; Cardona, L.; Casale, P.; Catteau, S.; David, L.; Dell'Amico, F.; et al. Toward a common approach for assessing the conservation status of marine turtle species within the European Marine Strategy Framework Directive. *Front. Mar. Sci.* **2022**, *9*, 790733. [CrossRef]
20. Jensen, M.P.; Allen, C.D.; Eguchi, T.; Bell, I.P.; LaCasella, E.L.; Hilton, W.A.; Hof, C.A.M.; Dutton, P.H. Environmental warming and feminization of one of the largest sea turtle populations in the world. *Curr. Biol.* **2018**, *28*, 154–159.e4. [CrossRef]
21. Clusa, M.; Carreras, C.; Pascual, M.; Demetropoulos, A.; Margaritoulis, D.; Rees, A.F.; Hamza, A.A.; Khalil, M.; Aureggi, M.; Levy, Y.; et al. Mitochondrial DNA reveals Pleistocenic colonisation of the Mediterranean by loggerhead turtles (*Caretta caretta*). *J. Exp. Mar. Bio. Ecol.* **2013**, *439*, 15–24. [CrossRef]
22. Garofalo, L.; Mingozzi, T.; Micò, A.; Novelletto, A. Loggerhead turtle (*Caretta caretta*) matrilines in the Mediterranean: Further evidence of genetic diversity and connectivity. *Mar. Biol.* **2009**, *156*, 2085–2095. [CrossRef]
23. Carpio Camargo, A.J.; Álvarez Gutiérrez, Y.; Jaramillo Véliz, J.; Sánchez Tortosa, F. Nesting failure of sea turtles in Ecuador—Causes of the loss of sea turtle nests: The role of the tide. *J. Coast. Conserv.* **2020**, *24*, 55. [CrossRef]
24. Patrício, A.R.; Hawkes, L.A.; Monsinjon, J.R.; Godley, B.J.; Fuentes, M.M.P.B. Climate change and marine turtles: Recent advances and future directions. *Endanger. Species Res.* **2021**, *44*, 363–395. [CrossRef]
25. Gammon, M.; Whiting, S.; Fossette, S. Vulnerability of sea turtle nesting sites to erosion and inundation: A decision support framework to maximize conservation. *Ecosphere* **2023**, *14*, e4529. [CrossRef]
26. Pike, D.A.; Roznik, E.A.; Bell, I. Nest inundation from sea-level rise threatens sea turtle population viability. *R. Soc. Open Sci.* **2015**, *2*, 4–6. [CrossRef]
27. Tanner, C.E.; Marco, A.; Martins, S.; Abella_perez, E.; Hawkes, L.A. Highly feminised sex ratio estimations for the world's third largest nesting aggregation of the loggerhead sea turtle. *Mar. Ecol. Prog. Ser.* **2019**, *621*, 209–219. [CrossRef]

28. Martins, S.; Silva, E.; Abella, E.; de Santos Loureiro, N.; Marco, A. Warmer incubation temperature influences sea turtle survival and nullifies the benefit of a female-biased sex ratio. *Clim. Chang.* **2020**, *163*, 689–704. [CrossRef]
29. Lopez, G.G.; Saliés, E.d.C.; Lara, P.H.; Tognin, F.; Marcovaldi, M.A.; Serafini, T.Z. Coastal development at sea turtles nesting ground: Efforts to establish a tool for supporting conservation and coastal management in northeastern Brazil. *Ocean Coast. Manag.* **2015**, *116*, 270–276. [CrossRef]
30. Matiddi, M.; Hochsheid, S.; Camedda, A.; Baini, M.; Cocumelli, C.; Serena, F.; Tomassetti, P.; Travaglini, A.; Marra, S.; Campani, T.; et al. Loggerhead sea turtles (*Caretta caretta*): A target species for monitoring litter ingested by marine organisms in the Mediterranean Sea. *Environ. Pollut.* **2017**, *230*, 199–209. [CrossRef] [PubMed]
31. Franchini, D.; Paci, S.; Ciccarelli, S.; Valastro, C.; Salvemini, P.; Di Bello, A. Clinical findings, management, imaging, and outcomes in Sea turtles with traumatic head injuries: A retrospective study of 29 *Caretta caretta*. *Animals* **2023**, *13*, 152. [CrossRef] [PubMed]
32. Anjos, M.R.; Pelegrini, L.S.; Barreiros, J.B.; Barcelos, L.M.D.; Barreiros, J.P. A healed propeller wound on a wild loggerhead turtle, *Caretta caretta* (Testudines, Cheloniidae), occupied by a pelagic crab, *Planes minutus*, in the Azores Islands, Portugal. *Herpetol. Notes* **2021**, *14*, 981–984.
33. Stelfox, M.; Hudgins, J.; Sweet, M. A review of ghost gear entanglement amongst marine mammals, reptiles and elasmobranchs. *Mar. Pollut. Bull.* **2016**, *111*, 6–17. [CrossRef] [PubMed]
34. Lucchetti, A.; Bargione, G.; Petetta, A.; Vasapollo, C.; Virgili, M. Reducing sea turtle bycatch in the mediterranean mixed demersal fisheries. *Front. Mar. Sci.* **2019**, *6*, 387. [CrossRef]
35. Micheli, F.; Halpern, B.S.; Walbridge, S.; Ciriaco, S.; Ferretti, F.; Fraschetti, S.; Lewison, R.; Nykjaer, L.; Rosenberg, A.A. Cumulative human impacts on Mediterranean and Black Sea marine ecosystems: Assessing current pressures and opportunities. *PLoS ONE* **2013**, *8*, e79889. [CrossRef]
36. Wang, J. Computationally efficient sibship and parentage assignment from multilocus marker data. *Genetics* **2012**, *191*, 183–194. [CrossRef]
37. Jones, O.R.; Wang, J. Colony: A program for parentage and sibship inference from multilocus genotype data. *Mol. Ecol. Resour.* **2010**, *10*, 551–555. [CrossRef]
38. Legislation Malta Subsidiary Legislation 549.44: Flora, Fauna, and Natural Habitats Protection Regulations. Available online: https://legislation.mt/eli/sl/549.44/eng/pdf (accessed on 8 November 2023).
39. ERA Natura 2000 Datasheets & Maps. Available online: https://era.org.mt/topic/natura-2000-datasheets-maps/ (accessed on 8 November 2023).
40. Abreu-Grobois, F.A.; Horrocks, J.A.; Formia, A.; Dutton, P.H.; LeRoux, R.A.; Velez-Zuazo, X.; Soares, L.S.; Meylan, A.B. New mtDNA D-loop primers which work for a variety of marine turtle species may increase the resolution of mixed stock analysis. In Proceedings of the 26th Annual Symposium on Sea Turtle Biology, Island of Crete, Greece, 3–8 April 2006; p. 179.
41. Shamblin, B.M.; Bolten, A.B.; Abreu-grobois, F.A.; Bjorndal, K.A.; Cardona, L.; Nairn, C.J.; Nielsen, J.T.; Dutton, P.H. Geographic patterns of genetic variation in a broadly distributed marine vertebrate: New insights into Loggerhead turtle stock structure from expanded mitochondrial DNA sequences. *PLoS ONE* **2014**, *9*, e85956. [CrossRef]
42. Bowen, B.W.; Bass, A.L.; Soares, L.; Toonen, R.J. Conservation implications of complex population structure: Lessons from the loggerhead turtle (*Caretta caretta*). *Mol. Ecol.* **2005**, *14*, 2389–2402. [CrossRef]
43. Fitzsimmons, N.N. Single paternity of clutches and sperm storage in the promiscuous green turtle (*Chelonia mydas*). *Mol. Ecol.* **1998**, *7*, 575–584. [CrossRef] [PubMed]
44. Moore, M.K.; Ball, R.M. Multiple paternity in loggerhead turtle (*Caretta caretta*) nests on Melbourne Beach, Florida: A microsatellite analysis. *Mol. Ecol.* **2002**, *11*, 281–288. [CrossRef] [PubMed]
45. Martin, A.P.; Pardini, A.T.; Noble, L.R.; Jones, C.S. Conservation of a dinucleotide simple sequence repeat locus in sharks. *Mol. Phylogenet. Evol.* **2002**, *23*, 205–213. [CrossRef] [PubMed]
46. Monzón-Argüello, C.; Muñoz, J.; Marco, A.; López-Jurado, L.F.; Rico, C. Twelve new polymorphic microsatellite markers from the loggerhead sea turtle (*Caretta caretta*) and cross-species amplification on other marine turtle species. *Conserv. Genet.* **2008**, *9*, 1045–1049. [CrossRef]
47. Shamblin, B.M.; Faircloth, B.C.; Dodd, M.; Wood-Jones, A.; Castleberry, S.B.; Carroll, J.P.; Nairn, C.J. Tetranucleotide microsatellites from the loggerhead sea turtle (*Caretta caretta*). *Mol. Ecol. Notes* **2007**, *7*, 784–787. [CrossRef]
48. Shamblin, B.M.; Faircloth, B.C.; Dodd, M.G.; Bagley, D.A.; Ehrhart, L.M.; Dutton, P.H.; Frey, A.; Nairn, C.J. Tetranucleotide markers from the loggerhead sea turtle (*Caretta caretta*) and their cross-amplification in other marine turtle species. *Conserv. Genet.* **2009**, *10*, 577–580. [CrossRef]
49. Kearse, M.; Moir, R.; Wilson, A.; Stones-Havas, S.; Cheung, M.; Sturrock, S.; Buxton, S.; Cooper, A.; Markowitz, S.; Duran, C.; et al. Geneious Basic: An integrated and extendable desktop software platform for the organization and analysis of sequence data. *Bioinformatics* **2012**, *28*, 1647–1649. [CrossRef]
50. NCBI. GenBank. Available online: http://www.ncbi.nlm.nih.gov/genbank/ (accessed on 8 November 2023).
51. Benson, D.A.; Cavanaugh, M.; Clark, K.; Karsch-Mizrachi, I.; Lipman, D.J.; Ostell, J.; Sayers, E.W. GenBank. *Nucleic Acids Res.* **2013**, *41*, 36–42. [CrossRef]
52. Amos, W.; Hoffman, J.I.; Frodsham, A.; Zhang, L.; Best, S.; Hill, A.V.S. Automated binning of microsatellite alleles: Problems and solutions. *Mol. Ecol. Notes* **2007**, *7*, 10–14. [CrossRef]

53. Excoffier, L.; Lischer, H.E.L. Arlequin suite ver 3.5: A new series of programs to perform population genetics analyses under Linux and Windows. *Mol. Ecol. Resour.* **2010**, *10*, 564–567. [CrossRef]
54. Van Oosterhout, C.; Hutchinson, W.F.; Wills, D.P.M.; Shipley, P. Micro-checker: Software for identifying and correcting genotyping errors in microsatellite data. *Mol. Ecol. Notes* **2004**, *4*, 535–538. [CrossRef]
55. Encalada, S.E.; Bjorndal, K.A.; Bolten, A.B.; Zurita, J.C.; Schroeder, B.; Possardt, E.; Sears, C.J.; Bowen, B.W. Population structure of loggerhead turtle (*Caretta caretta*) nesting colonies in the Atlantic and Mediterranean as inferred from mitochondrial DNA control region sequences. *Mar. Biol.* **1998**, *130*, 567–575. [CrossRef]
56. Splendiani, A.; Fioravanti, T.; Giovannotti, M.; D'Amore, A.; Furii, G.; Totaro, G.; Nisi Cerioni, P.; Lucchetti, A.; Caputo Barucchi, V. Mitochondrial DNA reveals the natal origin of *Caretta caretta* (Testudines: Cheloniidae) stranded or bycaught along the South-western Adriatic coasts. *Eur. Zool. J.* **2017**, *84*, 566–574. [CrossRef]
57. Garofalo, L.; Mastrogiacomo, A.; Casale, P.; Carlini, R.; Eleni, C.; Freggi, D.; Gelli, D.; Knittweis, L.; Mifsud, C.; Mingozzi, T.; et al. Genetic characterization of central Mediterranean stocks of the loggerhead turtle (*Caretta caretta*) using mitochondrial and nuclear markers, and conservation implications. *Aquat. Conserv. Mar. Freshw. Ecosyst.* **2013**, *23*, 868–884. [CrossRef]
58. Loisier, A.; Savelli, M.P.; Arnal, V.; Claro, F.; Gambaiani, D.; Sénégas, J.B.; Cesarini, C.; Sacchi, J.; Miaud, C.; Montgelard, C. Genetic composition, origin and conservation of loggerhead sea turtles (*Caretta caretta*) frequenting the French Mediterranean coasts. *Mar. Biol.* **2021**, *168*, 52. [CrossRef]
59. Kaska, A.; Koç, G.; Sözbilen, D.; Salih, D.; Glidan, A.; Alati, A. Increased sample size provides novel insights into population structure of Mediterranean loggerhead sea turtles. *Conserv. Genet. Resour.* **2023**, *15*, 67–75. [CrossRef]
60. Clusa, M.; Carreras, C.; Pascual, M.; Gaughran, S.J.; Piovano, S.; Giacoma, C.; Fernández, G.; Levy, Y.; Tomás, J.; Raga, J.A.; et al. Fine-scale distribution of juvenile Atlantic and Mediterranean loggerhead turtles (*Caretta caretta*) in the Mediterranean Sea. *Mar. Biol.* **2014**, *161*, 509–519. [CrossRef]
61. Vecchioni, L.; Arculeo, M.; Vamberger, M.; Marrone, F. Current status of and threats to Sicilian turtles. *Diversity* **2022**, *14*, 798. [CrossRef]
62. Carreras, C.; Pascual, M.; Cardona, L.; Aguilar, A.; Margaritoulis, D.; Rees, A.; Turkozan, O.; Levy, Y.; Gasith, A.; Aureggi, M.; et al. The genetic structure of the loggerhead sea turtle (*Caretta caretta*) in the Mediterranean as revealed by nuclear and mitochondrial DNA and its conservation implications. *Conserv. Genet.* **2007**, *8*, 761–775. [CrossRef]
63. Margaritoulis, D.; Argano, R.; Baran, I.; Bentivegna, F.; Bradai, M.N.; Camiñas, J.A.; Casale, P.; De Metrio, G.; Demetropoulos, A.; Gerosa, G.; et al. Loggerhead turtles in the Mediterranean: Present knowledge and conservation perspectives. In *Loggerhead Sea Turtles*; Bolten, A.B., Ed.; Smithsonian Institution Press: Washington, DC, USA, 2003; pp. 175–198.
64. Hamann, M.; Shimada, T.; Duce, S.; Foster, A.; To, A.T.Y.; Limpus, C. Patterns of nesting behaviour and nesting success for green turtles at Raine Island, Australia. *Endanger. Species Res.* **2022**, *47*, 217–229. [CrossRef]
65. Banan, I.; Turkozan, O. Nesting Activity of the Loggerhead Turtle, *Caretta caretta*, on Fethiye Beach, Turkey, in 1994. *Chelonian Conserv. Biol.* **1996**, *2*, 93–96.
66. Silva, E.; Marco, A.; Pérez, H.; Abella, E.; Patino-martinez, J.; Martins, S.; Almeida, C. Light pollution affects nesting behavior of loggerhead turtles and predation risk of nests and hatchlings. *J. Photochem. Photobiol. B Biol.* **2017**, *173*, 240–249. [CrossRef] [PubMed]
67. Lasala, J.A.; Hughes, C.; Wyneken, J. Female loggerhead sea turtles (*Caretta caretta* L.) rarely remate during nesting season. *Ecol. Evol.* **2020**, *10*, 163–174. [CrossRef] [PubMed]
68. Bowen, B.W.; Karl, S.A. Population genetics and phylogeography of sea turtles. *Mol. Ecol.* **2007**, *16*, 4886–4907. [CrossRef] [PubMed]
69. Purnama, D.; Zamani, N.P.; Farajallah, A. Microsatellite DNA Analysis on the polyandry of Green sea turtle *Chelonia mydas*. *Hayati J. Biosci.* **2013**, *20*, 182–186. [CrossRef]
70. Alfaro-Núñez, A.; Jensen, M.P.; Abreu-Grobois, F.A. Does polyandry really pay off? The effects of multiple mating and number of fathers on morphological traits and survival in clutches of nesting green turtles at Tortuguero. *PeerJ* **2015**, *3*, e880. [CrossRef]
71. Gaos, A.R.; Lewison, R.L.; Liles, M.J.; Henriquez, A.; Chavarría, S.; Yañez, I.L.; Stewart, K.; Frey, A.; Jones, T.T.; Dutton, P.H. Prevalence of polygyny in a critically endangered marine turtle population. *J. Exp. Mar. Biol. Ecol.* **2018**, *506*, 91–99. [CrossRef]
72. Lasala, J.A.; Harrison, J.S.; Williams, K.L.; Rostal, D.C. Strong male-biased operational sex ratio in a breeding population of loggerhead turtles (*Caretta caretta*) inferred by paternal genotype reconstruction analysis. *Ecol. Evol.* **2013**, *3*, 4736–4747. [CrossRef]
73. Harry, J.L.; Briscoe, D.A. Multiple paternity in the loggerhead turtle (*Caretta caretta*). *J. Hered.* **1988**, *79*, 96–99. [CrossRef]
74. Bollmer, A.J.L.; Irwin, M.E.; Rieder, J.P.; Parker, P.G.; Bollmer, J.L.; Irwin, M.E.; Rieder, J.P.; Parker, P.G. Multiple paternity in Loggerhead turtle clutches. *Copeia* **1999**, 475–478. [CrossRef]
75. Tedeschi, J.N.; Mitchell, N.J.; Berry, O.; Whiting, S.; Meekan, M.; Kennington, W.J. Reconstructed paternal genotypes reveal variable rates of multiple paternity at three rookeries of loggerhead sea turtles (*Caretta caretta*) in Western Australia. *Aust. J. Zool.* **2014**, *62*, 454–462. [CrossRef]
76. Sari, F.; Koseler, A.; Kaska, Y. First observation of multiple paternity in loggerhead sea turtles, *Caretta caretta*, nesting on Dalyan Beach, Turkey. *J. Exp. Mar. Bio. Ecol.* **2017**, *488*, 60–71. [CrossRef]
77. Zbinden, J.A.; Largiadèr, C.R.; Leippert, F.; Margaritoulis, D.; Arlettaz, R. High frequency of multiple paternity in the largest rookery of Mediterranean loggerhead sea turtles. *Mol. Ecol.* **2007**, *16*, 3703–3711. [CrossRef] [PubMed]
78. Lee, P.L.M.; Schofield, G.; Haughey, R.I.; Mazaris, A.D.; Hays, G.C. A review of patterns of multiple paternity across sea turtle rookeries. *Adv. Mar. Biol.* **2018**, *79*, 1–31. [CrossRef] [PubMed]

79. Caspers, B.A.; Krause, E.T.; Hendrix, R.; Kopp, M.; Rupp, O.; Rosentreter, K.; Steinfartz, S. The more the better—Polyandry and genetic similarity are positively linked to reproductive success in a natural population of terrestrial salamanders (*Salamandra salamandra*). *Mol. Ecol.* **2014**, *23*, 239–250. [CrossRef] [PubMed]
80. Uller, T.; Olsson, M. Multiple copulations in natural populations of lizards: Evidence for the fertility assurance hypothesis. *Behaviour* **2005**, *142*, 45–56. [CrossRef]
81. Slatyer, R.A.; Mautz, B.S.; Backwell, P.R.Y.; Jennions, M.D. Estimating genetic benefits of polyandry from experimental studies: A meta-analysis. *Biol. Rev.* **2012**, *87*, 1–33. [CrossRef] [PubMed]
82. Casale, P.; Freggi, D.; Maffucci, F.; Hochscheid, S. Adult sex ratios of loggerhead sea turtles (*Caretta caretta*) in two Mediterranean foraging grounds. *Sci. Mar.* **2014**, *78*, 303–309. [CrossRef]
83. Pearse, D.E.; Avise, J.C. Turtle mating systems: Behavior, sperm storage, and genetic paternity. *J. Hered.* **2001**, *92*, 206–211. [CrossRef]
84. Pearse, D.E.; Janzen, F.J.; Avise, J.C. Multiple paternity, sperm storage, and reproductive success of female and male painted turtles (*Chrysemys picta*) in nature. *Behav. Ecol. Sociobiol.* **2002**, *51*, 164–171. [CrossRef]
85. Sakaoka, K.; Sakai, F.; Yoshii, M.; Okamoto, H.; Nagasawa, K. Estimation of sperm storage duration in captive loggerhead turtles (*Caretta caretta*). *J. Exp. Mar. Bio. Ecol.* **2013**, *439*, 136–142. [CrossRef]
86. Blasi, M.F.; Hochscheid, S.; Bardelli, R.; Bruno, C.; Melodia, C.; Salzeri, P.; De Rosa, P.; Madonia, P. First report on two loggerhead turtle (*Caretta caretta*) nests in the Aeolian Archipelago (Southern Italy). *Acta Herpetol.* **2022**, *17*, 37–43. [CrossRef]
87. Schofield, G.; Dickson, L.C.D.; Westover, L.; Dujon, A.M.; Katselidis, K.A. COVID-19 disruption reveals mass-tourism pressure on nearshore sea turtle distributions and access to optimal breeding habitat. *Evol. Appl.* **2021**, *14*, 2516–2526. [CrossRef] [PubMed]
88. Martins, S.; Patrício, R.; Clarke, L.J.; Santos, N.D.; Marco, A. High variability in nest site selection in a loggerhead turtle rookery, in Boa Vista Island, Cabo Verde. *J. Exp. Mar. Biol. Ecol.* **2022**, *556*, 151798. [CrossRef]
89. Vella, N.; Vella, A. Phylogeographic analyses of the Shortfin mako, *Isurus oxyrinchus* Rafinesque, 1810 (Chondrichthyes: Lamniformes) from the Central Mediterranean Sea, a critically endangered species in the region. *Fishes* **2023**, *8*, 520. [CrossRef]
90. Vella, N.; Vella, A. Population genetics of the deep-sea bluntnose sixgill shark, *Hexanchus griseus*, revealing spatial genetic heterogeneity. *Mar. Genom.* **2017**, *36*, 25–32. [CrossRef]
91. Vella, A.; Vella, N.; Karakulak, F.S.; Oray, I.; Garcia-Tiscar, S.; de Stephanis, R. Population genetics of Atlantic bluefin tuna, *Thunnus thynnus* (Linnaeus, 1758), in the Mediterranean: Implications for its conservation management. *J. Appl. Ichthyol.* **2016**, *32*, 523–531. [CrossRef]

animals

MDPI

Article

Finless Porpoise Bycatch and Stranding along the Shandong Peninsula, China, Based on Public Reports from 2000 to 2018

Tao Zuo [1,2,*], Yongtao Li [1], Zhaolong Cheng [1], Jun Wang [1,2], Jianqiang Sun [1], Wei Yuan [1] and Mingxiang Niu [1,2]

1 Yellow Sea Fisheries Research Institute, Chinese Academy of Fishery Sciences, Qingdao 266071, China
2 Laboratory for Marine Ecology and Environmental Science, Pilot National Laboratory for Marine Science and Technology (Qingdao), Qingdao 266237, China
* Correspondence: zuotao@ysfri.ac.cn

Simple Summary: The East Asian finless porpoise is a common whale along Chinese coasts. The coast of the Shandong Peninsula is a substantial habitat for the species. However, there is relatively insufficient knowledge of and protection measures for the species population in this region. Based on public literature, media, and internet social websites, we collected and analyzed over two hundred reports on the stranding and accidental catching of finless porpoises along the coast of the Shandong Peninsula from 2000 to 2018. We found that bycatch and stranding incidents occurred widely across the peninsula throughout all seasons. The extensive use of fishing gear was the principal cause of porpoise mortalities and injuries along Shandong Peninsula. We recommend a comprehensive network consisting of an incident reporting system, fishing gear regulations, and scientific monitoring plans to protect the cetacean species in this region.

Abstract: The Shandong Peninsula is located on the western coast of the Pacific and is adjacent to the Bohai Sea (BS) and the Yellow Sea (YS) to the east. The East Asian finless porpoise *Neophocaena asiaeorientalis sunameri*, a subspecies of the narrow-ridged finless porpoise *N. asiaeorientalis*, is the dominant cetacean resident along the Shandong Peninsula. However, there is insufficient monitoring data to determine the status of the cetacean species in this region. Based on the publicly available literature, media, and internet social website, this study investigated the spatial–temporal distribution of porpoise stranding and bycatch along the coast of the Shandong Peninsula. Data on over five hundred porpoises from two hundred reports between 2000 and 2018 were compiled and analyzed. Results showed that the bycatch and stranding of porpoises occurred widely across the peninsula throughout all months and increased rapidly between 2010 and 2017. The incidents were more frequent in the area where the BS and YS converged during the spring and early summer than in other seasons. The mean body length of bycaught porpoises was smaller than that of those found stranded. Fishing activities could be the principal cause of local finless porpoise incidents. However, limited data hindered a quantitative evaluation of the living conditions of finless porpoises in this area. Establishing a comprehensive monitoring system, which includes standardized reporting, rescue operations, and scientific research, is essential to finless porpoise protection along the Shandong Peninsula.

Keywords: finless porpoise; *Neophocaena asiaeorientalis sunameri*; stranding; bycatch; Shandong Peninsula

Citation: Zuo, T.; Li, Y.; Cheng, Z.; Wang, J.; Sun, J.; Yuan, W.; Niu, M. Finless Porpoise Bycatch and Stranding along the Shandong Peninsula, China, Based on Public Reports from 2000 to 2018. *Animals* **2023**, *13*, 3868. https://doi.org/10.3390/ani13243868

Academic Editor: Chunwang Li

Received: 7 November 2023
Revised: 6 December 2023
Accepted: 14 December 2023
Published: 15 December 2023

1. Introduction

The East Asian Finless Porpoise *Neophocaena asiaeorientalis sunameri* is a small-toothed cetacean with a wide distribution. It inhabits the coastal waters of the Taiwan Strait, the East China Sea, the Bohai Sea (BS), and the Yellow Sea (YS) in China, as well as in the coastal waters of Korea and Japan [1,2]. The species was believed to be closely related to the flagship Yangtze Finless Porpoise *N.* a. *asiaeorientalis* species in the Yangtze River, as two subspecies of the narrow-ridged finless porpoise *N. asiaeorientalis* [3]. However, recent phylogenetic and genomic studies have suggested that these two subspecies may be distinct and independent species [4]. The narrow-ridged finless porpoise plays a vital role in monitoring of the general health status of coastal ecosystems and maintaining ecological balance and biodiversity. However, the marine finless porpoise, like its subspecies in the Yangtze River [5], faces multiple endangering factors that put its existence at risk as a vulnerable species [6–9]. These factors include bycatch, vessel strikes, habitat loss, degradation, and food shortages. The population of this species has significantly declined on the west coast of Korea [10] and in the Inland Sea of Japan [11] over the past few decades. As a result, its conservation status was upgraded from Vulnerable (VU) to Endangered (EN) by the IUCN/SSC in 2017. In some local waters, it is at a high risk of becoming "Critically Endangered" (CR) [12].

The Shandong Peninsula is situated in northern China, on the western shore of the Pacific Ocean, and shares a border with the BS and YS in the east. It also faces the Korean Peninsula and the Japanese archipelago across a vast stretch of sea. It has over two hundred bays and ten estuaries of seagoing rivers including the Yellow River, the largest river in northern China. There are numerous traditional spawning, nursing, and feeding grounds for the different economically significant fish species [13] and whales [14–17] in these habitats.

N. a. sunameri is the predominant marine mammal in the coastal waters of the Shandong Peninsula [15]. However, anthropogenic activities, such as coastal engineering, petroleum exploitation, and aquaculture, have significantly increased the pressure on the finless porpoise in this area. The finless porpoise is inherently vulnerable to hazards due to its small size and coastal distribution [6]. Local fishermen have reported a decrease in sightings of the porpoise, suggesting a potential decline in its population in the area [18].

Monitoring such highly mobile and cryptic cetaceans in the wild is challenging and costly, especially over large-scale oceanic areas. Public reporting networks for the stranding and bycatch of aquatic mammals could be a supplemental method for cetacean ecological and preservation research. These networks could offer valuable scientific information on the status of cetaceans, especially for species with limited data [19–22]. Long-term reporting data have the same scientific reference value as at-sea scientific surveys, which could provide insights into cetacean diversity, distribution, and migration [19,20]. Similar reporting networks have been established in Korean waters [23], Hong Kong, Taiwan (http://tcsn.whale.org.tw accessed on 9 January 2019), and Hainan Island (http://www.cetacean.csdb.cn accessed on 9 January 2019) in the southern region of the China Seas [6,24–26]. However, the marine cetacean along the Shandong Peninsula is still data-deficient, except for a few studies on its species distribution in the 1970s–1980s [14,27–29] and in recent years [18,30,31]. They have not received the same level of conservation attention as their freshwater subspecies, the Yangtze finless porpoise. Regional information on finless porpoise bycatch and stranding is scattered in the media, literature, and other public platforms, covering various topics such as physiology and phylogenetics [32–37]. Therefore, constructing a reporting network is necessary for incidental events of standing and bycatch, which will be valuable for subsequent research and conservation efforts for the finless porpoise.

This study investigates stranding and bycatch incidents involving finless porpoises over the past twenty years along the Shandong Peninsula. Data from various sources, including the media, literature, and social websites, were analyzed to achieve two main objectives: Firstly, this study aims to understand the conservation status of finless porpoises regarding incidents of stranding, bycatch, and injury. The second aim is to establish an organized incident-reporting network for marine cetaceans and to provide insights for future decision making on the conservation and management of finless porpoises in the area.

2. Materials and Methods

2.1. Study Area

The Shandong Peninsula is located on the western coast of the Pacific Ocean, within the coordinates of 119°16′–122°42.3′ E and 35°05′–37°50′ N (Figure 1). It has a coastline of approximately 3345 km and consists of seven coastal districts: Bingzhou, Dongying, Weifang, Yantai, Weihai, Qingdao, and Rizhao. The region is characterized by a consistent eastward coastal current in the BS, and a southward current in the YS, following 30 m isobaths along the peninsula [38]. During winter, the Bohai Strait brings northward saline YS Warm Water into the BS [39]. The region is known for its important fishing grounds, including the Laizhou Bay fishing ground, Yan(tai)-Wei(hai) fishing ground, and Wei(hai)-Qing(dao) fishing ground, due to the confluence of multiple water systems [13]. For statistical analysis, the coastal areas were divided into three geographical sections: SBS (Laizhou Bay and Yellow River Estuary in the southern part of the BS), NYS (the northern YS from Penglai to Rongcheng), and SYS (the southern YS from Rongcheng to Qingdao) for analysis and comparison. The study defined four seasons as follows: Spring (March to May), Summer (June to August), Autumn (September to November), and Winter (December to February).

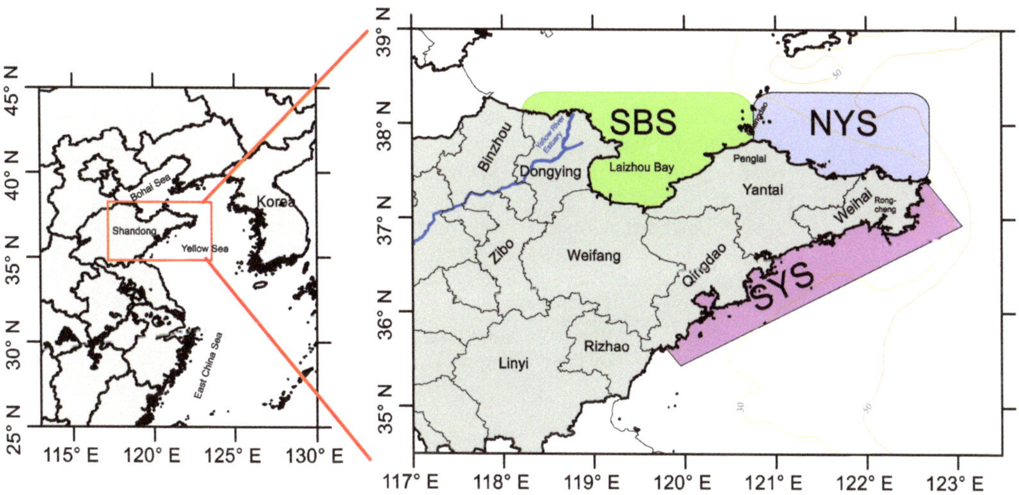

Figure 1. Map of the study area showing the regions analyzed.

2.2. Data Collection

In this study, stranding applied to an individual porpoise, whether alive or dead, who had been beached or washed up on the shore [25]. Bycatch applied to an individual, alive or dead, entangled in fishing gear [25]. Rescue referred to animals injured when discovered, but later successfully returned to the sea [25]. We collected extensive data from various sources, including the published literature, interviews with local fishermen, media reports, and social websites. The media consisted of newspapers, broadcasting, and television. The literature consisted of published articles in academic journals and dissertations. Social websites encompassed internet forums, blogs, and social platforms, as well as online reports on official websites.

Our Microsoft Access database only contained records verified by experts or confirmed with detailed descriptions, photos, or videos. Each record entry in the database included the porpoise number, sighting date, location of discovery, and physical characteristics, such as length, weight, and sex. We also recorded the life status of the porpoise (alive, dead, injured, or decomposed) and, if available, possible causes for injury or death. Missing information was registered as "undetermined" in the database. Events without the exact geographical coordinates were categorized based on the smallest relevant administrative unit. In cases where there were multiple reports of an event, we adopted the original reporting or discovering date. A mass event in this study referred to two or more porpoises caught or stranded at the same place and on the same date.

2.3. Data Analysis

In the following statistical analysis, we focused on data from 2000 to 2018. The locations of events were digitized on a map to visualize the spatial distribution of stranding and bycatch. The encounter rates of stranding and bycatch were calculated as the number of individuals per unit distance (ind./100 km) of the coastline [40]. The coastline length for each district was derived from the Shandong Administration of Surveying Mapping and Geo-Information.

The records of bycatch and stranding Incidents were categorized by year and month to identify annual and seasonal patterns. To explore the regional and seasonal differences in incident occurrence, an ANOVA with a statistical significance level of 0.05 was employed. Additionally, the body length distributions of stranding and bycatch were compared using the Kolmogorov–Smirnov test, with a statistical significance level of 0.05. The above analyses were conducted using STATISTICA 6.0 (StatSoft, Inc., Tulsa, OK, USA).

3. Results

3.1. Data Overview

Data on 606 finless porpoises were collected from 240 reports of target fishing, bycatch, and stranding between 1958 and 2018. Most of the information came from the published literature and the media, as shown in Table 1. The following analysis excluded the records of 66 targeted fishing porpoises before 1985 and two bycatches before 2000.

Table 1. Collection of incident events with individual number of finless porpoises along Shandong Peninsula.

	Media		Literature		Social Websites	
	Case Number	**Individual Number**	**Case Number**	**Individual Number**	**Case Number**	**Individual Number**
Stranding	111	123	4	33	48	47
Bycatch	37	61	12	237	16	28
Target-fished	0	0	7	66	0	0
Witnessed at sea	1	3	0	0	4	8

Between 2000 and 2018, about 326 bycatch and 203 stranding porpoises were recorded. However, only 150 bycatch and 165 stranding events had the exact location and date descriptions. There were about 14 mass events, consisting of 8 strandings (involving 19 individuals) and 6 bycatch incidents (involving 42 individuals).

3.2. Annual and Seasonal Variations

About 184 stranding and 185 bycatch porpoise individuals had information on the reporting year available. Figure 2 illustrates a significant increase in porpoise incidents since 2010. On average, there have been approximately 30 finless porpoise incidents per year over the past eighteen years along the Shandong Peninsula.

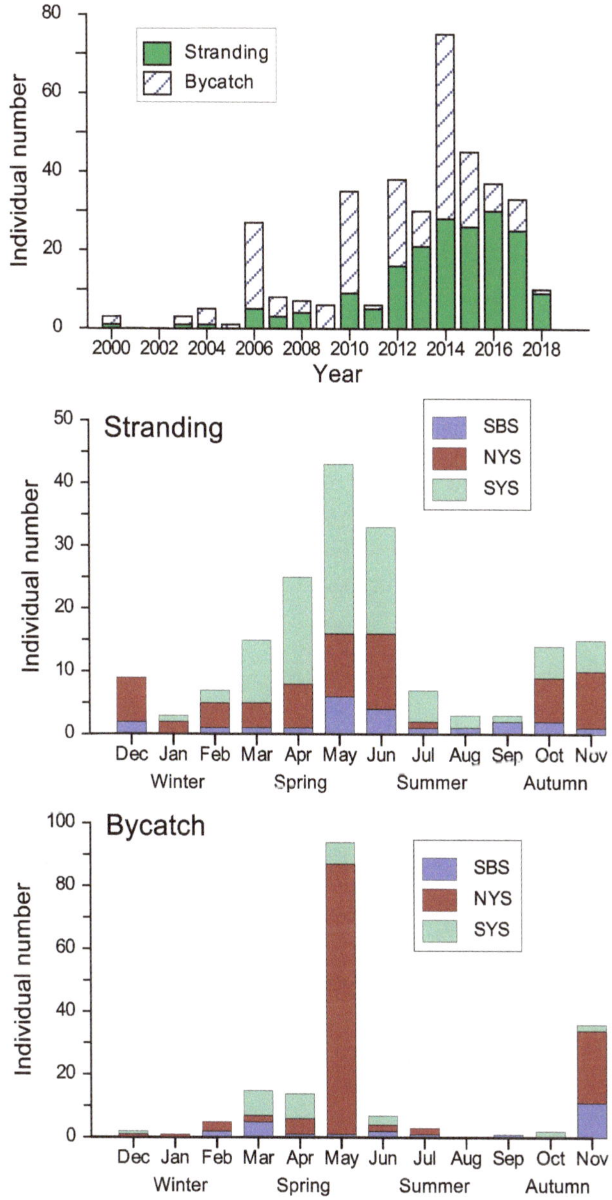

Figure 2. Annual and seasonal individual numbers of finless porpoise stranding and bycatch along the Shandong Peninsula.

Out of these incidents, exact dates were available for 177 strandings and 180 bycaught individuals. Both bycatch and stranding events occurred throughout the year, with a peak in May during the spring season (Figure 2). The maximum value of stranding occurred during spring and early summer (from April to June), primarily concentrated in SYS. The primary peak of bycatch in May was contributed mainly by Yantai and Weihai in NYS (Figures 2 and 3).

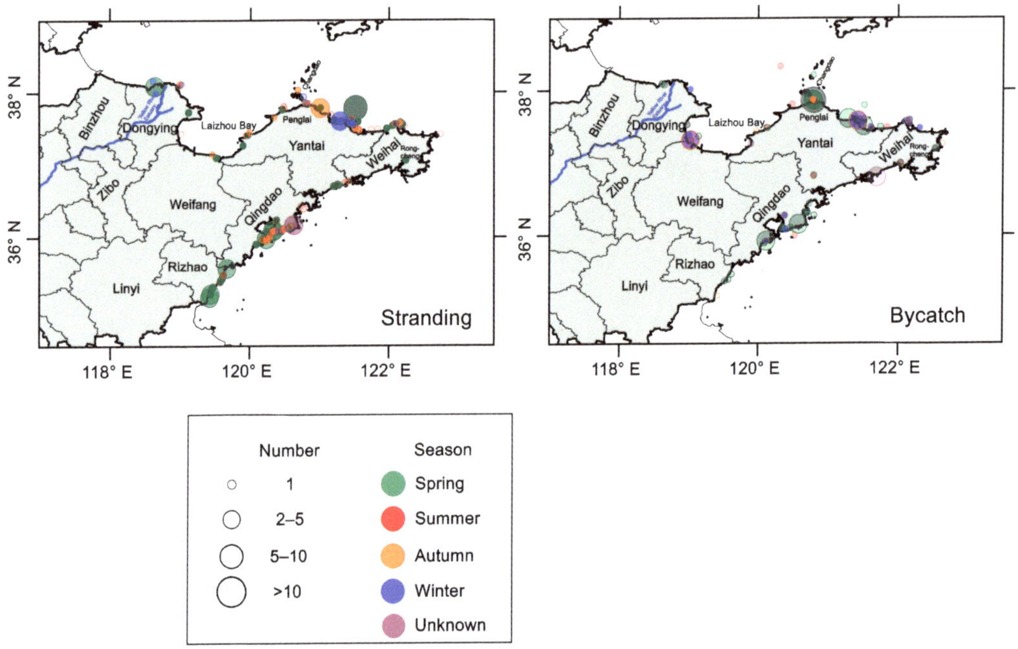

Figure 3. Distribution of finless porpoise stranding and bycatch from 2000 to 2018 along the Shandong Peninsula.

3.3. Geographic Distribution

Table 2 and Figure 3 show that stranding and bycatch occurred throughout the coasts of the Shandong Peninsula. NYS had the highest number of bycatch cases, while SYS had the highest stranding incidence. Both the bycatch and stranding encounter rates were the highest in NYS. Yantai had the most stranding and bycatch cases among the six districts. The highest stranding encounter rate occurred in Rizhao (25.1 ind. 100 km^{-1} shoreline), followed by Yantai and Qingdao. The highest bycatch encounter rate occurred in Yantai (17.9 ind. 100 km^{-1} shoreline), followed by Weihai and Weifang. However, there was no significant difference in either stranding or bycatch among three regions (the BS, NYS, and SYS) or six districts (ANOVA, $p > 0.05$).

Table 2. Finless porpoise number and encounter rate (individuals/100 km shoreline) of stranding and bycatch along Shandong Peninsula between 2000 and 2018.

		Stranding		Bycatch	
		Number	Encounter Rate	Number	Encounter Rate
District	Binzhou	0	-	0	-
	Dongying	10	2.4	15	3.6
	Weifang	2	1.4	12	8.6
	Yantai	70	7.6	151	16.6
	Weihai	34	3.0	124	10.9
	Qingdao	59	6.8	20	2.3
	Rizhao	26	26.1	4	4.0
Defined region	SBS	22	2.1	49	4.7
	NYS	83	10.6	216	26.9
	SYS	96	6.1	61	4.0

3.4. Status and Disposal

Out of the collected records, there were very few descriptions of the individual-level measurements of porpoises recorded (<5), apart from body length. In the study, the body-length records were available for 237 porpoises (102 strandings and 135 bycatches). Figure 4 shows that the body length of individual porpoises ranged from 50 cm to 260 cm. Over 85% of the recorded individuals had body lengths of between 100 cm and 200 cm. Additionally, over 63% of these individuals had body lengths of between 100 cm and 160 cm. The mean body length of stranded individuals was (137 ± 41 cm), which was significantly higher than that of bycatch individuals (127 ± 30 cm; Kolmogorov–Smirnov test, $p < 0.001$). Furthermore, there were significant differences in mean body length among the six districts (one-way analysis of variance, $F = 2.45$, $df = 6$, $p = 0.025$), with Qingdao having the highest value, followed by Rizhao, Dongying, Yantai, Weihai, and Weifang.

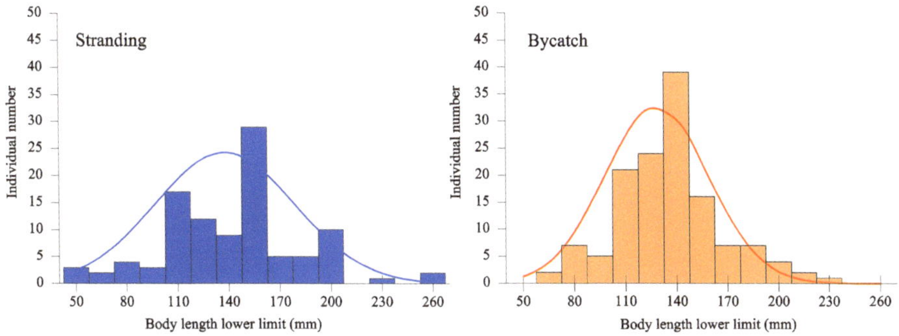

Figure 4. Body length-frequency distribution for stranded and bycatch finless porpoises between 2000 and 2018 along the Shandong Peninsula.

From the stranding records with a description of the cause of death, only four porpoises died of postpartum infection. Some stranding individuals had scars on their bodies, likely from fishing gear such as propellers, nets, or vessel collisions. Most of the bycatch porpoise's bodies had visible net marks or physical injuries (Figure 5). The bycatch porpoises were reported to be injured mainly by net cages in aquatic waters or entangled in fishing nets while at sea.

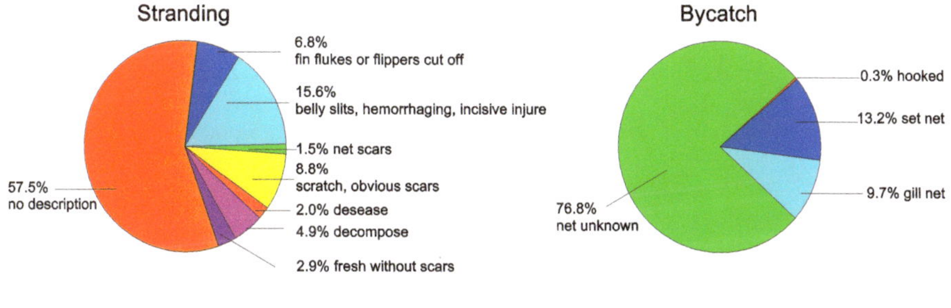

Figure 5. Mortality causes from the stranding and bycatch of finless porpoise along the Shandong Peninsula.

A total of 83 strandings and 277 bycatch individuals had descriptions of their carcass' disposal or rescue. Among them, only 66 individuals found were alive, and 42 were released (20 from stranding, 22 from bycatch). However, only ten individuals (four from stranding and the rest from bycatch) were in relatively good physical condition when released. The remaining dead individuals consisted of 59 strandings and 235 bycatches, of which approximately 74.8% were preserved or frozen in universities, aquaria, museums, and research institutes, 12.2% buried in situ, and 13% found for sale without authorization.

4. Discussion

4.1. Spatial and Temporal Variation

This study compiled the bycatch and stranding incidents of finless porpoises through an extensive review of public reports over the past two decades along the Shandong Peninsula. Despite some limitations in the data, they were still valuable for their insights into the temporal and spatial distribution of the porpoise population in the area. The results indicated that finless porpoises can be found year-round along the coasts of the Shandong Peninsula. Finless porpoise incidents occurred with relatively high frequency in the Laizhou Bay, Yan(tai)-Wei(hai), and Qing(dao)-Rong(cheng) coastal waters (Figure 2), which is consistent with previous studies in the 1980s [14,28]. These regions are known as highly productive fishing grounds [13] in zones where river runoff and oceanic waters are mixed between the YS and BS [38,39]. Moreover, they are also the substantial breeding areas for the finless porpoise [41].

As shown in Figures 2 and 3, porpoise incidents were relatively higher during the spring and early summer than in other seasons. These seasonal variations in finless porpoise occurrences could be related to factors such as prey movements, freshwater discharge, or monsoon rains [42]. The porpoise's prey, such as the fishes *Liza haematocheila*, *Sillago sihama*, and *Lateolabrax maculatus*, prefer to gather near the shore and reproduce at this time of the year [24]. Interestingly, a porpoise aggregation reported in June 2017 coincided with the spawning and schooling of *L. haematocheila* in Laoshan Bay, Qingdao. Additionally, there was a sub-peak in the bycatch and stranding of porpoises during the autumn season (Figure 2), which may be related to the migration of fish and continuous freshwater discharges. Historical records suggest that porpoises prefer to pursue highly migratory fish northward into the BS during early spring and southward into the NYS during fall and winter [14,15]. Schools of finless porpoises were observed swimming southeastward near Changdao Islands between the BS and NYS on 15 January 1999 [43]. Continuous freshwater discharges may be another factor promoting increases in porpoise levels. There were few cetacean records in local chronicles until the late 1800s, when the Yellow River emptied into the BS [17]. Local fishermen confirmed that finless porpoises almost disappeared near the Yellow River Estuary before implementing the Yellow River Water-Sediment Regulation project in 2002, especially during the severe drought period of 1987–2000.

4.2. Bycatch and Stranding

This study documented incidents of bycatch and stranding throughout the Shandong Peninsula (Figure 3). This continuous occurrence of incidents may be related to the extensive construction of aquaculture facilities and fishing enclosures (see Figure 6). Offshore aquaculture has been expanding rapidly along these coasts since the late 1990s, covering almost the entire traditional habitats of finless porpoises, from 10 m isobaths to 30 m isobaths. According to the China Fishery Statistics Yearbook of 2017, the total offshore aquaculture area in the Shandong Peninsula reached 3153 km^2, with offshore and ordinary cages accounting for 1,970,068 m^3 and 1,627,355 m^2, respectively. Additionally, this region hosts over 38,410 marine fishing vessels with an annual fish catch of 1.74 million tons (Figure 6), representing one-fifth of China's total marine catch. Such intensive aquaculture and fishing activities could lead to accidental entanglement and migration difficulties for finless porpoises, resulting in population declination and fragmentation of their distribution [7].

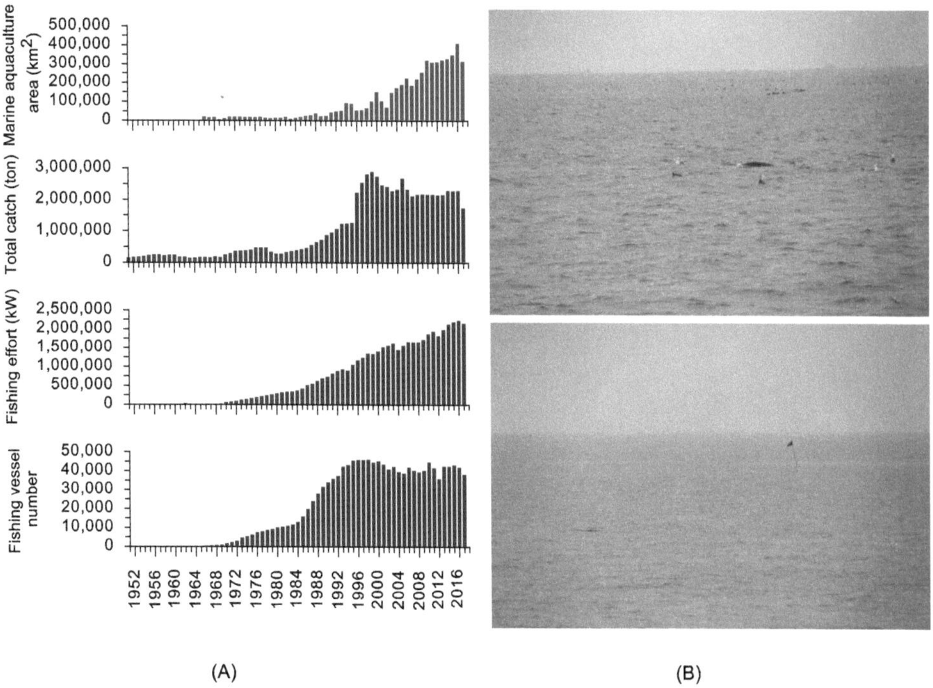

Figure 6. Annual trend of aquaculture area and fishing efforts along Shandong Peninsula (**A**) and a photo of a finless porpoise swimming close to nets in the aquaculture zone outside Rongcheng (**B**).

Fishery gear poses a significant threat to marine mammals worldwide, including the finless porpoise [21,22]. Evidence suggests that these porpoises experience incidental mortality directly or indirectly from fishing gear within their habitats [6–8,23]. Bycatch has resulted in an annual reduction in the porpoise population of over 15% in the Korean portion of YS [10]. In Ariake Sound and Tachibana Bay in Japan, gillnet entanglement alone has caused a population decline of 30–86% over three generations of porpoises [9]. Similarly, our results (Figure 5) indicate that the dominant causes of the porpoise mortalities and injuries along Shandong are interactions with gillnets, set nets, vessel propellers, and anti-predator facilities in aquaculture enclosures. In May 2014, about forty bycatch carcasses found at the Penglai port in Yantai showed evident scars from fishing gear [35]. However, the finless porpoise was the most commonly bycaught cetacean species along Chinese coasts, with the number of bycatch reaching more than 2000 individuals in 1994 [6]. This situation has not improved and may have worsened [24–26]. Due to the absence of a reliable regulatory and reporting network for the bycatch of marine cetaceans, the recorded number of porpoise incident events in this study is likely lower than the actual occurrence. In addition, similar to the situation in South Korea [44], most bycaught porpoises along the Shandong Peninsula were discarded at sea and rarely reported or taken back to the dock due to their low economic value and illegality, according to informal interviews with local fishermen. The individuals discarded or injured by collisions with vessels or propellers may drift to the beach via sea currents. This may explain the notable proportion of stranded individuals found with scars on their bodies (Figure 5).

Porpoise bycatch varied regionally and seasonally, as shown in Figures 2 and 3. The spring peak of bycatch may be due to intensified fishing efforts and the presence of finless porpoises closer to the shore during this season [23,24]. Mass bycatch events may be related to the local use of set nets, bottom driftnets in the NYS and SYS, and gillnets in the BS, respectively. Moreover, entanglement in fishing gear may have a more negative impact on younger individuals [41]. Our study found that bycaught porpoises tended to be smaller in body length than those found stranded (Figure 4). Similar observations were reported for humpback dolphins in Australia [22] and South Africa [45], which suggests that younger dolphins may engage in more risky behavior compared to older individuals, and be more at risk of bycatch.

Cetacean stranding can result from various factors, including changes in habitat, the availability of prey, and illness or mortality from natural causes, as well as currents and winds [46]. However, limited information is available on the reasons for porpoise stranding along the Shandong coasts. Based on a few necropsy reports, stranded individuals have shown symptoms of illness, parasite infection, or difficult births [32,37], as well as injuries from fishing gear.

4.3. Conservation and Management

It is still challenging to assess the threat level to the finless porpoise population around the Shandong Peninsula, owing to the absence of long-term scientific monitoring data. However, some prior cases and conversations with fishermen [28,47] suggest that the current finless porpoise population along the Shandong Peninsula could be significantly lower than in the past. For instance, in June 1959, over forty porpoises were found aggregating in a small bay in Dongying during an ebb tide [28]. Questionnaire surveys with local fishermen indicate that the current porpoise population may be less than 20% of its size in the early 1980s [18]. Recent surveys estimated a finless porpoise density ranging from 0.044 to around 0.115 ind. km^{-2} in Laizhou Bay [30] and 0.169 ind. km^{-2} in the eastern Shandong Peninsula [31]. These densities were much lower than those estimated in western Korean waters [10] and Japanese waters [48–51]. Given the relatively low density of porpoise populations along the Shandong Peninsula and the increasing trend of incidents, conservation efforts for porpoises in this area are necessary.

Fortunately, China joined the International Whaling Commission (IWC) in the 1980s and terminated whaling immediately. Furthermore, the finless porpoise is on the list of second-class National Protected Animals in China, and there is growing interest in its conservation from the public, authorities, scientists, and wildlife protection organizations in Shandong. Increasing reports of stranding and bycatch in recent years (Figure 3) may be attributed partly to increased public attention. Moreover, some conservation and rescue measures have been implemented to mitigate current threats to porpoises. For example, an aquarium in Penglai has rescued and rehabilitated more than 20 individuals, and released six injured individuals successfully into the sea between 2012 and 2014 [52]. In addition, enhanced conservation regulations have reduced the illegal trade of porpoises, as evidenced by the first prosecution in June 2018 [53]. Furthermore, since 2018, fishing restrictions in the BS and YS have been extended from May to August, which aligns with the finless porpoise's breeding season and the highest incidence of bycatch and stranding (Figure 3). These restrictions, by reducing fishing efforts, have proven effective in lessening fishery-related porpoise mortality in the Pearl River Estuary [54].

Based on this study, fishing activities pose a notable threat to the East Asian finless porpoise along the Shandong coastline. Since fishing efforts cannot be reduced significantly, it is essential to establish a local incident network for cetacean conservation. We recommend the following measures: (i) Implementing a routine and standardized reporting program under authorized laws and regulations. Improving spatial and temporal coverage of reported data, especially in traditional porpoise habitats and hotspots where incidents occur. (ii) Strengthening and integrating various stakeholders, including local fishery authorities, market regulations, aquariums, fishermen, trained volunteers, and mammal experts. (iii) Implementing local modifications and restrictions on fishery gear, such as installing mammal escape devices and acoustic deterrents to reduce unintentional entanglement. Our study identified specific fishing gear responsible for porpoise incidents, allowing for targeted modifications and restrictions based on the actual situation. (iv) Utilizing the data collected by the network and scientific monitoring projects to assess the population status of cetaceans and the effectiveness of these measures, which will facilitate improvements to the existing marine cetacean management and conservation system.

5. Conclusions

During the past two decades, public reports have indicated that finless porpoises could be found throughout the Shandong Peninsula. These porpoises have been stranded or bycaught unintentionally in fishing gear in all seasons and regions, with a higher occurrence in spring and early summer. The primary causes of accidental deaths and injuries of these porpoises are believed to be fishery facilities and activities. Smaller individuals are particularly affected by bycatch, which has a severe negative impact on them. Although there may be some bias in data from public reports, they still provide valuable insights into the current status of the data-limited finless porpoise along the Shandong Peninsula. Therefore, it is desirable to establish a comprehensive reporting and scientific monitoring network to protect this cetacean species.

Author Contributions: Conceptualization, T.Z. and Y.L.; methodology, T.Z.; software, T.Z.; validation, T.Z., Y.L., Z.C. and J.W.; formal analysis, T.Z.; investigation, T.Z., J.S. and M.N.; resources, W.Y. and M.N.; data curation, J.S.; writing—original draft preparation, T.Z.; writing—review and editing, T.Z.; supervision, J.W.; project administration, J.W. and Z.C.; funding acquisition, J.W. and Z.C. All authors have read and agreed to the published version of the manuscript.

Funding: This study was supported by the Shandong Provincial Natural Science Foundation (grant no. ZR2022QC099), the Bureau of Fisheries, The Ministry of Agriculture and Rural Affairs of the People's Republic of China (grant no. 125162002000210008), and the China National Offshore Oil Corporation Foundation (grant no. CF-MEEC/TR/2023-16).

Institutional Review Board Statement: Not applicable.

Informed Consent Statement: Not applicable.

Data Availability Statement: The data presented in this study are available on request from the corresponding author.

Conflicts of Interest: The authors declare no conflict of interest.

References

1. Gao, A. Morphological Differences and Genetic Variations among the Populations of *Neophocaena phocaenoides*. Ph.D. Dissertation, Nanjing Normal University, Nanjing, China, 1991. (In Chinese).
2. Gao, A.; Zhou, K. Geographical variation of external measurements and three subspecies of *Neophocaena phocaenoides* in Chinese waters. *Act. Theriol. Sin.* **1995**, *15*, 81–92. (In Chinese)
3. Jefferson, T.A.; Wang, J.Y. Revision of the taxonomy of finless porpoises (genus Neophocaena): The existence of two species. *J. Mar. Anim. Ecol.* **2011**, *4*, 3–16.
4. Zhou, X.; Guang, X.; Sun, D.; Xu, S.; Li, M.; Seim, I.; Jie, W.; Yang, L.; Zhu, Q.; Xu, J.; et al. Population genomics of finless porpoises reveal an incipient cetacean species adapted to freshwater. *Nat. Commun.* **2018**, *9*, 1276. [CrossRef] [PubMed]
5. Mei, Z.; Zhang, X.; Huang, S.L.; Zhao, X.; Hao, Y.; Zhang, L.; Qian, Z.; Zheng, J.; Wang, K.; Wang, D. The Yangtze finless porpoises: On an accelerating path to extinction? *Biol. Conserv.* **2014**, *172*, 117–123. [CrossRef]
6. Yang, G.; Zhou, K.Y.; Xu, X.R.; Leatherwood, S. A survey on the incidental catches of small cetaceans in coastal waters of China. *Chin. J. Appl. Ecol.* **1999**, *10*, 713–716. (In Chinese)
7. Reeves, R.R.; Wang, J.Y.; Leatherwood, S. The finless porpoise, *Neophocaena phocaenoides* (G. Cuvier 1829): A summary of current knowledge and recommendations for conservation action. *Asian Mar. Biol.* **1997**, *14*, 111–143.
8. Shirakihara, K.; Shirakihara, M. Finless porpoise bycatch in Ariake Sound and Tachibana Bay, Japan. *Endang. Species Res.* **2013**, *21*, 255–262. [CrossRef]
9. Hashimoto, M.; Shirakihara, K.; Shirakihara, M. Effects of bycatch on the population viability of the narrow-ridged finless porpoises in Ariake Sound and Tachibana Bay, Japan. *Endang. Species Res.* **2015**, *27*, 87–94. [CrossRef]
10. Park, K.J.; Sohn, H.; An, Y.R.; Kim, H.W.; An, D.H. A New Abundance Estimate for the Finless Porpoise *Neophocaena asiaeorientalis* on the West Coast of Korea: An Indication of Population Decline. *Fish Aquat. Sci.* **2015**, *18*, 411–416. [CrossRef]
11. Kasuya, T.; Yamamoto, Y.; Iwatsuki, T. Abundance decline in the finless porpoise population in the Inland Sea of Japan. *Raffles Bull. Zool.* **2002**, *50* (Suppl. S10), 57–65.
12. Wang, J.Y.; Reeves, R. *Neophocaena asiaeorientalis*. The IUCN Red List of Threatened Species 2017: e.T41754A50381766. Available online: https://www.iucnredlist.org/species/41754/50381766 (accessed on 9 January 2019).

13. Chen, D.G. *Fishery Ecology in the Yellow Sea and Bohai Sea*; China Ocean Press: Beijing, China, 1991; pp. 2–140. (In Chinese)
14. Dong, J.; Wang, G.; Xiao, Z. Migration and population difference of the finless porpoise in China. *Mar. Sci.* **1993**, *17*, 42–45. (In Chinese)
15. Wang, P.L. *Chinese Cetaceans*; Chemical Industry Press: Beijing, China, 2012; pp. 27–376. (In Chinese)
16. Yang, X.Y.; Sha, D.Y. On the research of whales. *Yindu J.* **2012**, *33*, 118–124. (In Chinese)
17. Sha, D.Y. Research on Whale Resources in the Local Chronicles of the Qing Dynasty. Master's Thesis, Ocean University of China, Qingdao, China, 2012. (In Chinese).
18. Cheng, Z.; Pine, M.K.; Li, Y.; Zuo, T.; Niu, M.; Wan, X.; Zhao, X.; Wang, K.; Wang, J. Using local ecological knowledge to determine ecological status and threats of the East Asian finless porpoise, *Neophocaena asiaeorientalis sunameri*, in south Bohai Sea, China. *Ocean. Coast. Manag.* **2021**, *203*, 105516. [CrossRef]
19. Pyenson, N.D. The high fidelity of the cetacean stranding record: Insights into measuring diversity by integrating taphonomy and macroecology. *Proc. Biol. Sci.* **2011**, *278*, 3608–3616. [CrossRef] [PubMed]
20. Peltier, H.; Dabin, W.; Daniel, P.; Van Canneyt, O.; Dorémus, G.; Huon, M.; Ridoux, V. The significance of stranding data as indicators of cetacean populations at sea: Modelling the drift of cetacean carcasses. *Ecol. Indic.* **2012**, *18*, 278–290. [CrossRef]
21. Reeves, R.R.; Mcclellan, K.; Werner, T.B. Marine mammal bycatch in gillnet and other entangling net fisheries, 1990 to 2011. *Endang. Species Res.* **2013**, *20*, 71–97. [CrossRef]
22. Meager, J.J.; Sumpton, W.D. Bycatch and strandings program as ecological indicator for data-limited cetaceans. *Ecol. Indic.* **2016**, *60*, 987–995. [CrossRef]
23. Song, K. Bycatch of cetaceans in Korea fisheries in the East Sea. *Fish. Res.* **2018**, *197*, 7–9. [CrossRef]
24. Wang, Y.; Li, W.; Van Waerebeek, K. Strandings, bycatches and injuries of aquatic mammals in China, 2000–2006, as reviewed from official documents: A compelling argument for a nationwide strandings programme. *Mar. Policy* **2015**, *51*, 242–250. [CrossRef]
25. Zhao, L.; Zhu, Q.; Miao, X.; Xu, M.; Wu, F.; Dai, Y.; Tao, C.; Mou, J.; Wang, X. An overview of cetacean strandings, bycatches and rescues along the western coast of the Taiwan Strait, China: 2010–2015. *Act. Oceanol. Sin.* **2017**, *36*, 31–36. [CrossRef]
26. Liu, M.; Lin, M.; Zhang, P.; Xue, T.; Li, S. An overview of cetacean stranding around Hainan Island in the South China Sea, 1978–2016: Implications for research, conservation and management. *Mar. Policy* **2019**, *101*, 147–153. [CrossRef]
27. Wang, P.L.; Liu, X.S. A study on finless porpoise from Bohai Sea and Yellow Sea. *Chin. Fish. Sci.* **1989**, *8*, 17–22. (In Chinese)
28. Wang, P.L. Surveys on the small toothed whale in the Bohai Sea and Yellow Sea. *Chin. J. Zool.* **1979**, *14*, 31–34. (In Chinese)
29. Wang, P.L. The distribution, ecology and resource protection of finless porpoise in the coastal oceans of China. *Trans. Liaoning Zool. Soc.* **1984**, *6*, 105–110. (In Chinese)
30. Zuo, T.; Sun, J.Q.; Shi, Y.Q.; Wang, J. Primary survey of finless porpoise population in the Bohai Sea. *Act. Theriol. Sin.* **2018**, *38*, 551–561. (In Chinese)
31. Li, Y.; Cheng, Z.; Zuo, T.; Niu, M.; Chen, R.; Wang, J. Distribution and Abundance of the East Asian Finless Porpoise in the coastal waters of Shandong Peninsula, Yellow Sea, China. *Fishes* **2023**, *8*, 410. [CrossRef]
32. Song, Z.X.; Luo, Q. Pathological report of a rescue stranding finless porpoise in Huanghai. *Chin. J. Vet. Med.* **2014**, *50*, 37–40. (In Chinese)
33. Lu, Z.; Xu, S.; Song, N.; Gao, T.X.; Tian, J.S.; Han, J.B. Analysis of the diet of finless porpoise (*Neophocaena asiaeorientalis sunameri*) based on prey morphological characters and DNA barcoding. *Conserv. Genet. Resour.* **2016**, *8*, 523–531. [CrossRef]
34. Zeng, X.; Huang, S.; Qian, Z.; Hao, Y.; Wang, D.; Ji, J.; Nabi, G. Characterization of milk composition in narrow-ridged finless porpoises (*Neophocaena asiaeorientalis*) at different lactation stages. *Mar. Mamm. Sci.* **2017**, *33*, 529–540. [CrossRef]
35. Zeng, X.; Ji, J.; Hao, Y.; Wang, D.; Hao, Y. Topographical distribution of blubber in finless porpoises (*Neophocaena asiaeorientalis sunameri*): A result from adapting to living in coastal waters. *Zool. Stud.* **2015**, *54*, 32. [CrossRef]
36. Wan, X.L.; McLaughlin, R.W.; Zheng, J.S.; Hao, Y.J.; Fan, F.; Tian, R.M.; Wang, D. Microbial communities in different regions of the gastrointestinal tract in East Asian finless porpoises (*Neophocaena asiaeorientalis sunameri*). *Sci. Rep.* **2018**, *8*, 14142. [CrossRef] [PubMed]
37. Wan, X.L.; Zheng, J.S.; Li, W.X.; Zeng, X.Y.; Wang, J.W.; Hao, Y.J.; Wang, D. Parasitic infections in the East Asian finless porpoise *Neophocaena asiaeorientalis sunameri* living off the Chinese Yellow/Bohai sea coast. *Dis. Aquat. Organ.* **2017**, *125*, 63. [CrossRef]
38. Zhang, Z.X.; Qiao, F.L.; Guo, J.S.; Yu, F. Seasonal variation of coastal water movement in the Southern Bohai Sea and water exchange between the Bohai Sea and the Yellow Sea. *Adv. Mar. Sci.* **2010**, *28*, 142–148. (In Chinese)
39. Chen, C.T.A. Chemical and physical fronts in the Bohai, Yellow and East China seas. *J. Mar. Syst.* **2009**, *78*, 394–410. [CrossRef]
40. Barbieri, M.M.; Raverty, S.; Hanson, M.; Venn-Watson Bradley, S.; Ford, J.K.; Gaydos, J.K. Spatial and temporal analysis of killer whale (*Orcinus orca*) strandings in the North Pacific Ocean and the benefits of a coordinated stranding response protocol. *Mar. Mamm. Sci.* **2013**, *29*, 448–462. [CrossRef]
41. Li, X.; Liu, Y.; Tzika, A.C.; Zhu, Q.; Van Doninck, K.; Milinkovitch, M.C. Analysis of global and local population stratification of finless porpoises *Neophocaena phocaenoides* in Chinese waters. *Mar. Biol.* **2011**, *158*, 1791–1804. [CrossRef]
42. Jefferson, T.A.; Hung, S.K.; Law, L.; Torey, M.; Tregenza, N. Distribution and abundance of finless porpoise in Hong Kong and adjacent waters of China. *Raffles Bull. Zool.* **2002**, *50* (Suppl. S10), 43–55.
43. Sun, Y.M. Thousands of files porpoise across the boundary of Bohai and Yellow Seas. *Ye Sheng Dong Wu* **1999**, *20*, 44. (In Chinese)
44. Lee, Y.R.; An, Y.R.; Park, K.J.; Sohn, H.; An, D.H.; Su, A.K. Age and reproduction of the finless porpoises, *Neophocaena asiaeorientalis*, in the Yellow Sea, Korea. *Anim. Cells Syst.* **2013**, *17*, 366–373. [CrossRef]

45. Atkins, S.; Cliff, G.; Pillay, N. Humpback dolphin bycatch in the shark nets in KwaZulu-Natal, South Africa. *Bio. Conserv.* **2013**, *159*, 442–449. [CrossRef]
46. Saavedra, C.; Pierce, G.J.; Gago, J.; Jusufovski, D.; Begoña Santos, M. Factors driving patterns and trends in strandings of small cetaceans. *Mar. Biol.* **2017**, *164*, 165. [CrossRef]
47. Meng, F.; Wang, Z.X.; Li, W. Capture, transfer and aquaculture of *Turiops truncatus* and *Neomeris phocaeuoides*. *Chin. J. Zool.* **1981**, *16*, 26–29. (In Chinese)
48. Yoshida, H. Population structure of finless porpoises (*Neophocaena phocaenoides*) in coastal waters of Japan. *Raffles Bull. Zool.* **2002**, *50* (Suppl. S10), 35–42.
49. Amano, M.; Nakahara, F.; Hayano, A.; Shirakihara, K. Abundance estimate of finless porpoises off the Pacific coast of eastern Japan based on aerial surveys. *Mamm. Study* **2003**, *28*, 103–110. [CrossRef]
50. Shirakihara, K.; Shirakihara, M.; Yamamoto, Y. Distribution and abundance of finless porpoise in the Inland Sea of Japan. *Mar. Biol.* **2007**, *150*, 1025–1032. [CrossRef]
51. Shirakihara, K.; Nakahara, F.; Shinohara, M.; Shirakihara, M.; Hiramatsu, K.; Irie, T. Abundance decline in the narrow-ridged finless porpoise population off the Pacific coast of eastern Japan. *Popul. Ecol.* **2019**, *61*, 325–332. [CrossRef]
52. Yin, E.M. Finless porpoise rescue and conservation. In Proceedings of the 1st Cross-Strait Symposium on Aquarium Development, China Wildlife Conservation Association, Zhuhai, China, 15 October 2014; pp. 25–28.
53. Na, M. First Case of illegal trade of Finless Porpoise-seller and buyer convicted. *Inside Outs. Court.* **2018**, *290*, 24–29. (In Chinese)
54. Lin, W.; Karczmarski, L.; Li, J.; Chan, S.C.Y.; Guo, L.; Wu, Y. Differential population dynamics of a coastal porpoise correspond to the fishing effort in a large estuarine system. *Aquat. Conserv. Mar. Feshw. Ecosyst.* **2019**, *29*, 223–234. [CrossRef]

animals

MDPI

Article

Predicting the Potential Distribution of the Szechwan Rat Snake (*Euprepiophis perlacea*) and Its Response to Climate Change in the Yingjing Area of the Giant Panda National Park

Xinqiang Song [1,2,†], Ying Jiang [1,3,†], Li Zhao [1,4,5], Long Jin [1,3,4,5], Chengzhi Yan [1,4,5] and Wenbo Liao [1,4,5,*]

[1] Key Laboratory of Southwest China Wildlife Resources Conservation (Ministry of Education), China West Normal University, Nanchong 637009, China
[2] Daxiangling Provincial Nature Reserve, Ya'an 625200, China
[3] School of Ecology and Nature Conservation, Beijing Forestry University, Beijing 100083, China
[4] Key Laboratory of Artificial Propagation and Utilization in Anurans of Nanchong City, China West Normal University, Nanchong 637009, China
[5] College of Panda, China West Normal University, Nanchong 637009, China
* Correspondence: liaobo_0_0@126.com
† These authors contributed equally to this work.

Simple Summary: This study employs maxent models to investigate the impact of climate change on the potential distribution of the Szechwan rat snake (*Euprepiophis perlacea*) in the Yingjing Area of the Giant Panda National Park. The analysis reveals the significant effects of distance from the streams and slope of the geographic distribution of *E. perlacea*. Furthermore, this study indicates a non-significant reduction in the potential distribution area for the species by the 2050s, which is attributed to reduced anthropogenic activities and habitat fragmentation within the Yingjing Area of the Giant Panda National Park. These findings validate the development of conservation guidelines for *E. perlacea*. More research is needed to identify suitable protected areas and wildlife corridors for *E. perlacea* across its distribution regions, thus enhancing its conservation prospects.

Abstract: Climate change is a significant driver of changes in the distribution patterns of species and poses a threat to biodiversity, potentially resulting in species extinctions. Investigating the potential distribution of rare and endangered species is crucial for understanding their responses to climate change and for the conservation of biodiversity and ecosystem management. The Szechwan rat snake (*Euprepiophis perlacea*) is an endemic and endangered species co-distributed with giant pandas, and studying its potential distribution contributes to a better understanding of the distribution pattern of endangered species. In this study, we confirmed seven presence points of this species in the Yingjing Area of the Giant Panda National Park, and selected eleven key factors to predict the potential distribution of *E. perlacea* under current and future scenarios using MaxEnt models. Our study consistently achieved AUC values exceeding 0.79, meeting the precision requirements of the models. The results indicated that the high potential distribution area of *E. perlacea* is mainly located near Yunwu mountain and the giant panda rewilding and reintroduction base, accounting for approximately 12% of the protected area. Moreover, we identified the primary environmental factors influencing the distribution of *E. perlacea* as the distance from streams and the slope degree, with their contribution rates exceeding 41% and 31%, respectively. In comparison to the current scenario, the potential habitat range for *E. perlacea* did not show an overall reduction in the context of future climate scenarios. To ensure the long-term preservation of *E. perlacea*, it is advisable to validate its actual distribution based on the models' results. Particular attention should be given to safeguarding its core distribution areas and raising awareness among residents within the potential distribution range about the conservation of *E. perlacea*.

Keywords: climate change; *E. perlacea*; environmental variables; distribution; conservation

Citation: Song, X.; Jiang, Y.; Zhao, L.; Jin, L.; Yan, C.; Liao, W. Predicting the Potential Distribution of the Szechwan Rat Snake (*Euprepiophis perlacea*) and Its Response to Climate Change in the Yingjing Area of the Giant Panda National Park. *Animals* **2023**, *13*, 3828. https://doi.org/10.3390/ani13243828

Academic Editor: Chunwang Li

Received: 31 October 2023
Revised: 4 December 2023
Accepted: 8 December 2023
Published: 12 December 2023

1. Introduction

The Earth is currently experiencing the sixth mass extinction event in biodiversity, with over 500 terrestrial animal species teetering on the brink of extinction [1,2]. Amid the myriad of factors contributing to the decline in species populations and extinctions, climate change resulting from human activities has already emerged as a primary and enduring threat to biodiversity [3–6]. A substantial body of research evidence underscores the profound effects of climate change on various facets of species, encompassing their distribution, phenology, morphology, physiology, and population dynamics [7–12]. As an illustration, the annual phenology traits of a Mediterranean snake species (*Vipera aspis*) are undergoing changes due to climate shifts [13]. Species exhibit adaptability to climate change through adaptive evolution and alterations in their distribution ranges. A wealth of evidence suggests that numerous species are undergoing migration toward polar regions and higher altitudes [14–16]. Considering the limited dispersal ability of ectothermic organisms and their pronounced temperature dependence, certain studies anticipate a substantial reduction in the potential distribution area of snakes [17,18]. Simultaneously, other research projects anticipate an expansion in the potential range of numerous temperate reptiles in the Northern Hemisphere due to climate change [19–22]. Notably, some species may be migrating to habitats beyond protected areas, resulting in their disappearance within these designated zones, thereby posing a challenge to the effective conservation of these species.

Climate change is instigating shifts in the geographical distribution, suitable habitats, and phenological patterns of wildlife species [23,24]. These shifts are causing them to move beyond the boundaries and functional zones of nature reserves primarily dedicated to wildlife conservation [25,26]. Consequently, the protective functions of these sanctuaries are at risk of being compromised. Previous studies indicate that the current and future network of protected areas in Morocco may not be sufficient to prevent amphibian and reptile species loss. New protected areas should be considered to protect species identified as highly vulnerable to climate change [27,28]. Meanwhile, a study on snakes in the Brazilian Atlantic Forest hotspot indicates that, by 2080, 73.6% of oviparous species and 67.6% of viviparous species could lose at least half of their original range [17]. The research finds that existing protected areas in the Atlantic Forest Hotspot have limited capacity to safeguard snakes presently, maintaining this precarious protection in the future [17]. Collectively, all the evidence underscores the significant impact of climate change on the biodiversity conservation functions of Nature Reserves, imposing substantial pressures and challenges for preserving biodiversity in these protected areas in the future. Therefore, the analysis and identification of the primary impacts of climate change on species distribution are critically important for understanding and addressing the climate change risks in Nature Reserves.

Rare and endangered species often exhibit limited geographic ranges and specific habitat requirements, rendering them less adaptable to climate change and elevating the risk of habitat loss and heightened extinction vulnerability [29]. Examining the influence of climate change on the distribution of these endangered and rare species holds paramount significance. Species Distribution Models (SDMs) constitute a class of models that utilize species occurrence data and environmental variables to project the potential distribution of species [30,31]. They have arisen as invaluable tools in extensive research on climate change prediction and have become indispensable for investigating how species' geographic distributions respond to climate change, and for assessing the effectiveness of protected areas under global change [32,33]. Among the array of species distribution models at our disposal, the MaxEnt model, grounded in the maximum entropy principle, distinguishes itself through its precision in outcome prediction and robust spatiotemporal extrapolation capabilities [34]. The MaxEnt model is characterized by several advantages, including its low sensitivity to collinearity among environmental variables, its robustness in situations with limited data points, and its capacity to model intricate relationships among variables [35,36]. Consequently, it has gained widespread and frequent application in recent years within the realms of conservation biology, invasion biology, and other fields that pertain to the impact of global climate change on species distribution and genetic diversity [37]. Of particular significance is the model's extensive utility in the domain of conservation biology, where it is extensively employed for the assessment of habitat suitability, exploration of species' ecological niche shifts, and the prognostication of global climate change impacts on the geographical distribution of specific species [38].

The Szechwan rat snake (*Euprepiophis perlacea*), is a species endemic to western China. It holds a critical status as a protected wildlife species in Sichuan Province, being classified as "Critically Endangered (CR)" in the China Red List of Endangered Animals [39]. Furthermore, it is listed as "Endangered (EN)" by the International Union for Conservation of Nature [40]. Since Stejneger named a male specimen collected in Ya'an in 1929 [41], no further discoveries were made for the next half-century, casting doubts on the authenticity of its species. Not until the 1980s did Deng collect a specimen labeled as number 3 in the Wolong Nature Reserve in Wenchuan County and the Hailuogou Forest Glacier Park in Luding County [42]. Zhao compared the differences in hemipenes between the *E. perlacea* and the Mandarin rat snake (*Euprepiophis mandarinus*), discussing the validity of the *E. perlacea* [39]. Chen et al. confirmed the validity of this species through molecular biology research [43]. In 2017, significant breakthroughs were achieved in the artificial breeding of *E. perlacea* [44]. However, this does not signify that *E. perlacea* has completely escaped a precarious situation. Whether *E. perlacea* remains endangered still depends on the stability of its habitat and the stability of the wild population. The sole investigation into the distribution of this species revealed a significant overlap in the distribution range between the *E. perlacea* and the giant pandas (*Ailuropoda melanoleuca*) [45]. The Szechwan rat snake primarily inhabits moist mountainous areas at elevations ranging from 1500 m to 2600 m, with habitats often found under deciduous broad-leaved forests, farmland, grasslands, and shrub thickets [45].

As an indicator species for the ecological environment, the presence of the Szechwan rat snake can be considered strong evidence of the excellent ecological conditions in giant panda habitats. However, despite being endangered and rare, the Szechwan rat snake, like most reptiles, has not received proactive protection from humans. Due to its small size and non-venomous nature, the Szechwan rat snake has limited migration and self-defense capabilities [43–45]. With the acceleration of habitat development, the Szechwan rat snake is frequently threatened by human activities. During the field investigation conducted in the Yingjing Area of the Giant Panda National Park, it was observed that many local residents, including farmers living around the habitat, adopt a hostile attitude towards non-venomous snakes, including the Szechwan rat snake. There was even a distressing incident where one Szechwan rat snake was found beaten to death. Therefore, the immediate priority is to predict the potential distribution areas of the Szechwan rat snake in the current and future scenarios within this protected area. This is essential for initiating habitat conservation and restoration efforts, aiming to better protect the natural population of the Szechwan rat snake in the wild. Moreover, it extends a range of recommendations and references that contribute to the broader cause of biodiversity conservation within the Yingjing Area of the Giant Panda National Park.

2. Materials and Methods

2.1. Study Site

The study was carried out in Yingjing Area of Giant Panda National Park, China which lies between 102°19′–102°55′ longitude and 29°28′–29°56′ latitude with a total geographical area of 836 km^2 (Figure 1) [46]. The altitudinal range within the Yingjing Area of the Giant Panda National Park is relatively extensive, with the highest elevation cresting at 3481 m, and the lowest point descending to 1150 m. This locale predominantly features a subtropical monsoon mountainous climate, boasting an annual mean temperature of 16.3 °C, with precipitation levels exceeding 1000 mm and an average frost period of approximately 60 days [47]. The spectrum of vegetation types encompasses evergreen broadleaf forests, soft (and hard) broadleaf forests, mixed coniferous and broadleaf forests, temperate coniferous forests, as well as subalpine shrublands and meadows [46]. This region is crisscrossed by meandering valleys, abounding in abundant water sources, thereby nurturing a lavish tapestry of biodiversity.

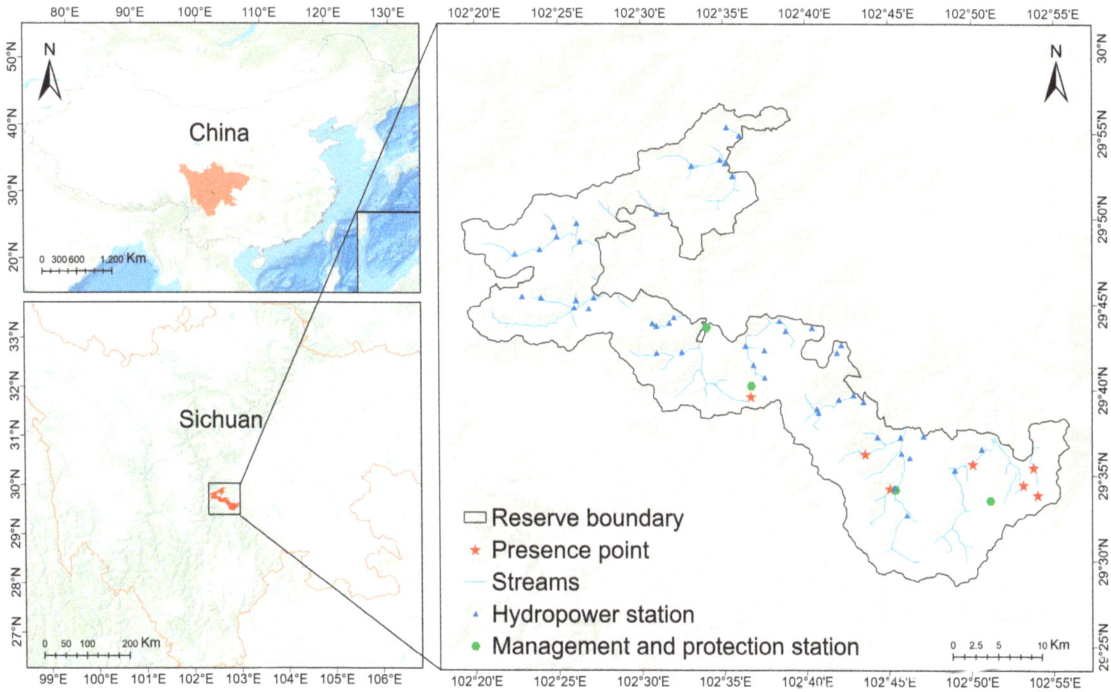

Figure 1. Distribution of presence points of *E. perlacea*.

The aquatic system within the protected zone is part of the primary tributary of the Qingyi River, known as the Yingjing River basin. Dominant tributaries in this region include the Jing River and the Xiangling River [48]. Within the safeguarded region, the Jing River gives rise to two primary branches: the Heishi River and the Baishi River. The Baishi River spans 34.05 km with a watershed area of 188.8 km^2, while the Heishi River extends for 31.2 km, covering a watershed area of 219.85 km^2 [48]. Originating from the northern foothill of Beihou Mountain, situated on the western slope of Xiaoxiangling, is the Xiangling River. Referred to as Huanglian Gou, the upper reaches extend over a length of 25.2 km, encompassing a watershed area of 166.35 km^2. The natural elevation drop is 2090 m, with a mouth flow measuring 11.5 m^3/s [48].

2.2. Input Data

Drawing on the ecological behaviors of the Szechwan rat snake and considering the topography, terrain, and vegetation distribution within the conservation area, this study established a total of 15 effective transects near three conservation stations: Yunwu Mountain, Niba Mountain, and the Giant panda rewilding and reintroduction base (Figure S1). Survey teams, consisting of 2 to 4 individuals per group, conducted meticulous searches along the transects at a speed of 1–2 km/h, covering areas within a 5 m radius on both sides [46]. They recorded discovery locations, habitat types, and species images, utilizing GPS to track their trajectories. Additionally, images of the Szechwan rat snake were provided to conservation area staff or residents for identification, thereby obtaining potential distribution information. Finally, seven occurrence points were successfully obtained.

In the present study, we exclusively utilized seven occurrence points to construct the model. These presence points were meticulously geo-referenced during primary ground surveys using GPS technology. To ensure data quality, the positional accuracy of the occurrence points was verified through Google Earth [49]. Duplicate points were identified and subsequently removed, resulting in the retention of only one point within each 1×1 km^2 grid, a measure taken to mitigate potential sampling bias [49]. Such bias could disproportionately favor the climatic conditions of locations with concentrated sampling efforts [49]. Given the limited number of presence points, which fell below 20, we applied the $1.5 \times$ Interquartile Range (1.5 IQR) method to identify and address potential outliers in the climate data [49]. This approach was based on the environmental data acquired from the WorldClim website at a spatial resolution of 30 arc seconds.

19 bioclimatic variables, encompassing both current (1970–2000) and three future (2050s) scenarios, were obtained from the WorldClim database (http://www.worldclim.org/, accessed on 17 Augst 2023) at a spatial resolution of 30 arc seconds. For the future scenarios, the bioclimatic data for the 2050s represents the mean values from 2041 to 2060. To estimate future climate change, we utilized predictions from general circulation models (GCMs) based on the Shared Socio-economic Pathways (SSPs) scenarios, introduced as part of the Coupled Model Intercomparison Project Phase 6 (CMIP6) by the IPCC [50]. The GCM we chose was the BCC-CSM2-MR climate system model which was developed by the National Climate Center [51]. Three scenarios, SSP 1-2.6, SSP 2-4.5, and SSP 5-8.5, were chosen in this study.

Additionally, we employed the WGS84 projection and acquired a nationwide 30 m $\times$ 30 m digital elevation model (DEM) dataset from the National Aeronautics and Space Administration [52]. With this dataset, we conducted calculations to determine slope, aspect, and the Euclidean distance from each grid cell to the nearest stream (distance from stream) based on the DEM data [53]. The soil water regime data were sourced from the Harmonized World Soil Database version 1.2 [54]. The vegetation data correspond to the Normalized Difference Vegetation Index (NDVI), which was produced using MODIS/Terra (MOD13C1 and MOD13C2). To ensure synchronization with the survey's timeframe, we opted for the July 2022 dataset. All environmental data underwent a thorough cross-check for resolution accuracy and were adjusted to a 30 arc second pixel resolution.

2.3. Ecological Niche Modeling (ENM)

All modeling was conducted using MAXENT Version 3.4.3, given the reliance on presence points exclusively and the limited sample size [49,55]. MAXENT is specifically designed to effectively accommodate small samples. The model was constructed using the Jackknife method [56]. To validate the model's robustness, we executed seven replicated model runs, employing a threshold rule of the 10th percentile for training presence and a cross-validation technique to partition the samples into replicate folds, with the remaining data serving as the test dataset, while keeping all other parameters at their default values [49]. For instance, in the default state, 10,000 points were randomly chosen as background points. Furthermore, recognizing the potential for model instability when dealing with a multitude of factors, we conducted an initial simulation using 25 environmental variables. The aim was to identify the factors with a relatively significant contribution. Subsequently, a secondary simulation was conducted. Three methods, namely True Skill Statistic (TSS), Cohen's Kappa (Kappa), and Area Under the Curve of Receiver Operator Characteristic (ROC) curves (AUC), were employed to assess the predictive accuracy of the model [53]. Utilizing ArcGIS 10.8's transformation tool, we converted the average results of the seven model predictions from ASCII-encoded files to a raster format [49]. These results were subsequently reclassified using manual grading techniques. Considering the maximum training sensitivity plus specificity threshold (range from 0.46 to 0.54) and the natural intermittent classification method (Jenks), we opted to categorize the distribution potential of the model into five classes: very low (0–0.2), low (0.2–0.4), medium (0.4–0.6), high (0.6–0.8), and very high (0.8–1) [49,52].

3. Results

The mean AUC values for the model's predictions of *E. perlacea* under different climatic scenarios were consistently greater than 0.79 (Table 1). The model exhibited a strong performance, meeting the accuracy requirements, and demonstrated relatively good stability over ten repeated runs. However, the mean TSS and Kappa values were comparatively low (Table S1). We attributed this to the relatively small size of presence points and the use of background points instead of true absence points [57]. Additionally, it has been demonstrated in studies that in cases with a limited number of presence points, the AUC value is the optimal metric to evaluate the effectiveness of the model [58]. Consequently, the MaxEnt model utilized in this study maintains high accuracy and reliability in predicting the distribution range of *E. perlacea* under different climate scenarios.

Table 1. AUC value of MaxEnt model under different scenarios.

Different Scenarios	AUC Value *
Current	0.83 ± 0.16
2050s SSP 1-2.6	0.81 ± 0.21
2050s SSP 2-4.5	0.80 ± 0.21
2050s SSP 5-8.5	0.79 ± 0.21

* The evaluation criteria for AUC are: Excellent, 0.9–1; Good, 0.8–0.9; Fair, 0.7–0.8; Poor, 0.6–0.7; Failed, 0.5–0.6.

The results revealed that the distribution of *E. perlacea* was influenced by 11 environmental factors, each with varying degrees of contribution (Table 2). Notably, the Euclidean distance to streams and the slope gradient exhibited relatively high contributions, each exceeding 41% and 31%, respectively. The mean diurnal range (Bio2) and isothermality (Bio3) follow in terms of their relative contributions; the remaining variables make minor contributions. It is evident that the 11 bioclimatic variables displayed differences in their contributions to the modeling of *E. perlacea*. Additionally, *E. perlacea* was predominantly found in low-lying mountainous regions, often associated with the presence of streams.

Table 2. Relative contribution of environmental variables under different scenarios.

Environmental Variable	Current	2050s		
		SSP 1-2.6	SSP 2-4.5	SSP 5-8.5
Distance from stream	41.90	48.80	48.50	49.50
Slope degree	32.00	37.50	38.80	40.40
NDVI	0.40		0.30	0.30
Soil water regime	0.10	0.40	0.40	0.50
Bio2 (mean diurnal range)	4.30	8.30	12.10	7.70
Bio3 (isothermality)	19.90	0.60		1.10
Bio7 (temperature annual range)				0.10
Bio14 (precipitation of driest month)	1.40			0.10
Bio15 (precipitation seasonality)	0.10	3.90		0.20
Bio17 (precipitation of driest quarter)		0.40		0.30
Bio19 (precipitation of coldest quarter)		0.20		

The results obtained from the MaxEnt model revealed that the potential distribution area of *E. perlacea* was predominantly situated in the eastern region of the Yingjing Area of the Giant Panda National Park, particularly in areas with streams near Yunwu mountain and the Giant panda rewilding and reintroduction base (Figure 2). In comparison to the present, there was no conspicuous trend of significant reduction in the potential distribution area of *E. perlacea* for the 2050s (Table 3).

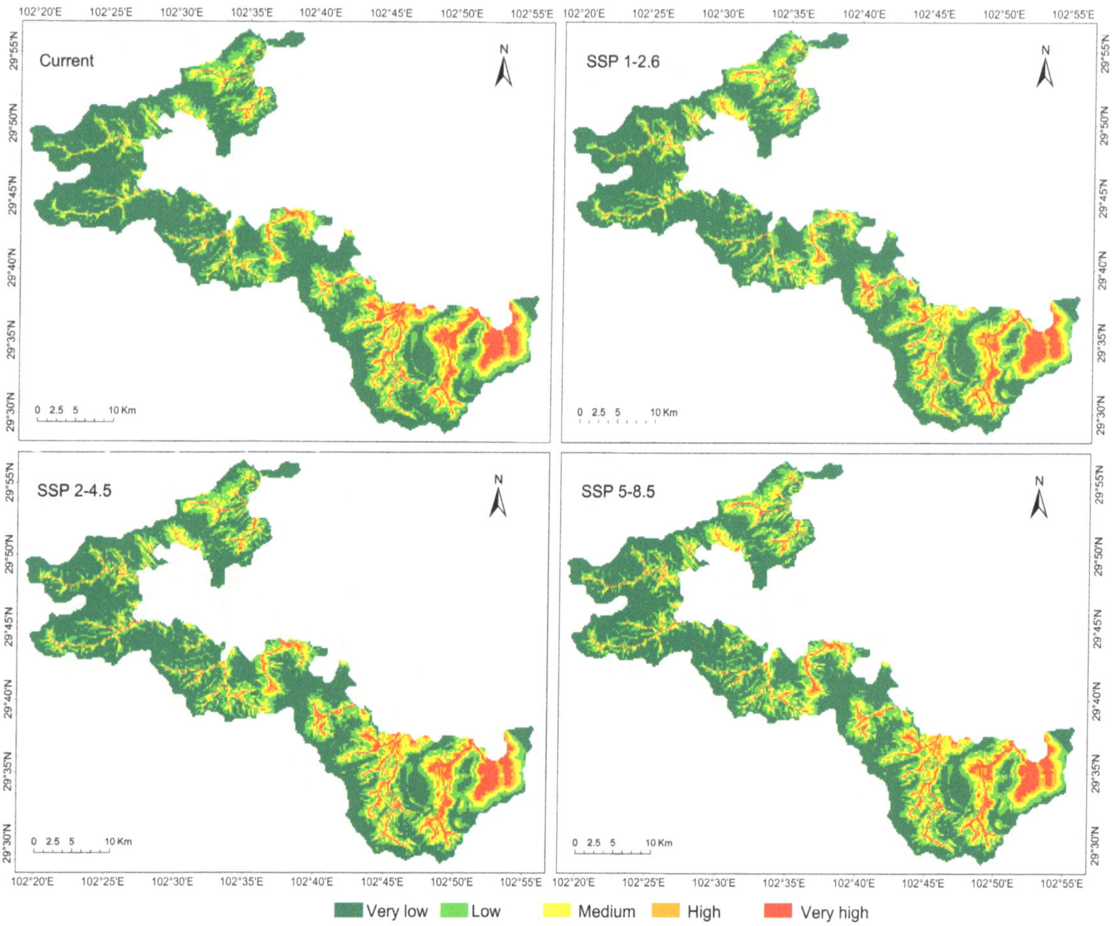

Figure 2. The potential distribution of *E. perlacea* in Yingjing Area of the Giant Panda National Park under different scenarios.

Table 3. Percentage of potential areas for *E. perlacea* in different scenarios in Yingjing Area of Giant Panda National Park.

Distribution Potential	Current	2050s		
		SSP 1-2.6	SSP 2-4.5	SSP 5-8.5
very low	53.59	51.88	50.45	49.56
low	23.30	24.06	24.95	25.40
medium	10.64	12.02	11.95	12.27
high	6.90	6.91	7.14	7.21
very high	5.58	5.14	5.50	5.56

4. Discussion

Our study on the potential distribution of *E. perlacea* in the Yingjing Area of the Giant Panda National Park provides novel insights into the significant effect of the distance from streams, the slope degree and the mean diurnal range on distribution. We found that individuals lived in low-lying mountainous regions associated with the rivers. Moreover, this species preferred the potential distribution area situated in the eastern region of the

National Park, and we found no trend in significant reductions in distribution area for the 2050s, suggesting no significant effect of climate change on the potential distribution of this species in the Yingjing Area of the Giant Panda National Park.

The MaxEnt model was selected to develop predictive models despite the constraints of a limited sample size within a smaller area of 836 km^2 for forecasting the potential distribution of *E. perlacea*. The MaxEnt model is notably valuable for assessing the current and potential distributions of rare, threatened, and poorly known species, particularly among various snake species. Consequently, it has gained widespread acceptance as an essential tool in systematic conservation planning and management [59]. Applying the MaxEnt model has yielded notable outcomes, such as a comprehensive understanding of 39 species of New World coral snakes spanning North, Central, and South America [60]. Additionally, the model has facilitated predictions regarding the potential distribution for many snake species, accompanied by an evaluation of the conservation status of existing protected areas for the species based on potential distribution assessments [61–65]. Within the context of our study, it is evident that the MaxEnt models for *E. perlacea* consistently achieve AUC values exceeding 0.79, underscoring how our study has, for the first time, established reliable models for predicting the potential distribution of *E. perlacea*. It is essential to acknowledge that our study does not account for factors such as human interference, the influence of natural predators, or competition for food resources, which can exert varying degrees of influence on the potential distribution of *E. perlacea*. Consequently, the model's results predominantly portray a distribution that aligns with the fundamental ecological niche requirements of *E. perlacea* and may, to some extent, exceed the actual species distribution [66]. Moreover, the MaxEnt model solely predicts static ecological niches, overlooking the gradual adaptation of species to their environment throughout the evolutionary process [67]. The potential distribution of a species hinges on its intrinsic adaptability and dispersal ability, signifying that species distribution is genuinely a dynamic ecological niche [67,68]. Therefore, to achieve a thorough forecast of suitable habitats for *E. perlacea*, it is imperative to incorporate pertinent biological factors. Future research should address the challenges related to quantifying specific indicators and collecting additional data, incorporating various factors influencing species distribution to enhance the precision of species distribution modeling.

The model yielded valuable insights on the topographic factors which are influencing the potential distribution of *E. perlacea*. Notably, it identified the distance from streams and the slope degree as the most important predictors, aligning with previous findings that individuals preferred living near streams and low degrees of sloping [45,69,70]. In a study predicting the potential distribution of six snake species, it was also found that the primary explanatory variable for species distribution was proximity to streams and a low slope degree [69]. Meanwhile, in a separate study focusing on the Pygmy Rattlesnake (*Sistrurus miliarius*), a locally rare snake species, it was observed that *S. miliarius* primarily inhabited riverine or riparian habitats associated with the lower Tennessee river valley [70]. Interestingly, our investigation found that areas close to rivers and lowlands were also suitable for *E. perlacea*, and were often connected to riverine or riparian regions. These findings underscore the potential significance of riparian corridors as suitable habitats for snakes. However, it is essential to acknowledge that riparian ecosystems are susceptible to anthropogenic disturbances, including hydropower stations, pollution and development [71], which may adversely affect *E. perlacea* populations if suitable habitats are indeed positively associated with riparian ecosystems. To safeguard the critical habitat of *E. perlacea* throughout the Yingjing Area of the Giant Panda National Park, it is crucial to consider the preservation of current river water levels and the measures required to mitigate the loss of riparian habitats due to hydrological alterations.

Reptiles are poikilothermic organisms that experience daily fluctuations in body temperature, which is affected by environmental factors. Temperature is regarded as the most important environmental factor which affects the potential geographical distribution and survival of species [72,73]. For instance, the minimum temperature of the coldest month influences the potential geographical distribution of insects [74,75]. However, we did not find a significant effect of the minimum temperature of the coldest month on the potential geographical distribution of *E. perlacea*. Precipitation is also regarded as an important effect on potential geographical distribution because the change in the soil moisture in fields is associated with an increase in potato damage [76]. For instance, a previous study using the MaxEnt model to predict the potential geographical distribution has shown that precipitation seasonality affects the potential geographical distribution of *Phenacoccus solenopsis* in India [77]. In this study, precipitation did not affect the potential geographical distribution of *E. perlacea*. Previous studies have shown that climate change drives changes in the variation of temperature and food resources, and thus the population dynamics and potential geographical distribution [6,78–80]. Indeed, there is evidence that variation in temperature affects the potential geographical distribution of reptiles because extreme temperature leads to declining population density [81]. Meanwhile, significant variation in temperature in the Yingjing Area of the Giant Panda National Park will lead to a decrease in the population, thus declining the number of regions of potential distribution. Our results provided evidence that the mean diurnal range was one significant variable affecting the potential geographical distribution of *E. perlacea*.

There is increasing evidence that an increase in the emissions of greenhouse gases causes an increase in global average temperature and global climate warming [82]. As a result, global climate change leads to variation in the distribution of many species. For instance, the distribution range of *Xanthium italicum* will decrease in the future due to climate change [83]. Della et al. [84] have revealed that global climate change was positively associated with the distribution of four species. On the contrary, global climate change is negatively correlated with the distribution of insects, birds and mammals [49,85,86]. The fates of potential geographical distribution of species are decided by their capacity to cope with an increase in temperature and variation in precipitation. Indeed, variation in temperature and precipitation is related to the total number of highly suitable habitat areas, thus affecting potential geographical distributions. In this study, we found there was non-significant reduction in the potential distribution area for the 2050s, suggesting that temperature and precipitation seasonality cannot change the potential geographical distribution of *E. perlacea* in future.

The models predict that the potential habitats (medium, high, and very high) for *E. perlacea* do not exceed 26% (217 km^2) of the Yingjing Area of the Giant Panda National Park. Hence, the *E. perlacea*'s habitats in the Giant Panda National Park were far fewer than the average area for the wild animals in the area. While protected areas can effectively confirm the threats from human activities for endangered species, many species facing climate change scenarios possibly shift their distribution ranges outside the protected areas [87,88]. Meanwhile, habitat fragmentation is likely to amplify the constraints on species distribution range shifts by hampering young dispersal [89–91]. Those patterns underscore the need to establish valuable protected areas for endangered species under climate change [92]. For *E. perlacea*, this means that anthropogenic activities and habitat fragmentation under climate change are likely to affect their habitats and potential distributions. However, all distribution locations of *E. perlacea* in the Yingjing Area of the Giant Panda National Park have less anthropogenic activity and habitat fragmentation. Hence, we suggest that the conservation of *E. perlacea* in the other distribution regions should rely on building small protected areas and wildlife corridors.

5. Conclusions

We explored the consequences of climate change on the potential distribution of *E. perlacea*, an endangered species that is challenging to observe in the field. However, we cannot delimit the boundaries of protected areas due to insufficient data on a single species. Although we found that there are significant effects of distance from streams, slope degree and mean diurnal range on geographical distribution, more similar studies on building small protected areas and wildlife corridors in the other distribution regions of *E. perlacea* are needed to confirm the most suitable protected areas. At the same time, we found a non-significant reduction in the potential distribution area for the 2050s because less anthropogenic activity and habitat fragmentation in the Yingjing Area of the Giant Panda National Park benefits species conservation. Furthermore, the non-significant variation in distribution areas in *E. perlacea* experiencing climate change is likely to be associated with the refugia regions established through climate change. Nevertheless, there is a probability that other climatic and non-climatic factors, such as biotic interactions, dispersal abilities, and evolutionary adaptations, not considered in our models, influence the spatial distribution of this species. Our future plans thus involve integrating multiple models and considering various factors influencing species distribution. This aims to conduct a more in-depth analysis of *E. perlacea* habitat suitability, offering guidance for *E. perlacea* habitat protection and core reservation planning. Lastly, from a conservation standpoint, there is cause for concern as residents within the potential distribution range currently demonstrate inadequate awareness and protective measures for *E. perlacea*. Consequently, we strongly encourage personnel in conservation areas to not only enhance the protection of the native habitat of *E. perlacea* but also to engage in educational outreach to local residents.

Supplementary Materials: The following supporting information can be downloaded at: https://www.mdpi.com/article/10.3390/ani13243828/s1, Figure S1: Map depicting the distribution of management and protection stations and transects in Yingjing Area of Giant Panda National Park; Table S1: The basic information for Maxent result.

Author Contributions: Conceptualization, W.L.; methodology, X.S.; software, Y.J.; validation, X.S. and Y.J.; formal analysis, X.S. and Y.J.; investigation, X.S., L.J., L.Z. and C.Y.; resources, X.S. and W.L.; data curation, Y.J.; writing—original draft preparation, X.S. and Y.J.; writing—review and editing, W.L.; visualization, Y.J.; supervision, W.L.; project administration, X.S.; funding acquisition, X.S. and W.L. All authors have read and agreed to the published version of the manuscript.

Funding: This research was funded by the National Natural Science Foundation of China (32300358, 32370456) and the Key Project of Natural Sciences Foundation of Sichuan Province (22NSFSC0011).

Informed Consent Statement: Not applicable.

Data Availability Statement: The data presented in this study are available on request from the corresponding author. The data are not publicly available due to privacy or ethical restrictions.

Acknowledgments: We would like to thank C. Chen, X. Zhu, W.J. Shao, and Y.J. Yang for their assistance in data collection during the experiment.

Conflicts of Interest: The authors declare no conflict of interest.

References

1. Ceballos, G.; Ehrlich, P.R.; Dirzo, R. Biological annihilation via the ongoing sixth mass extinction signaled by vertebrate population losses and declines. *Proc. Natl. Acad. Sci. USA* **2017**, *114*, E6089–E6096. [CrossRef] [PubMed]
2. Barnosky, A.D.; Matzke, N.; Tomiya, S.; Wogan, G.O.U.; Swartz, B.; Quental, T.B.; Marshall, C.; McGuire, J.L.; Lindsey, E.L.; Maguire, K.C.; et al. Has the Earth's sixth mass extinction already arrived? *Nature* **2011**, *471*, 51–57. [CrossRef] [PubMed]
3. Pacifici, M.; Foden, W.B.; Visconti, P.; Watson, J.E.; Butchart, S.H.; Kovacs, K.M.; Scheffers, B.R.; Hole, D.G.; Martin, T.G.; Resit Akçakaya, H.; et al. Assessing species vulnerability to climate change. *Nat. Clim. Chang.* **2015**, *5*, 215–224. [CrossRef]
4. Pimm, S.L.; Jenkins, C.N.; Abell, R.; Brooks, T.M.; Gittleman, J.L.; Joppa, L.N.; Raven, P.H.; Roberts, C.M.; Sexton, J.O. The biodiversity of species and their rates of extinction, distribution, and protection. *Science* **2014**, *344*, 1246752. [CrossRef] [PubMed]
5. Thomas, C.D.; Cameron, A.; Green, R.E.; Bakkenes, M.; Beaumont, L.J.; Collingham, Y.C.; Erasmus, B.F.N.; de Siqueira, M.F.; Grainger, A.; Hannah, L.; et al. Extinction risk from climate change. *Nature* **2004**, *427*, 145–148. [CrossRef] [PubMed]
6. Obregon, R.L.; Scolaro, J.A.; IbargÜengoytÍa, N.R.; Medina, M. Thermal biology and locomotor performance in *Phymaturus calcogaster*: Are Patagonian lizards vulnerable to climate change? *Integr. Zool.* **2021**, *16*, 53–66. [CrossRef] [PubMed]
7. Gardner, J.L.; Peters, A.; Kearney, M.R.; Joseph, L.; Heinsohn, R. Declining body size: A third universal response to warming? *Trends Ecol. Evol.* **2011**, *26*, 285–291. [CrossRef]
8. Dillon, M.E.; Wang, G.; Huey, R.B. Global metabolic impacts of recent climate warming. *Nature* **2010**, *467*, 704–706. [CrossRef]
9. Parmesan, C. Ecological and evolutionary responses to recent climate change. *Annu. Rev. Ecol. Evol. Syst.* **2006**, *37*, 637–669. [CrossRef]

10. Perry, A.L.; Low, P.J.; Ellis, J.R.; Reynolds, J.D. Climate change and distribution shifts in marine fishes. *Science* **2005**, *308*, 1912–1915. [CrossRef]
11. Visser, M.E.; Both, C. Shifts in phenology due to global climate change: The need for a yardstick. *Proc. R. Soc. B Biol. Sci.* **2005**, *272*, 2561–2569. [CrossRef] [PubMed]
12. Butikofer, L.; Ji, W.; Sacchi, R.; Mangiacotti, M. Climate migrants' survival threatened by "C" shaped anthropic barriers. *Integr. Zool.* **2020**, *15*, 32–39. [CrossRef] [PubMed]
13. Rugiero, L.; Milana, G.; Petrozzi, F.; Capula, M.; Luiselli, L. Climate-change-related shifts in annual phenology of a temperate snake during the last 20 years. *Acta Oecologica* **2013**, *51*, 42–48. [CrossRef]
14. VanDerWal, J.; Murphy, H.T.; Kutt, A.S.; Perkins, G.C.; Bateman, B.L.; Perry, J.J.; Reside, A.E. Focus on poleward shifts in species' distribution underestimates the fingerprint of climate change. *Nat. Clim. Chang.* **2013**, *3*, 239–243. [CrossRef]
15. Lenoir, J.; Gégout, J.C.; Marquet, P.A.; de Ruffray, P.; Brisse, H. A significant upward shift in plant species optimum elevation during the 20th century. *Science* **2008**, *320*, 1768–1771. [CrossRef]
16. Parmesan, C.; Ryrholm, N.; Stefanescu, C.; Hill, J.K.; Thomas, C.D.; Descimon, H.; Huntley, B.; Kaila, L.; Kullberg, J.; Tammaru, T.; et al. Poleward shifts in geographical ranges of butterfly species associated with regional warming. *Nature* **1999**, *399*, 579–583. [CrossRef]
17. Lourenço-de-Moraes, R.; Lansac-Toha, F.M.; Schwind, L.T.F.; Arrieira, R.L.; Rosa, R.R.; Terribile, L.C.; Lemes, P.; Rangel, T.F.; Diniz-Filho, J.A.F.; Bastos, R.P.; et al. Climate change will decrease the range size of snake species under negligible protection in the Brazilian Atlantic Forest hotspot. *Sci. Rep.* **2019**, *9*, 8523. [CrossRef]
18. Zacarias, D.; Loyola, R. Climate change impacts on the distribution of venomous snakes and snakebite risk in Mozambique. *Clim. Chang.* **2019**, *152*, 195–207. [CrossRef]
19. Currie, D.J. Projected effects of climate change on patterns of vertebrate and tree species richness in the conterminous United States. *Ecosystems* **2001**, *4*, 216–225. [CrossRef]
20. Hansen, A.J.; Neilson, R.P.; Dale, V.H.; Flather, C.H.; Iverson, L.R.; Currie, D.J.; Shafer, S.; Cook, R.; Bartlein, P.J. Global change in forests: Responses of species, communities and biomes. *BioScience* **2001**, *51*, 765–779. [CrossRef]
21. Araújo, M.B.; Thuiller, W.; Pearson, R.G. Climate warming and the decline of amphibians and reptiles in Europe. *J. Biogeogr.* **2006**, *33*, 1712–1728. [CrossRef]
22. Archis, J.N.; Akcali, C.; Stuart, B.L.; Kikuchi, D.; Chunco, A.J. Is the future already here? The impact of climate change on the distribution of the eastern coral snake (*Micrurus fulvius*). *PeerJ* **2018**, *6*, e4647. [CrossRef]
23. Kamdem, M.M.; Ngakou, A.; Yanou, N.N.; Otomo, V.P. Habitat components and population density drive plant litter consumption by *Eudrilus eugeniae* (Oligochaeta) under tropical conditions. *Integr. Zool.* **2021**, *16*, 255–269. [CrossRef]
24. Kubisch, E.L.; Fernández, J.B.; Ibargüengoytía, N.R. Thermophysiological plasticity could buffer the effects of global warming on a Patagonian lizard. *J. Exp. Zool. Part A Ecol. Integr. Physiol.* **2023**, *339*, 590–601. [CrossRef]
25. McNeely, J.A. Today's protected areas: Supporting a more sustainable future for humanity. *Integr. Zool.* **2020**, *15*, 603–616. [CrossRef]
26. Sharnuud, R.; Ameca, E.I. Taxonomy, distribution, and contemporary exposure of terrestrial mammals to floods and human pressure across different areas for biodiversity conservation in China. *Integr. Zool.* **2023**, *00*, 1–10. [CrossRef]
27. Carvalho, S.B.; Brito, J.C.; Crespo, E.J.; Possingham, H.P. Incorporating evolutionary processes into conservation planning using species distribution data: A case study with the western Mediterranean herpetofauna. *Divers. Distrib.* **2011**, *17*, 408–421. [CrossRef]
28. Martínez-Freiría, F.; Argaz, H.; Fahd, S.; Brito, J.C. Climate change is predicted to negatively influence Moroccan endemic reptile richness. Implications for conservation in protected areas. *Naturwissenschaften* **2013**, *100*, 877–889. [CrossRef]
29. Li, X.; Tian, H.; Wang, Y.; Li, R.; Song, Z.; Zhang, F.; Xu, M.; Li, D. Vulnerability of 208 endemic or endangered species in China to the effects of climate change. *Reg. Environ. Chang.* **2013**, *13*, 843–852. [CrossRef]
30. Bond, N.; Thomson, J.; Reich, P.; Stein, J. Using species distribution models to infer potential climate change-induced range shifts of freshwater fish in south-eastern Australia. *Mar. Freshw. Res.* **2011**, *62*, 1043–1061. [CrossRef]
31. Wang, B.; Xu, Y.; Ran, J.H. Predicting suitable habitat of the Chinese monal (*Lophophorus lhuysii*) using ecological niche modeling in the Qionglai Mountains, China. *PeerJ* **2017**, *5*, e3477. [CrossRef] [PubMed]
32. Carvalho, J.S.; Graham, B.; Bocksberger, G.; Stewart, F.A.; Sunderland-Groves, J.; Tagg, N.; Todd, A.; Vosper, A.; Wenceslau, J.F.C.; Wessling, E.G.; et al. Predicting range shifts of African apes under global change scenarios. *Divers. Distrib.* **2021**, *27*, 1663–1679. [CrossRef]
33. Tang, J.; Swaisgood, R.R.; Owen, M.A.; Zhao, X.; Wei, W.; Hong, M.; Zhou, H.; Zhang, Z. Assessing the effectiveness of protected areas for panda conservation under future climate and land use change scenarios. *J. Environ. Manag.* **2023**, *342*, 118319. [CrossRef]
34. Elith, J.; Graham, C.H.; Anderson, R.P.; Dudík, M.; Ferrier, S.; Guisan, A.; Hijmans, J.R.; Huettmann, F.; Leathwick, R.J.; Lehmann, A.; et al. Novel methods improve prediction of species'distributions from occurrence data. *Ecography* **2006**, *29*, 129–151. [CrossRef]
35. Elith, J.; Phillips, S.J.; Hastie, T.; Dudík, M.; Chee, Y.E.; Yates, C.J. A statistical explanation of MaxEnt for ecologists. *Divers. Distrib.* **2011**, *17*, 43–57. [CrossRef]
36. Song, K.; Mi, C.R.; Yang, N.; Sun, L.; Sun, Y.H.; Xu, J.L. Improve the roles of nature reserves in conservation of endangered pheasant in a highly urbanized region. *Sci. Rep.* **2020**, *10*, 17673. [CrossRef]

37. Mays, H.L.; Hung, C.M.; Shaner, P.J.; Denvir, J.; Justice, M.; Yang, S.F.; Roth, T.L.; Oehler, D.A.; Fan, J.; Rekulapally, S.; et al. Genomic analysis of demographic history and ecological niche modeling in the endangered Sumatran rhinoceros *Dicerorhinus sumatrensis*. *Curr. Biol.* **2018**, *28*, 70–76. [CrossRef]
38. Qin, A.; Liu, B.; Guo, Q.; Bussmann, R.W.; Ma, F.; Jian, Z.; Xu, G.; Pei, S. Maxent modeling for predicting impacts of climate change on the potential distribution of *Thuja sutchuenensis* Franch., an extremely endangered conifer from southwestern China. *Glob. Ecol. Conserv.* **2017**, *10*, 139–146. [CrossRef]
39. Zhao, E. The validity of *Elaphe perlacea*-a rare endemic snake from Sichuan Province, China. *Asiat. Herpetol. Res.* **1990**, *3*, 101–103.
40. IUCN Red List of Threatened Species. Available online: https://www.iucnredlist.org/en (accessed on 17 August 2023).
41. Stejneger, L. A new snake from China. *Proc. Biol. Soc. Wash.* **1929**, *42*, 129–130.
42. Deng, Q.X.; Jiang, Y.M. The Szechwan rat snake in China. *J. China West Norm. Univ. (Nat. Sci.)* **1989**, *10*, 120–122.
43. Chen, X.; Jiang, K.; Guo, P.; Huang, S.; Rao, D.; Ding, L.; Takeuchi, H.; Che, J.; Zhang, Y.P.; Myers, E.A.; et al. Assessing species boundaries and the phylogenetic position of the rare Szechwan rat snake, *Euprepiophis perlaceus* (Serpentes: Colubridae), using coalescent-based methods. *Mol. Phylogenet. Evol.* **2014**, *70*, 130–136. [CrossRef] [PubMed]
44. Chen, Z.N.; Gan, S.X.; Ding, L.; Zhou, C.Q.; Mao, X.T.; Zhao, K.L. The experiment of the captive breeding for the *Euprepiophis perlacea*. *J. Sichuan For. Sci. Technol.* **2017**, *38*, 1003–5508.
45. Gan, S.; Ding, L.; Yang, H.; Zhou, C.; Zhao, K. A preliminary study of selecting key ecological factors of the suitable habitats for *Euprepiophis perlacer*. *J. Sichuan For. Sci. Technol.* **2017**, *38*, 40–44.
46. Shao, W.; Song, X.; Chen, C.; Zhao, L.; Jin, L.; Liao, W. Diversity and altitudinal distribution pattern of amphibians and reptiles in Yingjing Area of Giant Panda National Park. *Chin. J. Zool.* **2022**, *57*, 707–721.
47. Fu, J.; Sun, Z.; Liu, S.; Fu, Z.; Liu, Y.; Wang, X.; Zhao, J.; Cai, G. A survey of bird resources in Daxiangling Nature Reserve in Sichuan Province. *J. Sichuan For. Sci. Technol.* **2008**, *29*, 31–37.
48. Jia, W. Giant Panda Microhabitat Study in the Daxiangling Niba Mountain Corridor, Sichuan Province. Master Thesis, Guizhou Normal University, Guizhou, China, 10 March 2023.
49. Deb, C.R.; Jamir, N.S.; Kikon, Z.P. Distribution prediction model of a rare orchid species (*Vanda bicolor* Griff.) using small sample size. *Am. J. Plant Sci.* **2017**, *8*, 1388.
50. Popp, A.; Calvin, K.; Fujimori, S.; Havlik, P.; Humpenöder, F.; Stehfest, E.; Bodirsky, B.L.; Dietrich, J.P.; Doelmann, J.C.; Gusti, M.; et al. Land-use futures in the shared socio-economic pathways. *Glob. Environ. Chang.* **2017**, *42*, 331–345. [CrossRef]
51. Xin, X. Performance of BCC-CSM2-MR in simulating summer climate changes in East Asia. *EGU Gen. Assem. Conf. Abstr.* **2019**, *21*, 4711.
52. Gong, J.; Li, Y.; Wang, R.; Yu, C.; Fan, J.; Shi, K. MaxEnt modeling for predicting suitable habitats of snow leopard (*Panthera uncia*) in the mid-eastern Tianshan Mountains. *J. Resour. Ecol.* **2023**, *14*, 1075–1085.
53. Ye, W.J.; Yang, N.; Yang, B.; Li, Y.; Zhang, J.D.; Chen, D.M.; Zhou, C.Q.; Zhong, X.; Zhang, Z.J. Impacts of climate change on potential geographical distribution of golden pheasant (*Chrysolophus pictus*), an endemic species in China. *Chin. J. Ecol.* **2021**, *40*, 1783–1792.
54. Fischer, G.; Nachtergaele, F.; Prieler, S.; van Velthuizen, H.T.; Verelst, L.; Wiberg, D. *Global Agro-Ecological Zones Assessment for Agriculture (GAEZ 2008)*; IIASA: Laxenburg, Austria; FAO: Rome, Italy, 2008.
55. Hernandez, P.A.; Graham, C.H.; Master, L.L.; Albert, D.L. The effect of sample size and species characteristics on performance of different species distribution modeling methods. *Ecography* **2006**, *29*, 773–785. [CrossRef]
56. Pearson, R.G.; Raxworthy, C.J.; Nakamura, M.; Peterson, A.T. Predicting species distributions from small numbers of occurrence records: A test case using cryptic geckos in Madagascar. *J. Biogeogr.* **2007**, *34*, 102–117. [CrossRef]
57. Barbet-Massin, M.; Jiguet, F.; Albert, C.H.; Thuiller, W. Selecting pseudo-absences for species distribution models: How, where and how many? *Methods Ecol. Evol.* **2012**, *3*, 327–338. [CrossRef]
58. Allouche, O.; Tsoar, A.; Kadmon, R. Assessing the accuracy of species distribution models: Prevalence, kappa and the true skill statistic (TSS). *J. Appl. Ecol.* **2006**, *43*, 1223–1232. [CrossRef]
59. Qu, M.; Zhang, Y.; Gao, Z.; Zhang, Z.; Liu, Y.; Wan, S.; Lin, Q. The genetic basis of the leafy seadragon's unique camouflage morphology and avenues for its efficient conservation derived from habitat modeling. *Sci. China Life Sci.* **2023**, *66*, 1213–1230. [CrossRef] [PubMed]
60. Terribile, L.C.; Diniz-Filho, J.A.F. How many studies are necessary to compare niche-based models for geographic distributions? Inductive reasoning may fail at the end. *Braz. J. Biol.* **2010**, *70*, 263–269. [CrossRef]
61. Hosseinzadeh, M.S.; Ghezellou, P.; Kazemi, S.M. Predicting the potential distribution of the endemic snake *Spalerosophis microlepis* (Serpentes: Colubridae), in the Zagros Mountains, western Iran. *Salamandra* **2010**, *53*, 294–298.
62. Chefaoui, R.M.; Hosseinzadeh, M.S.; Mashayekhi, M.; Safaei-Mahroo, B.; Kazemi, S.M. Identifying suitable habitats and current conservation status of a rare and elusive reptile in Iran. *Amphib. Reptil.* **2018**, *39*, 355–362. [CrossRef]
63. Andrade-Díaz, M.S.; Sarquis, J.A.; Loiselle, B.A.; Giraudo, A.R.; Díaz-Gómez, J.M. Expansion of the agricultural frontier in the largest South American Dry Forest: Identifying priority conservation areas for snakes before everything is lost. *PLoS ONE* **2019**, *14*, e0221901. [CrossRef]
64. Kirk, D.A.; Karimi, S.; Maida, J.R.; Harvey, J.A.; Larsen, K.W.; Bishop, C.A. Using ecological niche models for population and range estimates of a threatened snake species (*Crotalus oreganus*) in Canada. *Diversity* **2021**, *13*, 467. [CrossRef]

65. Mizsei, E.; Szabolcs, M.; Szabó, L.O.R.Á.N.D.; Boros, Z.; Mersini, K.; Roussos, S.A.; Lengyel, S. Determining priority areas for an endangered cold-adapted snake on warming mountaintops. *Oryx* **2021**, *55*, 334–343. [CrossRef]
66. Phillips, S.J.; Anderson, R.P.; Schapire, R.E. Maximum entropy modeling of species geographic distributions. *Ecol. Model.* **2006**, *190*, 231–259. [CrossRef]
67. Pearman, P.B.; Guisan, A.; Broennimann, O.; Randin, C.F. Niche dynamics in space and time. *Trends Ecol. Evol.* **2008**, *23*, 149–158. [CrossRef]
68. Lin, L.H.; Zhu, X.M.; Du, Y.; Fang, M.C.; Ji, X. Global, regional, and cladistic patterns of variation in climatic niche breadths in terrestrial elapid snakes. *Curr. Zool.* **2019**, *65*, 1–9. [CrossRef]
69. Kalboussi, M.; Achour, H. Modelling the spatial distribution of snake species in northwestern Tunisia using maximum entropy (Maxent) and Geographic Information System (GIS). *J. For. Res.* **2018**, *29*, 233–245. [CrossRef]
70. Snyder, S.D.; Sutton, W.B.; Steen, D.A. Species distribution modeling reveals insights into the occurrence of a locally rare snake at the periphery of its geographic range. *Herpetologica* **2023**, *79*, 98–107.
71. Richardson, D.M.; Holmes, P.M.; Esler, K.J.; Galatowitsch, S.M.; Stromberg, J.C.; Kirkman, S.P.; Pysek, P.; Hobbs, R.J. Riparian vegetation: Degradation, alien plant invasions, and restoration prospects. *Divers. Distrib.* **2007**, *13*, 126–139. [CrossRef]
72. Bale, J.S.; Masters, G.J.; Hodkinson, I.D.; Awmack, C.; Bezemer, T.M.; Brown, V.K.; Butterfield, J.; Buse, A.; Coulson, J.C.; Farrar, J.; et al. Herbivory in global climate change research: Direct effects of rising temperature on insect herbivores. *Glob. Chang. Biol.* **2022**, *8*, 1–16. [CrossRef]
73. Kocmánková, E.; Trnka, M.; Juroch, J.; Dubrovský, M.; Semerádová, D.; Možný, M.; Žalud, Z. Impact of climate change on the occurrence and activity of harmful organisms. *Plant Prot. Sci.* **2009**, *45*, S48–S52. [CrossRef]
74. Dos Santos, L.A.; Mendes, M.F.; Krüger, A.P.; Blauth, M.L.; Gottschalk, M.S.; Garcia, F.R. Global potential distribution of *Drosophila suzukii* (Diptera, Drosophilidae). *PLoS ONE* **2017**, *12*, e0174318. [CrossRef]
75. Wei, J.; Peng, L.; He, Z.; Lu, Y.; Wang, F. Potential distribution of two invasive pineapple pests under climate change. *Pest Manag. Sci.* **2020**, *76*, 1652–1663. [CrossRef] [PubMed]
76. Staley, J.T.; Hodgson, C.J.; Mortimer, S.R.; Morecroft, M.D.; Masters, G.J.; Brown, V.K.; Taylor, M.E. Effects of summer rainfall manipulations on the abundance and vertical distribution of herbivorous soil macro-invertebrates. *Eur. J. Soil Biol.* **2007**, *43*, 189–198. [CrossRef]
77. Kumar, R.; Nagrare, V.S.; Nitharwal, M.; Swami, D.; Prasad, Y.G. Within-plant distribution of an invasive mealybug, *Phenacoccus solenopsis*, and associated losses in cotton. *Phytoparasitica* **2014**, *42*, 311–316. [CrossRef]
78. Wen, H.; Luo, T.; Wang, Y.; Wang, S.; Liu, T.; Xiao, N.; Zhou, J. Molecular phylogeny and historical biogeography of the cave fish genus Sinocyclocheilus (Cypriniformes: Cyprinidae) in southwest China. *Integr. Zool.* **2022**, *17*, 311–325. [CrossRef] [PubMed]
79. Yu, F.; Sun, Y.; Wang, T.; Skidmore, A.K.; Ding, C.; Ye, X. Linking the past and present to predict the distribution of Asian crested ibis (*Nipponia nippon*) under global changes. *Integr. Zool.* **2022**, *17*, 1095–1105. [CrossRef]
80. Tourinho, L.; Vale, M.M. Choosing among correlative, mechanistic, and hybrid models of species' niche and distribution. *Integr. Zool.* **2023**, *18*, 93–109. [CrossRef] [PubMed]
81. Muthoni, F.K. Modelling the Spatial Distribution of Snake Species under Changing Climate Scenario in Spain. Master's Thesis, University of Twente, Enschede, The Netherlands, 2010.
82. Muñoz, I.; Rigarlsford, G.; Canals, L.M.; King, H. Accounting for greenhouse gas emissions from the degradation of chemicals in the environment. *Int. J. Life Cycle Assess.* **2013**, *18*, 252–262. [CrossRef]
83. Zhang, Y.; Tang, J.; Ren, G.; Zhao, K.; Wang, X. Global potential distribution prediction of *Xanthium italicum* based on Maxent model. *Sci. Rep.* **2021**, *11*, 16545. [CrossRef]
84. Della, R.F.; Milanesi, P. Combining climate, land use change and dispersal to predict the distribution of endangered species with limited vagility. *J. Biogeogr.* **2020**, *47*, 1427–1438. [CrossRef]
85. Liu, S.; Xia, S.; Wu, D.; Behm, J.E.; Meng, Y.; Yuan, H.; Wen, P.; Hughes, A.C.; Yang, X. Understanding global and regional patterns of termite diversity and regional functional traits. *iScience* **2022**, *25*, 105538. [CrossRef] [PubMed]
86. Sentís, M.; Pacioni, C.; De Cuyper, A.; Janssens, G.P.J.; Lens, L.; Strubbe, D. Biophysical models accurately characterize the thermal energetics of a small invasive passerine bird. *iScience* **2023**, *26*, 107743. [CrossRef] [PubMed]
87. Kong, F.; Tang, L.; He, H.; Yang, F.; Tao, J.; Wang, W. Assessing the impact of climate change on the distribution of Osmanthus fragrans using Maxent. *Environ. Sci. Pollut. Res.* **2021**, *28*, 34655–34663. [CrossRef] [PubMed]
88. Thapa, A.; Wu, R.; Hu, Y.; Nie, Y.; Singh, P.B.; Khatiwada, J.R.; Yan, L.; Gu, X.; Wei, F. Predicting the potential distribution of the endangered red panda across its entire range using MaxEnt modeling. *Ecol. Evol.* **2018**, *8*, 10542–10554. [CrossRef] [PubMed]
89. De Kort, H.; Baguette, M.; Lenoir, J.; Stevens, V.M. Toward reliable habitat suitability and accessibility models in an era of multiple environmental stressors. *Ecol. Evol.* **2020**, *10*, 10937–10952. [CrossRef]
90. Yang, S.; Wang, X.; Hu, J. Mountain frog species losing out to climate change around the Sichuan Basin. *Sci. Total Environ.* **2022**, *806*, 150605. [CrossRef]

91. Bensch, S.; Caballero-López, V.; Cornwallis, C.K.; Sokolovskis, K. The evolutionary history of "suboptimal" migration routes. *iScience* **2023**, *26*, 108266. [CrossRef]
92. Fricke, E.C.; Ordonez, A.; Rogers, H.S.; Svenning, J.C. The effects of defaunation on plants' capacity to track climate change. *Science* **2022**, *375*, 210–214. [CrossRef]

animals
MDPI

Article

Russian Biodiversity Collections: A Professional Opinion Survey

Elena V. Alpeeva †, Natalia P. Sharova †, Konstantin S. Sharov * and Ekaterina A. Vorotelyak

Koltzov Institute of Developmental Biology of Russian Academy of Sciences, 26 Vavilov Street, 119334 Moscow, Russia; alpeeva_l@mail.ru (E.V.A.); npsharova@bk.ru (N.P.S.)
* Correspondence: const.sharov@mail.ru; Tel.: +7-499-135-7674; Fax: +7-499-135-8012
† These authors contributed equally to this work.

Simple Summary: The rate of human population growth continues to increase, with a growing adverse anthropogenic influence on the biosphere. This creates a completely new evolutionary challenge to animal species and new research tasks for bioconservation science. Biocollections are very helpful in bioconservation in this situation. We studied the professional opinions of bioconservation specialists on the future of creating/maintaining biocollections and biocollection networks in Russia. There is a significant degree of concordance among them about the necessity to unite biocollections in networks. This may provide several important opportunities: the ability to cross-check in research, the simplicity of access, redundancy in storing specimens, and effective data curation. We show the success and deliberate on the future potential of our scientific institution in developing and sustaining four large biocollections. These biocollections may become a basis for a national biodiversity centre.

Abstract: Biodiversity collections are important vehicles for protecting endangered wildlife in situations of adverse anthropogenic influence. In Russia, there are currently a number of institution- and museum-based biological collections, but there are no nation-wide centres of biodiversity collections. In this paper, we report on the results of our survey of 324 bioconservation, big-data, and ecology specialists from different regions of Russia in regard to the necessity to create several large national biodiversity centres of wildlife protection. The survey revealed specific goals that have to be fulfilled during the development of these centres for the protection and restoration of endangered wildlife species. The top three problems/tasks (topics) are the following: (1) the necessity to create large national centres for different types of specimens; (2) the full sequencing and creation of different "omic" (genomic, proteomic, transcriptomic, etc.) databases; (3) full digitisation of a biodiversity collection/centre. These goals may constitute a guideline for the future of biodiversity collections in Russia that would be targeted at protecting and restoring endangered species. With the due network service level, the translation of the website into English, and permission from the regulator (Ministry of Science and Higher Education of Russian Federation), it can also become an international project.

Keywords: biodiversity collection; biodiversity centre; biobank; biological conservation; extinction; wildlife restoration; endangered species

Citation: Alpeeva, E.V.; Sharova, N.P.; Sharov, K.S.; Vorotelyak, E.A. Russian Biodiversity Collections: A Professional Opinion Survey. *Animals* **2023**, *13*, 3777. https://doi.org/10.3390/ani13243777

Academic Editors: Clive J. C. Phillips and Chunwang Li

Received: 22 September 2023
Revised: 15 October 2023
Accepted: 22 November 2023
Published: 7 December 2023

1. Introduction

1.1. Problem Overview

Biodiversity collections are significant instruments of biological conservation, wildlife restoration, and research in situations of climatic change, deforestation, carbonisation, waste accumulation, including the plastification of water reservoirs, and other anthropogenic activities [1–3]. By the second decade of the twenty-first century, only small amounts of truly "wild nature" remained on the planet. Humans transform nature to a degree when artificial infrastructure is embedded in natural biotopes and biocoenoses and sometimes "melted" with them.

The adverse anthropogenic influence causes the shrinking of animal habitats, species extinction, and biodiversity loss [4,5]. Humans have been responsible for the extinction of around 900 vertebrate species since the sixteenth century (of them, more than half were eradicated in the twentieth to twenty-first centuries) [6–10].

Not only do *homo sapiens* influence ecosystems by themselves and cause species to become extinct, they also cause a variety of species to interfere with others. To date, humans have introduced around 6800 invasive animal species to different parts of the world: i.e., they have relocated them from one part of the planet to another, sometimes unintentionally and sometimes on purpose [9]. In many cases, invasive animal species drive native species out of a given area by supplanting them or feeding on them and, sometimes, even changing biocoenoses beyond recognition [7]. Invasive species may proliferate excessively in the absence of their natural enemies in new living territories. This may lead to a direct substantial hazard to biodiversity, which often causes panzoonotics (animal analogues of pandemics) and may dramatically affect wild species [11–14]. Additionally, panzoonotics may easily give rise to outbreaks of novel human emerging infectious diseases (EIDs). There are two well-known examples demonstrating this [15,16].

First, the growth of the human population has caused the even faster growth of rat and mouse populations, which now number approximately 8 billion each, just like *homo sapiens*, as well as the street pigeon (feral pigeon) population, which consists of up to 0.5 billion specimens [16]. Due to their proximity to humans, these rodents and birds have become frequent causes of outbreaks as intermediary hosts for EID agents.

Second, over the last three decades, *homo sapiens* have bred enormous populations of the domesticated chicken, *Gallus gallus domesticus*. Now, chickens number around 24 billion specimens on the planet, as *homo sapiens* use this species as one of their main foods. During the H5N1 avian influenza epidemics in China and Southeast Asia in 1997–2005, chickens were the main intermediary hosts for the virus, which switched to them from several wild bird species [17]. Chickens were not bred in incubators in this region. They lived freely in paled yards where wild geese, jays, common mynas, and crows had unrestricted access as they fed on the chickens' food. In turn, infected chickens began to infect wild birds. In some regions, huge colonies were wiped out. As a result, humans had to eradicate the largest part of the chicken population in Asia to stop the great bird panzootic.

Figure 1 demonstrates the dynamics of species extinction and invasion caused by anthropogenic influence.

The main ways in which humans influence the population sizes of animal species are as follows:

(1) The destruction of ecosystems and biogeocoenoses due to the founding of settlements, such as mega-cities in the extreme case, and their infrastructure (national electrical networks, roads, pipelines, etc.);
(2) Terraforming: cutting down forests, changing river beds, making water reservoirs, building artificial islands, etc., to generate spacious agricultural grounds or accommodate more people, including underwater lands and coastlines;
(3) The introduction of domesticated animals and plants (invasive species) into ecosystems and biogeocoenoses that cannot integrate them and, thus, are destroyed;
(4) The use of genetically modified husbandry species for food or agricultural forage;
(5) The induction of climate change due to the combustion of fossil fuels with the consequent greenhouse effect;
(6) The poorly deliberated use of alternative sources of energy;
(7) The wide use of wild food because of the augmented demand for traditional kitchens in the Far East, Southeast Asia, Latin America, and Africa;
(8) Chemical and biological pollution, including the carbonisation of the planet;
(9) The generation of huge masses of unprocessed waste; e.g., the plastification of the world's oceans creates a perfect basis for the growth of bacterial colonies and damages thermohaline circulation, which begets a malfunction in sustaining the climate of aquatic forms of life.

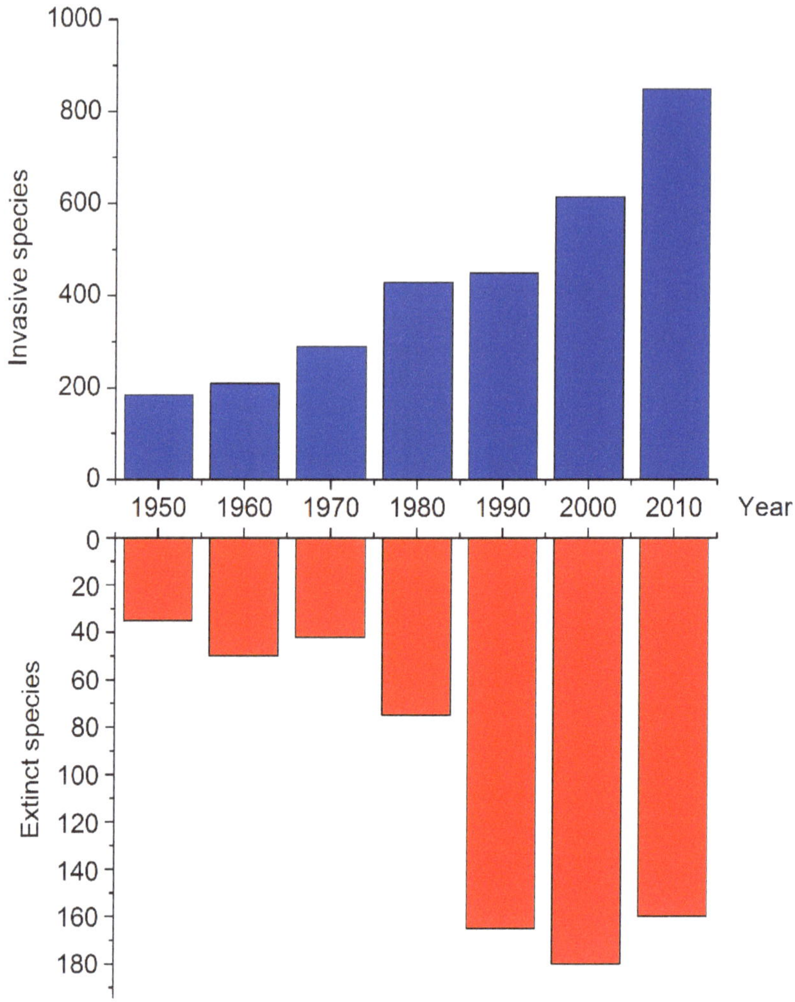

Figure 1. Vertebrate species' extinction and invasion caused by humans during the last seventy years. Based on the data provided in [18,19].

About 68 percent of the vertebrate animal biomass has disappeared since 1970 [11]. There are estimations that by 2100, *homo sapiens* may place 4 to 5 million species of the 8–9 million species remaining on the Earth now on the edge of extinction. This is sometimes called the "sixth great extinction of species" [10,12–14]. The line between "nature" and "culture" has never been as blurred as it is now [15–17].

The anthropogenic influence causes such a high level of species extinction mainly due to the dramatic influence on planetary processes and biotopes.

Biodiversity collections cannot diminish the rates of animal species extinction caused by anthropogenic factors, but they can provide ways to restore the populations of some species through the following mechanisms:

(1) The direct conservation of genetic material;
(2) Extensive scientific research that may lead to finding new means to restore species populations [20].

1.2. International Experience in Creating Biocollections

Biodiversity collections may be subdivided into the following collection types:

(1) Collections of lifeless forms suitable only for research, e.g., fossils, herbaria, entomology, ichthyology, herpetology, ornithology, and mammal collections;
(2) Collections of lifeless forms suitable for both research and biota restoration, e.g., seed collections;
(3) Collections of live forms, e.g., cell collections, microbial collections, cryogenic banks of biological material (biobanks);
(4) Information collections that do not contain biological samples but only information about them [20].

The collections of the first three types can (in real life—must) be supplemented with electronic databases for their accessibility to the scientific community, business entities, and other concerned parties [21]. A typical modern biocollection often possesses the instruments for "omic" investigations (genomics, proteomics, transcriptomics, etc.) and radioisotope, X-ray phase, X-ray structure, chromatography–mass spectrometry, and other analyses [22]. Photography, video recordings (e.g., recording of cell processes or recording of the growth of a specimen), and tomography are widely used in biocollections to image specimens [20]. Different types of information about a specimen may be sorted, analysed, rendered to a human-friendly view, and stored in digital form using network protocols [23].

There is a discrepancy in distinguishing between biocollections and biobanks. Some researchers suggest neglecting any linguistic or semantic differences between these two terms [24]. Others conceive of a biobank as a biocollection at the national scale [25]. The third group proposes considering human biological materials to belong to a biobank, whereas non-human species belong to a collection [26]. The fourth group supposes that a biobank can provide or lease biological material just like a bank lends money, while a biocollection only stores and/or processes samples and the corresponding data [27]. In our work, we follow the fourth approach. We define a biobank as a collection of animal or human biological materials that are stored, analysed, and researched and can be provided for various purposes on a free or commercial basis any time the material is needed. Therefore, in this understanding, the notion of a biobank is a part of the notion of a collection.

An important advantage of a biocollection is its accessibility to external researchers via a file-transfer protocol (FTP) route, the Internet, or private networks [28]. For instance, genetic biobanks can be integrated into international databases such as GenBank [29]. Big-data computing, artificial intelligence, and machine-learning algorithms play important roles in genotype studies and phenotype characterisation [30,31]. Several state-of-the-art collections use AI (artificial intelligence) technologies allowing the detection of correlations between morphology and phylogeny or between morphology and ecologic perturbations that have influenced biota in a given geographical zone [32].

To sum up, biocollections are very important today. With their use, one can meet the challenges of different scientific and other fields, including performing taxonomic analysis to model biological processes and the behavioural patterns of animals.

In countries where biocollections have been developed for a considerably long time (thirty–fifty years), large nation-wide biodiversity centres have appeared [33]. Figure 2 shows the number of documented biocollections and large biodiversity centres in several countries, which are significant for bioconservation, agriculture, medicine, veterinary, and scientific research [34–36]. Large centres are of national and international importance. They can contain millions of specimens [35]. Important international collections and centres contain information in digital form; many of them also store metadata [36]. Around one-third of North American collections provide external online access to them [36].

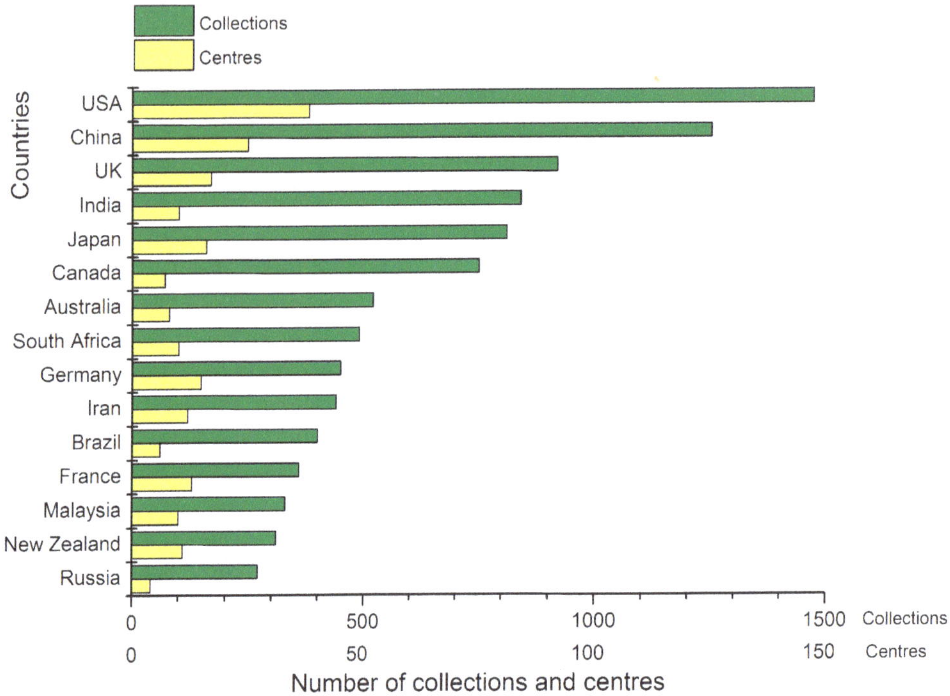

Figure 2. Number of documented biodiversity collections and centres important for bioconservation, agriculture, medicine, veterinary, and scientific research in fifteen chosen countries (as of 2021). Based on data provided in the studies [34–36].

A crucial step is the uniting of biodiversity collections and centres in networks. This may provide several important opportunities:

(1) The ability to cross-check in research;
(2) The simplicity of access;
(3) Redundancy in storing specimens;
(4) Effective data curation [37].

Data curation for many specimens is necessary even in the best biocollections, as, usually, information about a specimen may not be abundant [38]. In addition, progress in the development of technology, methods, and tools may eventually provide additional possibilities for analysis and effective data curation using appropriate algorithms [39].

The Pan American network (Extended Specimen Network, ESN) can serve as an example of a national network of biocollections. Its development commenced in 2011, and it includes the records of some 65 million specimens as of now [40]. By 2030, it is predicted to include the records of 1.2 billion specimens [40].

In the most advanced modern biocollections, one can trace a paradigmatic shift from the depositing and analysis of separate specimens to the storage of information about biotic associations [36]. Thus, the "omic" data about a species will be supplemented with meta-information about the following:

(a) Its habitat(s) (including geographic information system data);
(b) A typical ecological environment;
(c) Biogeocoenosis in which specimens of this species usually live;
(d) The behavioural patterns of animals;
(e) Organic and inorganic matter and forces (including climate and weather patterns in the habitat);

(f) Common food chains;
(g) Different organisms that form biotic associations with the species analysed (e.g., parasites, symbiotes, endophytes, epiphytes) [36].

Such an approach may open new possibilities in research. Studying population markers and characteristics can reshape our views on some evolutionary events, e.g., population bottlenecks, the contraction and expansion of habitats, transfer to another type of food, etc. [34]. More specifically, the genetic approach to studying speciation has developed largely due to success in creating a database of standardised genetic markers of different forms of life, such as Boldsystems [41]. Some local collections also provide sample information, e.g., the collection of PhMr. Tibor Weisz in Sarisske Museum Bardejov, Slovakia [42].

Table 1 compares three well-known international biodiversity centres. The comparison shows that biodiversity centres may have completely different tasks and technical specifications, but all of them have similar digitisation algorithms and enable easy online access to their databases for external research teams.

Table 1. Some pieces of information about three renowned international biodiversity centres.

Parameter	**Bioresource Centre Riken, Japan https://web.brc.riken.jp/en/, accessed on 15 July 2023**	**American Type Culture Collection (ATCC), USA https://www.lgcstandards-atcc.org/en/Products/Cells_and_Microorganisms/Cell_Lines.aspx, accessed on 21 July 2023**	**European Collection of Authenticated Cell Cultures (ECACC), United Kingdom https://www.phe-culturecollections.org.uk/collections/ecacc.aspx, accessed on 28 July 2023**
Type of funding	Government-sponsored	NGO	Government-sponsored
Number of biodiversity collections	5	5	2
Laboratory mice	9000	–	–
Plants	840,000	400 species (seeds)	–
Cell cultures	15,600	More than 20,000	Around 6700
DNA samples	3,900,000	+	–
Microorganisms	29,000	Around 20,000	+
Backup systems	+	+	+
Concordance to ISO 9001	?	ISO 9001:2015 [43] ISO 13485:2016 [44] ISO 17025:2017 [45] ISO 17034:2016 [46]	?
Digital catalogue of specimen	+	+	+
Genetic databases	–	–	–
Genome-editing possibilities	+	+	–
Sequencing possibilities	–	+	–
Other research possibilities	+	+	+
Inclusion in international projects	National Bioresource project	–	?
Inclusion in database networks	Asian collaborative networks	?	?
Educational options (undergraduate, graduate, and post-graduate studies)	+	+	+
Ethical committees	3	Adherence to common bioethical standards	?

+ Present; – absent; ? information regarding this biocollection is insufficient in open sources.

Biobanks, as a part of biocollections, are acquiring increasing importance for medical and veterinary purposes [47]. Their main medical and veterinary applications are transplantation, blood transfusion, and in vitro fertilisation procedures. In addition, they give an opportunity to perform genome-wide association studies (GWAS) [48]. An example of GWAS application is tracing the migration routes and interspecies genetic relations, i.e., studying the predisposition of a species to a given set of diseases [48].

We also witness a tendency to unite biobanks in networks. A prominent recent example is the Global Biobank Meta-Analysis Initiative, which combined digital data from twenty-four national cryobiobanks representing fifteen countries [49]. In this initiative, around 2.2 million genotype samples with approximately 70 million genetic variants have already been studied [50]. Thirteen diseases are currently under a comprehensive investigation in this network [37].

Important experience may be gained from the international biocollection networks that have been managed by US universities, scientific institutions, and societies for a long time, e.g., Network Integrated Biocollections Alliance (NIBA) [51] and Biodiversity Collections Network or Extended Specimen Network (BCN or ESN) [52]. They encompass more than one billion specimens. Both projects are rapidly developing. They offer a variety of research opportunities:

(1) Scientific discovery;
(2) Seamless data integration and attribution among different biocollections;
(3) The completion and improvement of digitised data;
(4) The ability to fill gaps in biodiversity data;
(5) The building and strengthening of international collaboration in bioconservation;
(6) The creation of an advanced specimen identification system;
(7) The development of new protocols for collecting underrepresented taxa;
(8) The provision of equal opportunities for small collections, either regional or personal;
(9) The provision of educational venues and capabilities;
(10) The strengthening of multidisciplinary work in bioconservation, biocollections, and big data.

The NIBA and ESN have caused a shift in our understanding of biocollections from separate specimens towards dynamic repositories of interconnected resources enriched by the study over time. They will help us to understand the organisms' growth, diversification, and their interaction with one another, as well as how climate change and the anthropogenic influence may affect biotic associations.

1.3. Biocollections in Russia

Russia can utilise the rich international experience in creating and maintaining biocollections obtained during the last thirty–forty years. Its variety of climatic zones, big area, and many endemic taxa make Russia important in the context of biocollections. Likewise, it is important to utilise rich data obtained during the Imperial and Soviet times of our history, as many discoveries were made at that time, and many specimens were collected, studied, described, and systematised [53].

In Russia, there are currently biocollections of the following types:

(a) Biocollections created by scientific institutions or universities as a result of their research work;
(b) Nation-wide collections subsidised by governmental authorities;
(c) Museums;
(d) Biobanks [53].

They are aimed at (1) storing and providing specimens for research and (2) preserving biodiversity in Russia [53,54].

Figure 3 summarises the information about the Russian documented biocollections that are supported by the Ministry of Science and Higher Education. There are 255 collections funded by the Ministry in total. The overall number of Russian documented biocollections has been estimated at around 280–300 [55,56].

The systematisation of Russian biocollections is at its very beginning. The information that the Ministry keeps on the biocollections contains merely the name of the collection; the institution that is responsible for managing the collection; the purpose of the collection's creation; the number of specimens; the list of standard operational steps necessary for maintaining the collection; and the list of key infrastructure components and equipment [57].

Despite the comparatively large number of biocollections, there are currently only four national biodiversity centres:

(1) National Collection of Pathogenic Microorganisms and Cell Cultures (State Research Centre for Applied Biotechnology and Microbiology) [58];
(2) National Collection of Pathogenic Microorganisms Causing Dangerous, Extremely Dangerous and Rare Diseases of Animals (Federal Research Centre for Virology and Microbiology) [59];
(3) National Collection of Industry-Related Microorganisms (State Research Institute of Genetics and Selection of Industrial Microorganisms of the National Research Centre Kurchatov Institute) [60];
(4) National Collection of Genetic Resources of Plants (N. I. Vavilov All-Russian Institute of Plant Genetic Resources) [61].

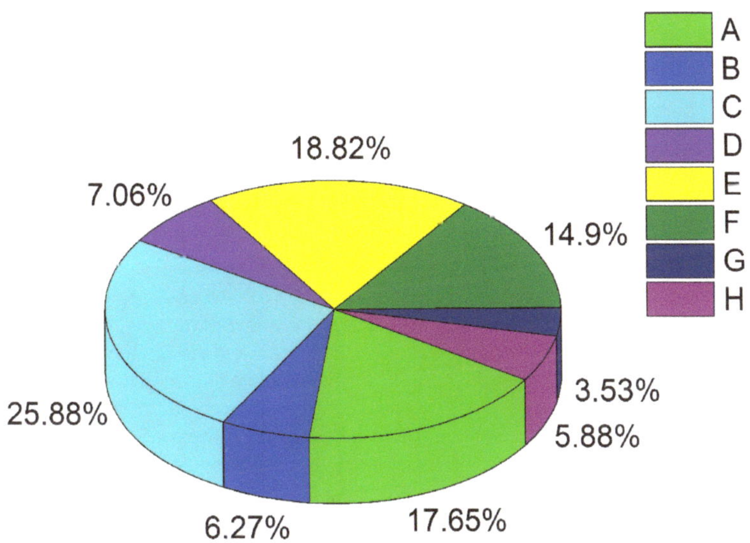

Figure 3. The number of documented biocollections in Russia as of May 2023 based on the data provided in [59,60]. A—microbial; B—cultures of humans and animal cells; C—agricultural plants; D—wild plants; E—herbaria; F—wild and laboratory animals; G—husbandry animals; H—biobanks.

No animal bioconservation centres are present in Russia at present.

Creating biocollection networks in Russia also seems to be in great demand. In addition to the benefits of the networks discussed above, they may significantly help in accompanying epidemiological studies, e.g., by accumulating necessary data about animal parasites, the carriers of zoonotic causative agents. This may serve to reduce the burden of zoonoses and mitigate the risk of zoonosis-related animal species extinction, as well as the risk of new human emerging infectious diseases resulting from an anthropozoonotic spillover. Therefore, this activity of biocollection networks may be of great international safety value.

1.4. Specific Goal

As we see from the above, in Russia, biocollections are developing, and further multifaceted work is required to expand them with the procurement of significant government and private funding. Thus, in the current work, we performed a professional opinion survey based on the opinions of Russian biologists (bioconservation specialists, big-data specialists, and ecologists). The goal was to learn their views on how to arrange, develop, and use biocollections in Russia in the most effective way.

2. Methods

2.1. Survey

To obtain a collective expert opinion on the problems of Russian biodiversity collections/centres and possible ways to solve them, we performed a survey. Our methodology is based upon the Child Health and Nutrition Research Initiative (CHNRI) approach [62–70]. This method allows one to detect and rank research and practical priorities in the field. In our case, this is the perspective on Russian biocollection development. The CHNRI method has been used in more than 100 published studies led by institutions and centres in the last decade [71].

Our CHNRI-based survey consisted of the following four steps:

(1) Invitation to participate in anonymous research/initial discovery of topics. Those experts who accepted the invitation to participate were asked to suggest up to three practical or research problems associated with biodiversity collections/centres in Russia. The initial invitation was dispatched to 324 persons through the social networks VKontakte, Facebook, and LinkedIn. The search for experts was carried out manually using these social networks. All of them are Russian specialists in ecology, evolution, genetics, and/or bioconservation. The selected specialists had to represent different provinces of Russia.

(2) Compilation. Identical answers were combined.

(3) Ranking. The ranking was determined in regard to five predefined criteria:

   (a) Criterion 1: Impact on the success of the research ("Do you think that the proposal will lead to new achievements in the research, facilitate the research, or open new investigation opportunities?");

   (b) Criterion 2: Impact on biodiversity conservation ("Do you think that the proposal will stimulate bioconservation or result in finding new, more effective ways to perform bioconservation or to retard biodiversity loss?");

   (c) Criterion 3: Impact on the promotion of "citizen science", i.e., on elevating community involvement ("Do you think that the proposal will increase the involvement of a general population in research or practice related to biocollections?");

   (d) Criterion 4: Potential for a paradigmatic shift in ecology and evolution ("Do you think that the proposal will or may facilitate expanding of our horizons in genetics, evolution, environmental science, or other branches of biology?");

   (e) Criterion 5: Potential for the transfer and applicability to different practical areas ("Do you think that the results of the proposal application will be useful and usable in other areas, e.g., medicine, pharmaceutics, or different fields of biology other than bioconservation?").

(4) Calculation. The CHNRI method involves completing a spreadsheet (e.g., rows list topics and columns criteria) with +1 (an expert supposes that a topic satisfies a criterion), 0 (an expert supposes the contrary), or 0.5 (an expert thinks that he/she has sufficient knowledge but is not inclined to answer "yes" or "no", though this option was generally discouraged), or the field in the spreadsheet may be left blank. The response "0.5" reduces the discriminatory power of the exercise and leads to the "regression to the mean" in the final distribution of the overall Research Priority Scores (RPSs) [71].

The survey was carried out from September 2022 to February 2023.

In addition, the CHNRI methodology permits the measurement of the level of agreement among the experts for every endpoint. The indicator Average Expert Agreement (AEA) is the average proportion of scorers that returned the most common answer for a research question [70]:

$$AEA = \frac{1}{5}\sum_{i=1}^{5}\frac{N_{i,Scorers\ that\ provided\ a\ most\ frequent\ response}}{N_{i,\ Total\ scorers}},$$

where i runs through the set of the chosen criteria. In our case, the number of criteria is 5.

A typical spreadsheet is shown in Table 2.

Table 2. An example of a completed spreadsheet that was offered for completion to 71 experts.

No. of Topic	Topic	Criterion 1	Criterion 2	Criterion 3	Criterion 4	Criterion 5
1	Topic 1	0	+1	+0.5	0	0
2	Topic 2	+1	+1	0	0	0
Etc.						

For each topic and each criterion, the mean was calculated based on all answers that were not blank.

2.2. *Software*

Origin 8.1 (OriginLab, Northampton, MA, USA) was used for calculations and visualisation.

3. Results

Seventy-one people (21.9% of the initial set) responded with their proposals of 1–3 relevant topics. Their main demographic characteristics are provided in Table 3.

Table 3. Main demographics and academic profiles of the respondents detected algorithmically based on the data provided in social network accounts (if any) or self-reported. Student distribution of the sample set was assumed.

Age	**Range: 26–71 Years *; Mean 36.3 ± 12.5 y.o.**
Gender	28 females (39.4%)
Bioconservation specialists	23 (32.4%)
Biocollection specialists	16 (22.5%)
Ecologists	18 (25.4%)
Evolutionists, geneticists	12 (16.9%)
Other biological specialties represented	2 (2.8%)
Academic training: • Bachelor's/Master's degree • Post-graduate student • PhD degree, not Professor • Professor	 11 (15.5%) 8 (11.3%) 46 (64.7%) 6 (8.5%)
Having considerable practical experience with starting/managing biocollections (one year or more)	53 (74.6%)
Average length of working experience in biocollections for those respondents who had practical experience	Mean 4.7 ± 0.8 years
Having published papers/books/grey literature in biocollections	2 (2.8%)

* Confidence interval 95%, $p = 0.05$.

We compiled the proposed topics in one list (188 topics). Identical or similar topics were merged, giving 104 topics (55.3% of the full list of topics). Only the 24 topics (23% of the set) that passed the 5% threshold, i.e., that were suggested by four experts (5%) or more, were retained.

Then, we returned the completed list of 24 topics to each of the 71 participants to rank the priorities by scoring the topics. Of the 71 participants, only 28 (39.4% of the intermediate set of participants) responded in the second round with their rankings.

Figure 4 provides scores for the top ten topics suggested by the experts. Table 4 describes the topics. Ranking was performed according to the RPS values and are shown in Table 4.

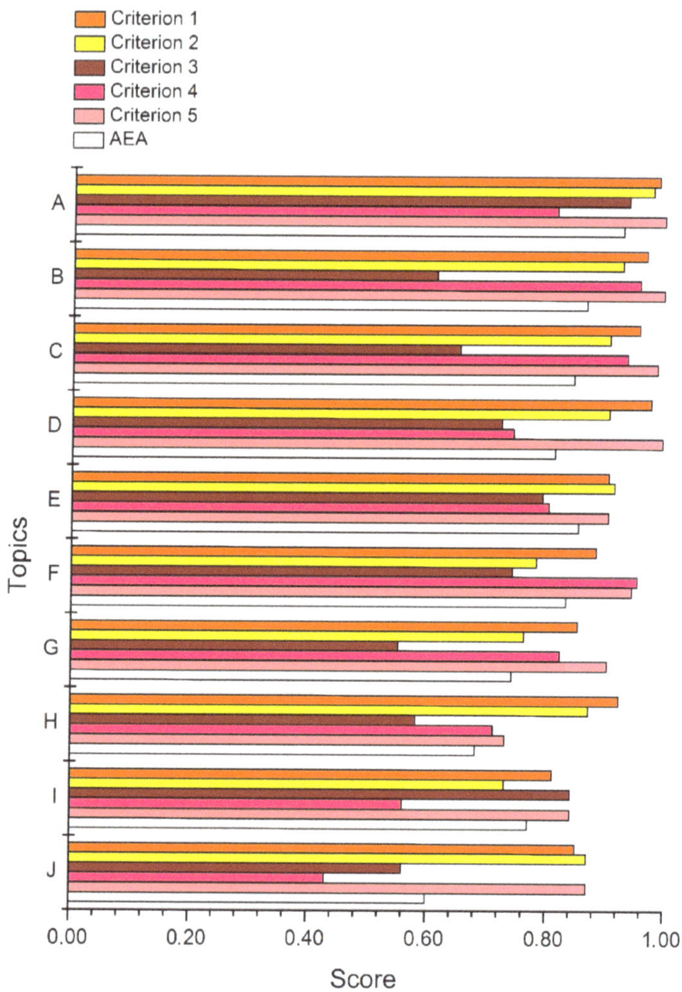

Figure 4. Expert scoring for the top ten proposed topics concerning the five predefined criteria along with their AEA values. AEA—Average Expert Agreement score (see Methods for details).

Table 4. The main topics that reflect problems of biodiversity collection/centre development in Russia.

No.	Designation	Topic	RPS *
1	A	Necessity to create large national centres of biological conservation.	0.936
2	B	Full sequencing and creation of different "omic" databases.	0.886
3	C	Full digitisation of a biodiversity collection/centre.	0.882
4	D	Free-of-charge and open access to a collection/centre database for external researchers.	0.864
5	E	The presence of rich metadata in a collection/centre database.	0.860
6	F	Utilising the big-data principle for a collection/centre database.	0.858
7	G	Standardisation of specimen and data curation for simplicity of a search.	0.776
8	H	Capability of collections/centres to perform their own research on the specimens.	0.762
9	I	Free-of-charge provision of specimens to concerned parties, specifically by biobanks.	0.756
10	J	Lack of bureaucratic barriers and minimum reporting in creating, managing, and using biodiversity collections/centres.	0.716
11	K	Modelling of restoration of biogeocoenoses in collections/centres.	0.682
12	L	Identification and systematisation of pathogens causing novel emerging infectious diseases. GIS (geographical information systems) approach for the rapid identification of new niduses and outbreaks.	0.656
13	M	Using a systemic approach to creating a collection/centre. Combining data about biological associations.	0.644
14	N	Simple procedures for business entities participating in funding of collections/centres with little paperwork.	0.622
15	O	Uniting collections/centres in national networks. Using cross-referencing in all national databases.	0.618
16	P	High standards for specimen and data protection in biobanks.	0.606
17	Q	Ensuring biological safety via thorough standardisation of genetically modified organisms that pose potential threats to human and biota.	0.588
18	R	Increasing the government funding of biocollection/centre development.	0.562
19	S	Inclusion of cultural- and region-specific metadata on agricultural, food, veterinary, or medicinal use of specimens.	0.512
20	T	International cooperation. Inclusion of national collections/centres in international projects, initiatives, and networks.	0.506
21	U	Relevant legal support of collection/centre functioning. Development of corresponding laws.	0.418
22	V	Simplifying the patent work in collaboration with collections/centres.	0.362
23	W	Integration of biodiversity collections/centres with educational programmes.	0.306
24	X	Ability to perform work on genetic modification of specimens.	0.248

* RPS—Research Priority Score (see Methods for details).

4. Discussion

We see that the higher the rank of a topic, the greater the agreement between experts about its importance. The Pearson correlation coefficient between the RPS (Research Priority Score) and AEA (Average Expert Agreement) is 0.938 ($p \leq 5.9 \times 10^{-5}$) for the top ten topics and 0.584 ($p \leq 7.1 \times 10^{-2}$) for the full set of 24 topics. This means that disagreement between the experts is considerable for the 14 lowest-ranked topics, whereas the top 10 topics seem to be important to most of the experts. For example, 92% of the 28 experts who returned completed questionnaires agree with each other that there is a strict necessity to form nation-wide centres of biological diversity (topic A). Moreover, merely 51% of the experts agree with each other about the importance of ensuring the ability to perform genetic modifications of specimens in biocollections/centres (topic X).

In addition, the variance in the responses also grows as we proceed down the table of topics (Figure 4, Table 3); i.e., the discrepancy in expert opinions increases.

The top ten goals/tasks can be subdivided into three groups: (1) management-related (topics A and H); (2) science-related and technical (topics B, C, E, F, and G); and (3) policy-related ones (topics D, I, and J). Interestingly, the policy-related topics seemed important enough for the experts to place them in the top-ten group. This may be accounted for by the huge amount of paperwork that biocollection specialists ought to do in Russia now and the presence of many bureaucratic procedures that hinder them from performing scientific research and managing collections. This is in line with our previous research [72–78].

Although the respondents have considerable experience in working with biocollections, only 2.8 percent have published the results of their research thus far. This fact may indicate that the development of biocollections and biocollection networks is at the initial stage in Russia, and there are broad prospects for further elaboration. In addition, there are currently no uniform standards of academic writing in this field in Russia. The third reason may be insufficient governmental financing of biocollection-related investigations and the scarcity of research grants.

Thus, the surveyed Russian experts were unanimous in their opinion that the most important task for developing biodiversity collections in the country is creating large, networked, nationally important biodiversity centres, including biobanks, with free and easy access to their databases.

From the results of our research, we can deduce the main vectors for the future development of biocollections in Russia.

Many collections and samples were lost in Russia in the 1990s. The fundamental work on the preservation and development of Russian biodiversity collections practically stopped during those years and continued only due to the enthusiasm of their custodians, only resuming in 2015 when the Academician Nikolai A. Kolchanov, the former head of the Institute of Cytology and Genetics of the Siberian branch of the Russian Academy of Sciences (RAS), commenced this important activity. He attracted the attention of the authorities to the problem of the continuing catastrophic loss of national strategic resources, including animal biodiversity. The first funds for these initiatives were granted by governmental bodies only in 2016–2017. As a result, the remaining collections were renovated, rebuilt, grouped according to the objects (plants, animals, microorganisms, human materials, and cell cultures), and rated. The scientific teams responsible for the development of biocollections were granted the possibility to expand the lists of necessary equipment and material assets.

The majority of the collections were registered as the core facility centres of the corresponding organisations on the specific web resource http://ckp-rf.ru (accessed on 1 December 2023) aimed at collecting data on their operation (e.g., annual information on money spent for maintenance, money received and gained, orders, users) and providing statistics [79]. It was also designed to be a platform for the interaction between users and providers for making arrangements for the paid and unpaid use of biocollection specimens. Unfortunately, so far, that goal has not been fully fulfilled, and the work is continuing. In addition, standard operating procedures (SOPs) to maintain and expand collections and the cost calculation method of the SOPs were developed. The important tasks for the collections that were included in the new programme are expansion, the creation of specimen catalogues, and the provision of better access to users.

The N.K. Koltzov Institute of Developmental Biology RAS (IDB RAS) houses four biodiversity collections:

(1) Cell Culture Collection (CCC);
(2) Collection of the Tissues of Wild Animals;
(3) Drosophila Genetic Lines Collection;
(4) Coccinellidae Polymorphous Species Collection.

These collections have existed for many years. They are managed by the corresponding laboratories of IDB RAS, which used collection samples in its routine work and took care of them.

The IDB RAS CCC is a flagship biodiversity collection of our institution and one of the very few collections of this kind in Russia.

The first and oldest one, known as the Cell Culture Collection of Vertebrates, was founded by the Honoured Scientist Georgy P. Pinaev in the Institute of Cytology in St. Petersburg in 1978 and gathered nine collections around the Soviet Union [79,80]. Nowadays, it offers about 150 cell cultures for users and has about 800 hybridoma and patented cell cultures, which are stored according to patent requirements. The CCC, on the contrary, is a "young" cell collection, officially founded and registered in 2016, and includes more than 200 different animal and human cell cultures now (around 8000 samples). Among them are primary cultures isolated from different tissues and organs of humans and animals (laboratory and domestic animals, such as mice, rats, rabbits, pigs, monkeys, cats, and dogs, and wild animals, such as different species of hamsters and mole voles), genetically modified cells (including immortalised ones), and induced pluripotent stem cells (iPSCs) [81].

The IDB RAS CCC is a quickly developing and expanding collection. Its researchers implement the basic rules of the world's biobanking knowledge in their routine work.

Currently, the most important tasks of the CCC are the characterisation of cell cultures (including mycoplasma tests, STR-profiling and karyotype investigations if possible, doubling time measurement, and immunocytochemistry for basic markers) and the provision of characterised cell culture samples (together with the corresponding cell culture passports) to different researchers in scientific institutions and business companies. The price of the samples provided by the CCC is very low compared with the price of the samples from known international collections, such as the American Type Culture Collection or the European Collection of Authenticated Cell Cultures (including logistics). This makes CCC materials very accessible for users and thus serves the goal of the development of Russian science and inventions in the fields of pharmaceutics, biomedicine, veterinary medicine, and gene technology.

5. Conclusions

We performed a survey of Russian specialists engaged in biodiversity collections. There is a substantial degree of concordance among these specialists that we need to create several nation-wide biodiversity centres. These centres will contain not only genetic material but also a variety of specimens and data, including metadata. They have to be united in a national electronic biodiversity network that will allow free access to the data stored in the centres to scientists, bioconservation researchers, and practitioners. They may significantly contribute to endangered species restoration in Russia in the future.

We have shown the success of our scientific institution in developing and managing biocollections and have deliberated upon their potential. These biocollections may become a basis for the creation of a national biodiversity centre.

Author Contributions: Conceptualisation—E.V.A. and K.S.S.; methodology—E.V.A. and K.S.S.; software—K.S.S.; validation—E.A.V., E.V.A., N.P.S. and K.S.S.; formal analysis—K.S.S.; data mining—E.V.A. and K.S.S.; investigation—E.A.V., E.V.A., N.P.S. and K.S.S.; resources—E.A.V. and N.P.S.; data curation—E.V.A. and K.S.S.; writing original draft—E.V.A. and K.S.S.; review and editing—E.A.V., E.V.A., N.P.S. and K.S.S.; visualisation—E.V.A. and K.S.S.; supervision—N.P.S. and E.A.V.; project administration—E.A.V.; funding acquisition—E.A.V. All authors have read and agreed to the published version of the manuscript.

Funding: This work was supported by the Ministry of Science and Higher Education of the Russian Federation under Agreement No. 075-15-2021-1063 from 28 September 2021.

Institutional Review Board Statement: Not applicable. All results are anonymised, and no identifying information about the respondents is disclosed in the research. Since no experiments were carried out on biological organisms, no special ethical permission from the affiliated institution was required.

Informed Consent Statement: Informed consent was obtained from all persons involved in the study to participate and publish this paper.

Data Availability Statement: Available from the authors upon reasonable request.

Conflicts of Interest: The authors declare no conflict of interest.

References

1. La Salle, J.; Williams, K.J.; Moritz, C. Biodiversity analysis in the digital era. *Philos. Trans. R. Soc. Lond. B Biol. Sci.* **2016**, *371*, 20150337. [CrossRef] [PubMed]
2. Meineke, E.K.; Davies, T.J.; Daru, B.H.; Davis, C.C. Biological collections for understanding biodiversity in the Anthropocene. *Philos. Trans. R. Soc. Lond. B Biol. Sci.* **2018**, *374*, 20170386. [CrossRef]
3. Walls, R.L.; Deck, J.; Guralnick, R.; Baskauf, S.; Beaman, R.; Blum, S.; Bowers, S.; Buttigieg, P.L.; Davies, N.; Endresen, D.; et al. Semantics in support of biodiversity knowledge discovery: An introduction to the biological collections ontology and related ontologies. *PLoS ONE* **2014**, *9*, e89606. [CrossRef] [PubMed]
4. Boguslavsky, D.V.; Sharova, N.P.; Sharov, K.S. Evolutionary Challenges to Humanity Caused by Uncontrolled Carbon Emissions: The Stockholm Paradigm. *Int. J. Environ. Res. Public Health* **2022**, *19*, 16920. [CrossRef] [PubMed]
5. Rudan, I. *Evil Air: Thoughts on Health and Disease in the 21st Century*; Naklada Ljevak: Zagreb, Croatia, 2018.
6. Dirzo, R.; Young, H.S.; Galetti, M.; Ceballos, G.; Isaac, N.J.; Collen, B. Defaunation in the Anthropocene. *Science* **2014**, *345*, 401–406. [CrossRef] [PubMed]
7. Seddon, P.J.; King, M. Creating proxies of extinct species: The bioethics of de-extinction. *Emerg. Top. Life Sci.* **2019**, *3*, 731–735. [CrossRef] [PubMed]
8. Perry, G.H. How human behavior can impact the evolution of genetically-mediated behavior in wild non-human species. *J. Comp. Physiol. A Neuroethol. Sens. Neural Behav. Physiol.* **2020**, *206*, 337–342. [CrossRef]
9. Trask, A.; Canessa, S.; Moehrenschlager, A.; Newland, S.; Medina, S.; Ewen, J. Extinct-in-the-wild species' last stand. *Science* **2020**, *369*, 516. [CrossRef]
10. Cowie, R.H.; Bouchet, P.; Fontaine, B. The Sixth Mass Extinction: Fact, fiction or speculation? *Biol. Rev. Camb. Philos. Soc.* **2022**, *97*, 640–663. [CrossRef]
11. Parvinen, K. Evolutionary suicide. *Acta Biotheor.* **2005**, *53*, 241–264. [CrossRef]
12. Dirzo, R.; Ceballos, G.; Ehrlich, P.R. Circling the drain: The extinction crisis and the future of humanity. *Philos. Trans. R. Soc. Lond. B Biol. Sci.* **2022**, *377*, 20210378. [CrossRef]
13. Mora, C.; Tittensor, D.P.; Adl, S.; Simpson, A.G.B.; Worm, B. How Many Species Are There on Earth and in the Ocean? *PLoS Biol.* **2011**, *9*, e1001127. [CrossRef]
14. Lewis, O.T. Climate change, species-area curves and the extinction crisis. *Philos. Trans. R. Soc. Lond. B Biol. Sci.* **2006**, *361*, 163–171. [CrossRef]
15. Sassin, W.; Donskikh, O.A.; Gnes, A.; Komissarov, S.A.; Liu, D. *Evolutionary Environments: Homo Sapiens—An Endangered Species?* Studia: Innsbruck, Austria, 2018.
16. Boguslavsky, D.V.; Sharov, K.S.; Sharova, N.P. Using Alternative Sources of Energy for Decarbonization: A Piece of Cake, but How to Cook This Cake? *Int. J. Environ. Res. Public Health* **2022**, *19*, 16286. [CrossRef]
17. Glaubrecht, M. *Das Ende der Evolution: Der Mensch und Die Vernichtung der Arten*; C. Bertelsmann Verlag: München, Germany, 2019.
18. Kolbert, E. *The Sixth Extinction: An Unnatural History*; Henry Holt and Co.: New York, NY, USA, 2014.
19. Smil, V. *Cycles of Life: Civilization and the Biosphere*; W. H. Freeman & Co.: New York, NY, USA, 1997.
20. National Academies of Sciences, Engineering, and Medicine. *Biological Collections: Ensuring Critical Research and Education for the 21st Century*; National Academies Press: Washington, DC, USA, 2020.
21. Van Insberghe, D.; Arevalo, P.; Chien, D.; Polz, M.F. How can microbial population genomics inform community ecology? *Philos. Trans. R. Soc. Lond. B Biol. Sci.* **2020**, *375*, 20190253. [CrossRef]
22. Fukushima, C.; West, R.; Pape, T.; Penev, L.; Schulman, L.; Cardoso, P. Wildlife collection for scientific purposes. *Conserv. Biol.* **2021**, *35*, 5–11. [CrossRef]
23. Camacho, D.M.; Collins, K.M.; Powers, R.K.; Costello, J.C.; Collins, J.J. Next-Generation Machine Learning for Biological Networks. *Cell* **2018**, *173*, 1581–1592. [CrossRef]
24. Coppola, L.; Cianflone, A.; Grimaldi, A.M.; Incoronato, M.; Bevilacqua, P.; Messina, F.; Baselice, S.; Soricelli, A.; Mirabelli, P.; Salvatore, M. Biobanking in health care: Evolution and future directions. *J. Transl. Med.* **2019**, *17*, 172. [CrossRef] [PubMed]
25. Hewitt, R.; Watson, P. Defining biobank. *Biopreserv. Biobank.* **2013**, *11*, 309–315. [CrossRef] [PubMed]
26. Day, J.G.; Stacey, G.N. Biobanking. *Mol. Biotechnol.* **2008**, *40*, 202–213. [CrossRef] [PubMed]
27. Manders, P.; Peters, T.M.A.; Siezen, A.E.; van Rooij, I.A.L.M.; Snijder, R.; Swinkels, D.W.; Zielhuis, G.A. A Stepwise Procedure to Define a Data Collection Framework for a Clinical Biobank. *Biopreserv. Biobank.* **2018**, *16*, 138–147. [CrossRef]
28. Shickle, D.; Griffin, M.; El-Arifi, K. Inter- and intra-biobank networks: Classification of biobanks. *Pathobiology* **2010**, *77*, 181–190. [CrossRef]
29. GenBank Overview. Available online: https://www.ncbi.nlm.nih.gov/genbank (accessed on 10 June 2023).
30. Zisis, K. Biobanking with Big Data: A Need for Developing "Big Data Metrics". *Biopreserv. Biobank.* **2016**, *14*, 450–451. [CrossRef]
31. Adams, C.I.M.; Knapp, M.; Gemmell, N.J.; Jeunen, G.J.; Bunce, M.; Lamare, M.D.; Taylor, H.R. Beyond Biodiversity: Can Environmental DNA (eDNA) Cut It as a Population Genetics Tool? *Genes* **2019**, *10*, 192. [CrossRef] [PubMed]
32. Li, X.; Han, J.; Zhang, S.; Chen, K.; Zhao, L.; He, Y.; Liu, S. Artificial Intelligence for Screening Chinese Electronic Medical Record and Biobank Information. *Biopreserv. Biobank.* **2021**, *19*, 386–393. [CrossRef]
33. Green, D.W.; Watson, J.A.; Jung, H.S.; Watson, G.S. Natural History Collections as Inspiration for Technology. *Bioessays* **2019**, *41*, e1700238. [CrossRef]

34. Kamau, E.C. *Global Transformations in the Use of Biodiversity for Research and Development: Post Nagoya Protocol Implementation Amid Unresolved and Arising Issues*; Springer: Cham, Switzerland, 2022.
35. Underkoffler, S.C.; Adams, H.R. *Wildlife Biodiversity Conservation Multidisciplinary and Forensic Approaches*; Springer: Cham, Switzerland, 2021.
36. Tupasela, A. *Populations as Brands: Marketing National Resources for Global Data Markets*; Springer: Cham, Switzerland; Palgrave Macmillan: London, UK, 2022.
37. Zhou, W.; Kanai, M.; Wu, K.H.H.; Rasheed, H.; Tsuo, K.; Hirbo, J.B.; Wang, Y.; Bhattacharya, A.; Zhao, H.; Namba, S.; et al. Global Biobank Meta-analysis Initiative: Powering genetic discovery across human disease. *Cell Genom.* **2022**, *2*, 100192. [CrossRef] [PubMed]
38. Chalmers, D. Genetic research and biobanks. *Methods Mol. Biol.* **2011**, *675*, 1–37. [CrossRef]
39. Alaimo, J.T.; Saunders, C.J. Big Data Strikes Again: Future Utilization of the UK Biobank as a Resource for Clinical Laboratories. *Clin. Chem.* **2021**, *67*, 932–934. [CrossRef] [PubMed]
40. Lendemer, J.; Thiers, B.; Monfils, A.K.; Zaspel, J.; Ellwood, E.R.; Bentley, A.; LeVan, K.; Bates, J.; Jennings, D.; Contreras, D.; et al. The Extended Specimen Network: A Strategy to Enhance US Biodiversity Collections, Promote Research and Education. *Bioscience* **2020**, *70*, 23–30. [CrossRef]
41. Barcode of Life Data System. Available online: http://www.boldsystems.org (accessed on 20 April 2023).
42. Hromada, M.; Čanády, A.; Mikula, P.; Peterson, A.T.; Tryjanowski, P. Old natural history collections for new millennium–birds and mammals in the collection of PhMr. Tibor Weisz in Sarisske Museum Bardejov, Slovakia. *Folia Oecol. Acta Univers. Presov.* **2015**, *7*, 115–141.
43. *ISO 9001:2015*; Quality Management Systems—Requirements. International Organization for Standardization: Geneva, Switzerland, 2015.
44. *ISO 13485:2016*; Medical Devices—Quality Management Systems—Requirements for Regulatory Purposes. International Organization for Standardization: Geneva, Switzerland, 2016.
45. *ISO 17025:2017*; General Requirements for the Competence of Testing and Calibration Laboratories. International Organization for Standardization: Geneva, Switzerland, 2017.
46. *ISO 17034:2016*; General Requirements for the Competence of Reference Material Producers. International Organization for Standardization: Geneva, Switzerland, 2016.
47. Paskal, W.; Paskal, A.M.; Dębski, T.; Gryziak, M.; Jaworowski, J. Aspects of Modern Biobank Activity—Comprehensive Review. *Pathol. Oncol. Res.* **2018**, *24*, 771–785. [CrossRef] [PubMed]
48. Ko, S.; German, C.A.; Jensen, A.; Shen, J.; Wang, A.; Mehrotra, D.V.; Sun, Y.V.; Sinsheimer, J.S.; Zhou, H.; Zhou, J.J. GWAS of longitudinal trajectories at biobank scale. *Am. J. Hum. Genet.* **2022**, *109*, 433–445. [CrossRef] [PubMed]
49. Global Biobank Meta-Analysis Initiative. Available online: https://www.globalbiobankmeta.org (accessed on 22 May 2023).
50. Kanai, M.; Elzur, R.; Zhou, W.; Daly, M.J.; Finucane, H.K.; Wu, K.-H.H.; Rasheed, H.; Tsuo, K.; Hirbo, J.B.; Wang, Y.; et al. Meta-analysis fine-mapping is often miscalibrated at single-variant resolution. *Cell Genom.* **2022**, *2*, 100210. [CrossRef] [PubMed]
51. Network Integrated Biocollections Alliance (NIBA). Available online: https://www.idigbio.org/content/network-integrated-biocollections-alliance-niba (accessed on 6 October 2023).
52. Biodiversity Collections Network. Available online: https://bcon.aibs.org (accessed on 6 October 2023).
53. Pavlinov, I. Biodiversity and biocollections: Problem of correspondence. In *Aspects of Biodiversity*; Pavlinov, I.W., Kalyakin, M.V., Sysoev, A.V., Eds.; KMK Sci Press: Moscow, Russia, 2016; Volume 54, pp. 733–786.
54. All on Board! How They Collect Livestock of a New Arc. Available online: https://www.rscf.ru/news/media/vse_na_bort_kak_sobirayut_zhivnost_novogo_kovchega (accessed on 25 May 2023).
55. Mokhov, A.A. To the question about bioresource centres and biocollections. *Bus. Manag. Law* **2021**, *5*, 26–30.
56. Ruda, M.G. Biological collections as an element of research infrastructure. *Bull. Kutafin Univ.* **2022**, *9*, 146–153. [CrossRef]
57. Davydov, D.S.; Koshechkin, K.A.; Movsesynts, A.A. Basic approaches to data management for administration of biological collections. *Bioprep. Prev. Diagn. Treat.* **2017**, *17*, 216–221.
58. State Research Centre for Applied Biotechnology and Microbiology. Available online: https://www.obolensk.org/en (accessed on 11 April 2023).
59. Federal Research Center for Virology and Microbiology. Available online: https://ficvim.ru/en (accessed on 18 April 2023).
60. State Research Institute of Genetics and Selection of Industrial Microorganisms of the National Research Center Kurchatov Institute. Available online: http://eng.genetika.ru (accessed on 19 April 2023).
61. N. I. Vavilov All-Russian Institute of Plant Genetic Resources. Available online: http://www.vir.nw.ru/en (accessed on 19 April 2023).
62. Agafonov, V.B.; Vypkhanova, G.V.; Zhavoronkova, N.G.; Mokhov, A.A. National law about bioresource centres and biological collections: A concept of development. *Bull. Kutafin Univ.* **2021**, *8*, 103–109.
63. Kalinin, R.S.; Goleva, O.V.; Illarionov, R.A.; Tsay, V.V.; Mukomolova, A.L.; Konstantinova, Y.E.; Markin, I.V.; Krylov, A.V.; Rogozina, N.V.; Bekhtereva, M.K.; et al. Development of a biobank in the structure of scientific and diagnostic and treatment institutions and prospects for interregional integration. *Cardiovasc. Ther. Prophyl.* **2022**, *21*, 3401. [CrossRef]
64. Rudan, I.; Yoshida, S.; Chan, K.Y.; Cousens, S.; Sridhar, D.; Bahl, R.; Martines, J. Setting health research priorities using the CHNRI method: I. Involving funders. *J. Glob. Health* **2016**, *6*, 010301. [CrossRef] [PubMed]

65. Yoshida, S.; Cousens, S.; Wazny, K.; Chan, K.Y. Setting health research priorities using the CHNRI method: II. Involving researchers. *J. Glob. Health* **2016**, *6*, 010302. [CrossRef] [PubMed]
66. Yoshida, S.; Wazny, K.; Cousens, S.; Chan, K.Y. Setting health research priorities using the CHNRI method: III. Involving stakeholders. *J. Glob. Health* **2016**, *6*, 010303. [CrossRef] [PubMed]
67. Rudan, I. Setting health research priorities using the CHNRI method: IV. Key conceptual advances. *J. Glob. Health* **2016**, *6*, 010501. [CrossRef] [PubMed]
68. Rudan, I.; Yoshida, S.; Wazny, K.; Chan, K.Y.; Cousens, S. Setting health research priorities using the CHNRI method: V. Quantitative properties of human collective knowledge. *J. Glob. Health* **2016**, *6*, 010502. [CrossRef] [PubMed]
69. Yoshida, S.; Rudan, I.; Cousens, S. Setting health research priorities using the CHNRI method: VI. Quantitative properties of human collective opinion. *J. Glob. Health* **2016**, *6*, 010503. [CrossRef]
70. Rudan, I.; Yoshida, S.; Chan, K.Y.; Sridhar, D.; Wazny, K.; Nair, H.; Sheikh, A.; Tomlinson, M.; Lawn, J.E.; Bhutta, Z.A.; et al. Setting health research priorities using the CHNRI method: VII. A review of the first 50 applications of the CHNRI method. *J. Glob. Health* **2017**, *7*, 011004. [CrossRef] [PubMed]
71. Polašek, O.; Wazny, K.; Adeloye, D.; Song, P.; Chan, K.Y.; Bojude, D.A.; Ali, S.; Bastien, S.; Becerra-Posada, F.; Borrescio-Higa, F.; et al. Research Priorities to Address the Burden of COVID-19 in Low- and Middle-Income Countries. *J. Glob. Health* **2022**, *12*, 09003. [CrossRef] [PubMed]
72. Legach, E.I.; Sharov, K.S. (Eds.) *SARS-CoV-2 and Coronacrisis: Epidemiological Challenges, Social Policies and Administrative Strategies*; Springer: Singapore, 2021.
73. Sharov, K.S. Adaptation to SARS-CoV-2 under stress: Role of distorted information. *Eur. J. Clin. Investig.* **2020**, *50*, e13294. [CrossRef]
74. Sharov, K.S. SARS-CoV-2 spread in different biosocial strata in Russia in 2020: Groups of risk and victimised groups. *J. Glob. Health* **2021**, *11*, 03006. [CrossRef]
75. Sharov, K.S. *Gender as a Political Instrument Forming New Boundaries by Ethnic and Religious Diasporas in European Union*; Springer: Singapore, 2022.
76. Sharov, K.S. Adaptation of a Russian population to SARS-CoV-2: Asymptomatic course, comorbidities, mortality, and other respiratory viruses—A reply to Fear versus Data. *Int. J. Antimicrob. Agents* **2020**, *56*, 106093. [CrossRef]
77. Boguslavsky, D.V.; Sharova, N.P.; Sharov, K.S. Public Policy Measures to Increase Anti-SARS-CoV-2 Vaccination Rate in Russia. *Int. J. Environ. Res. Public Health* **2022**, *19*, 3387. [CrossRef] [PubMed]
78. Boguslavsky, D.V.; Sharov, K.S.; Sharova, N.P. Counteracting conspiracy ideas as a measure of increasing propensity for COVID-19 vaccine uptake in Russian society. *J. Glob. Health* **2022**, *12*, 03013. [CrossRef] [PubMed]
79. Kalyuzhniy, K.A. Gosudarstvennaya podderzhka infrastrukturi kollektivnogo polzovaniya: Est' li rezultat? *Upr. Nauk. I Nauk.* **2021**, *16*, 416–440. [CrossRef]
80. Poljanskaya, G.G. Creation and 40 years of working of Collection of Cell Cultures of Vertebrates—Central bank of the Russian collection cell cultures. *Genes Cells* **2019**, *14*, 46.
81. Alpeeva, E.V.; Vorotelyak, E.A. World biobanking practice and its implementation in the operation of the Cell Culture Collection of IDB RAS. In *Kollekcii Kultur Kletok Tsheloveka i Zhivotnikh: Sovremennie Vizovi i Setevie Resheniya (Thezis of the All-Russian School-Conference, the 22th–23th of June, 2022)*; POLITEKH-PRESS: St. Petersburgh, Russia, 2022; pp. 19–21. [CrossRef]

animals

MDPI

Article

Estimating the Spatial Distribution and Future Conservation Requirements of the Spotted Seal in the North Pacific

Leyu Yang [1,2], Hongfei Zhuang [2], Shenghao Liu [2], Bailin Cong [2], Wenhao Huang [2,3], Tingting Li [2], Kaiyu Liu [1,2] and Linlin Zhao [2,*]

[1] School of Advanced Manufacturing, Fuzhou University, Jinjiang 362200, China; yangleyu123456789@163.com (L.Y.); liukaiyu@fio.org.cn (K.L.)
[2] Key Laboratory of Marine Eco-Environmental Science and Technology, First Institute of Oceanography, Ministry of Natural Resources, Qingdao 266061, China; zhuanghf@fio.org.cn (H.Z.); shliu@fio.org.cn (S.L.); biolin@fio.org.cn (B.C.); hwh@fio.org.cn (W.H.); ytlitingting@163.com (T.L.)
[3] College of Environmental Science and Engineering, Ocean University of China, Qingdao 266100, China
* Correspondence: zhaolinlin@fio.org.cn

Simple Summary: To understand the impact of climate change on marine mammals, we focused on the spotted seal population in the North Pacific. This ice-breeding species exhibits distinct variations across different regions. Our study aimed to quantify their ecological niches and conduct a conservation gap analysis. We found clear niche divergence among three populations and observed habitat contraction driven by climate change, potentially leading to breeding habitat loss in certain areas. Unfortunately, existing marine protected areas do not adequately cover most spotted seal habitats. By incorporating local adaptation into species distribution modeling, our research provides valuable insights for designing effective conservation policies to protect the different geographical populations of spotted seals in the face of climate change. This study highlights the importance of considering local adaptation in conservation and management strategies for marine mammal species.

Abstract: Local adaptation has been increasingly involved in the designation of species conservation strategies to response to climate change. Marine mammals, as apex predators, are climatechange sensitive, and their spatial distribution and conservation requirements are critically significant for designing protection strategies. In this study, we focused on an ice-breeding marine mammal, the spotted seal (*Phoca largha*), which exhibits distinct morphological and genetic variations across its range. Our objectives were to quantify the ecological niches of three spotted seal populations, construct the species-level model and population-level models that represent different regions in the Bering population (BDPS), Okhotsk population (ODPS) and southern population (SDPS), and conduct a conservation gap analysis. Our findings unequivocally demonstrated a clear niche divergence among the three populations. We predicted habitat contraction for the BDPS and ODPS driven by climate change; in particular, the spotted seals inhabiting Liaodong Bay may face breeding habitat loss. However, most spotted seal habitats are not represented in existing marine protected areas. Drawing upon these outcomes, we propose appropriate conservation policies to effectively protect the habitat of the different geographical populations of spotted seals. Our research addresses the importance of incorporating local adaptation into species distribution modeling to inform conservation and management strategies.

Keywords: spotted seal; climate change; ecological niche divergence; species distribution models; conservation gap analysis

Citation: Yang, L.; Zhuang, H.; Liu, S.; Cong, B.; Huang, W.; Li, T.; Liu, K.; Zhao, L. Estimating the Spatial Distribution and Future Conservation Requirements of the Spotted Seal in the North Pacific. *Animals* **2023**, *13*, 3260. https://doi.org/10.3390/ani13203260

Academic Editor: Chunwang Li

Received: 6 September 2023
Revised: 4 October 2023
Accepted: 7 October 2023
Published: 19 October 2023

1. Introduction

Climate change poses a significant peril to global biodiversity in the 21st century, unleashing a series of profound and unpredictable changes on our planet [1]. In comparison to terrestrial communities, marine communities stand as more vulnerable sentinels, acutely attuned to the environmental changes wrought by this climatic change [2]. The effects of climate change on the marine environment have changed the life history and distribution landscape of marine species, with the potential to cause habitat destruction and even species extinction [3]. Indeed, mounting evidence attests to the rapid reconfiguration of species distributions along depth or latitudinal gradients in response to changing climates [4,5]. Marine mammals, serving as highly susceptible apex predators within marine ecosystems, assume a pivotal role in maintaining biodiversity and regulating ecosystem processes [6]. Consequently, obtaining comprehensive and precise knowledge of the current distributions of marine mammals, as well as reliable projections under future climate change scenarios, is of paramount importance in developing effective strategies for resource management and conservation.

Species distribution models (SDMs) are useful tools in this endeavor, as they can predict potential changes in species habitats by determining statistical relationships between species occurrence data and environmental predictors, and can also be used to forecast how suitable areas may vary under different climate change scenarios [7]. Traditionally, SDMs have been constructed at the species level based on the "niche conservatism" hypothesis, which suggests that individuals of the same species have similar niche spaces and exhibit consistent responses to climate change across their range [8,9]. Nevertheless, over an extensive evolutionary timeframe, species with a wide range may gradually adapt to local conditions, leading to niche divergence [10]. Recent research highlights a growing recognition of the importance of integrating local adaptation or intraspecific variation into climate responses, with an increasing number of studies emphasizing this crucial aspect [11,12]. By modeling habitat suitability below the species level, more accurate estimations of species ranges and climate change projections can be attained. Notably, within the realm of species conservation, the consideration of climate change responses within phylogeographic lineages has gained currency for certain taxonomic groups in terrestrial communities [13,14], and marine communities [15,16].

Here, we examined how predictions of climate change responses can differ when considering local adaptation for spotted seals. The International Union for Conservation of Nature's Red List of Threatened Species ranked spotted seals (*Phoca largha*) as "least concern", indicating a low risk of extinction [17]. However, in China, spotted seals have the highest protection level among rare and endangered species (class-I protection) due to increasing threats from climate change and habitat loss [18]. North Pacific spotted seals have eight specialized breeding habitats and limited mobility, which contribute to their vulnerability to climate change [19]. Based on morphological and genetic differences, spotted seals were divided into three distinct geographical populations [20]: the Bering population (BDPS), the Okhotsk population (ODPS) and the Southern population (SDPS). Throughout their extensive evolutionary narratives, the three populations have taken up residence within divergent ecological domains, thereby hinting at the plausible presence of localized adaptation. While the precise degree of distributional overlap and genetic interchange amongst these populations remains shrouded in obscurity, the likelihood of their existence cannot be discounted [21,22]. Neglecting to account for such local adaptations when employing SDMs in conservation or management decisions may result in erroneous characterizations of species' responses to environmental changes throughout their ranges, thus misdirecting conservation efforts [23]. Therefore, to further estimate the impacts of climate change on spotted seals at a higher taxonomic resolution below the species level has become an urgent issue in their long-term conservation and management.

In this study, we quantified realized niches (i.e., the portion of the fundamental niche currently used by the species), developed SDMs and made future predictions to examine how climate change might influence spotted seals by constructing species-level versus population-level models. We sought to address the following hypotheses: (1) conspicuous disparities exist in spatial niches among the three distinct populations; (2) the three populations respond differently to climate change, and population-level SDMs are more reliable than species-level results; (3) the potential distribution of spotted seals under future climatic conditions will shift to higher latitudes; and (4) there are large gaps in spotted seal conservation outside protected areas that are not effectively protected. Our results emphasize the value of developing SDMs below the species level and serve as a useful guide for designing climate-adapted conservation and management strategies for spotted seals within more precise taxonomic units.

2. Materials and Methods

2.1. Study Area and Spotted Seal Occurrence Data

This study focused on the distribution range of spotted seals in the North Pacific, with the main study area located in the temperate and cold temperate coastal and littoral regions (90° E—240° W, 0° N—80° N; Figure 1). The spotted seal occurrence data were collected from the literature, the Global Biodiversity Information Facility (GBIF; http://www.gbif.org/, accessed on 21 December 2022) [24], and the Ocean Biogeographic Information System (https://obis.org/, accessed on 4 December 2022) [25] (Supplementary Table S1). To minimize sampling bias, we spatially thinned the occurrence data to match the resolution of the environmental data (5′ × 5′, approximately 9.2 km × 9.2 km) using the R package "spThin" [26], and only one random spotted seal distribution record was used in each raster. Following this data-filtering procedure, 1990 records were kept in order to construct the SDM at the species level (hereafter "species model"). Of these, 680 records belonged to the BDPS that was used to construct the SDM at the population level (hereafter "BDPS model"), 853 records belonged to the ODPS that was used to construct the ODPS model, and 457 records belonged to the SDPS that was used to construct the SDPS model.

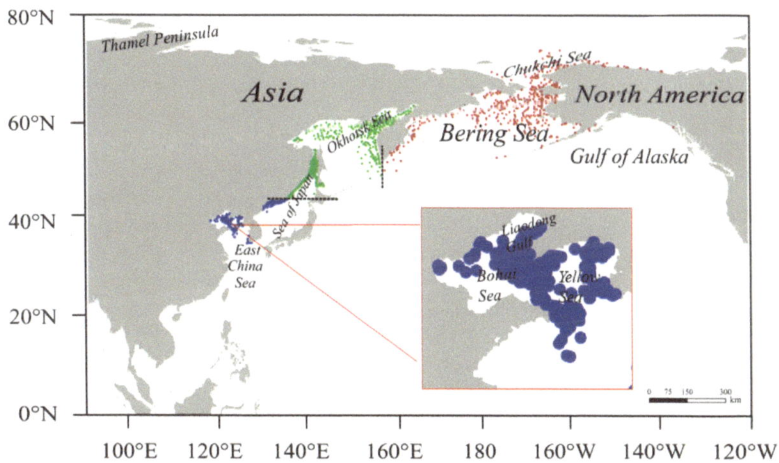

Figure 1. Map of the study area and occurrence records of spotted seal. Red dots represent the occurrence records of the Bering population (BDPS), green dots represent those of the Okhotsk population (ODPS), and blue dots those of the southern population (SDPS). The dotted black lines are drawn along 43° N latitude and 157° E longitude, which are considered the boundaries between the SDPS and ODPS and between the ODPS and BDPS.

2.2. Environmental Predictor Variables

Habitat surroundings have a significant influence on the distribution of spotted seals, and considering a combination of bioenvironmental relevance and data availability, 10 environmental variables that may influence the distribution of spotted seals were selected for this study (Table 1). Current and future environmental data were downloaded from online datasets: the water depth and distance to shore were downloaded from the Global Marine Environment Datasets (https://gmed.auckland.ac.nz/, accessed on 15 April 2023) [27], and the remaining predictors were downloaded from the Ocean Raster for Analysis of Climate and Environment (https://bio-oracle.org/, accessed on 6 April 2023) [28]. Considering the correlation between predictors, we completed Pearson's correlation factor analysis between environmental layers using the R package usdm [29], retaining only environmental variables with correlation factor values < |0.7| [30] (Supplementary Figure S1). Finally, seven predictors, including water depth (Dep), distance to shore (DTS), chlorophyll concentration (Chl), current velocity (CV), ice thickness (IT), salinity, and water temperature (Tmean), were retained for modeling analysis.

Table 1. Ten environmental variables initially selected for this study and their units, spatial resolution and sources.

Environment Variable	Unit	Spatial Resolution	Source
water depth	m	5 arc minutes	https://gmed.aucklandac.nz/, accessed on 15 April 2023
distance to shore	km	5 arc minutes	https://gmed.auckland.ac.nz/, accessed on 15 April 2023
calcite	$mol \cdot m^{-3}$	5 arc minutes	https://bio-oracle.org/, accessed on 6 April 2023
chlorophyll concentration	$mg \cdot m^{-3}$	5 arc minutes	https://bio-oracle.org/, accessed on 6 April 2023
currents velocity	$m \cdot s^{-1}$	5 arc minutes	https://bio-oracle.org/, accessed on 6 April 2023
dissolved oxygen	$mol \cdot m^{-3}$	5 arc minutes	https://bio-oracle.org/, accessed on 6 April 2023
sea ice concentration	fraction	5 arc minutes	https://bio-oracle.org/, accessed on 6 April 2023
ice thickness	m	5 arc minutes	https://bio-oracle.org/, accessed on 6 April 2023
salinity	PSS	5 arc minutes	https://bio-oracle.org/, accessed on 6 April 2023
water temperature	°C	5 arc minutes	https://bio-oracle.org/, accessed on 6 April 2023

To project the future habitat suitability of spotted seals, we considered four representative concentration pathway (RCP) scenarios (i.e., RCP 2.6, RCP 4.5, RCP 6.0 and RCP 8.5), and two periods (i.e., 2050s: the average for 2040s–2050s, and 2100s: the average for 2090–2100). We obtained the corresponding projections of future marine environmental layers from Bio-ORACLE v2.0. This study assumed no change in water depth and distance to shore [31].

2.3. Estimates of Niche Divergence

To assess whether the three spotted seal populations occupy different niche spaces, we used n-dimensional hypervolume to characterize their realized niche [32]. For this, we first performed principal component analysis (PCA) on seven selected environmental variables and retained the top four principal components, which cumulatively explained 83.6% of the total variance (Supplementary Figure S2). Then, we calculated the principal component retention values corresponding to each population using a Gaussian approach with the R package hypervolume [33]. Finally, the niche divergence between populations can be assessed by overlapping the hypervolume of each population using the R package BAT [34]. Total niche divergence (βTotal) was divided into the following two processes: niche shift (spatial replacement between hypervolumes) and niche contraction/expansion (net difference between hypervolumes). The βTotal ranged from 0 to 1, indicating the two hypervolumes of complete overlap to complete separation [11].

2.4. SDMs Establishment and Projection

We conducted SDM analysis based on the "biomod 2" package in the R platform (version 4.2.3) [35]. This package contained 10 modeling algorithms: generalized additive mode, generalized linear model (GLM), generalized boosting model (GBM), random fores, surface range envelope (SRE), artificial neural network (ANN), flexible discriminant analysis (FDA), classification tree analysis (CTA), multiple adaptive regression splines (MARS), and maximum entropy (Maxent). Since true absence data were lacking, we randomly simulated the same number of pseudo-absence records as that of presence records in the environmental conditions [36]. The dataset was divided into five groups during the modeling process, with an equal number of records in each group, four of which were used for model training and the remaining one for model testing. To evaluate the predictive performance of each model, the 5-fold cross-validation process was repeated 10 times. We used the TSS (the true skill statistic) and AUC (the area under the ROC curve) values to assess the accuracy of the models.

This study selected TSS > 0.8 and AUC > 0.9 as model selection standards [37,38] and used a weighted-average algorithm to build an integrated model for reducing the uncertainty of individual models. To better explain habitat suitability, we transformed continuous habitat suitability predictions into a binary map by maximizing the probability threshold of the TSS [39]. We applied a randomized method to measure Pearson correlations between all predictor and assessment variables [40] to assess the relative importance of each variable in predicting species distributions. Finally, we built two levels of species and population ensemble models to predict the potential distribution of habitat for the whole species and three geographical populations (the BDPS, ODPS and SDPS) under current and future (2050s, 2100s) climate scenarios under RCP 2.6 and RCP 8.5.

2.5. Protection Gap Analysis

The Global Marine Protected Areas layers were sourced from the World Database on Protected Areas (https://www.protectedplanet.net/, accessed on 4 May 2023) [41], while data on protected areas in the Yellow and Bohai Seas of China were sourced from a published article [42]. First, we overlaid the layers of existing protected areas and the range of spotted seals to analyze the proportion of the existing protected area covered in the spotted seal distribution range and the uncovered spatial area in QGIS 3.28.6 software (https://www.qgis.org/en/site/, accessed on 5 April 2023). We then conducted a conservation gap analysis of the uncovered spatial areas to identify uncovered habitat areas. Finally, by integrating current and future climate change scenarios, we projected trends in the range of spotted seals and further identified conservation gaps for these species under future climate scenarios.

3. Results

3.1. Niche Divergence among the Three Populations

According to the results of niche divergence studies of different geographical populations, the BDPS has the widest ecological range. The four-dimensional hypervolume for the BDPS, ODPS, and SDPS, respectively, was 1642.20, 353.46, and 145.16. The niche divergence between two populations was very high, with values of 0.81 (BDPS:ODPS), 0.92 (BDPS:SDPS), and 0.86 (ODPS:SDPS) shown by the paired comparison of hypervolumes. Contraction/expansion accounted for more than 85% of the niche divergence between the BDPS and the ODPS or SDPS, whereas niche transitions had a much smaller role (15%). The main cause of the niche divergence between the ODPS and SDPS was contraction/expansion (>65%), with niche shift accounting for a little part (35%) (Table 2).

Table 2. Total niche differentiation (βTotal) between populations and the proportion of niche shift and niche contraction/expansion. BDPS, Bering distinct population segment; ODPS, Okhotsk distinct population segment; SDPS, Southern distinct population segment.

Populations Pair	βTotal	Niche Shift	Niche Contraction/Expansion
BDPS-ODPS	0.81	0.12(15%)	0.69(85%)
BDPS-SDPS	0.92	0.04(4%)	0.88(96%)
ODPS-SDPS	0.86	0.27(32%)	0.59(68%)

When the three populations' general niches were compared, it was clear that PCA1 was the primary focus of niche divergence (Figure 2a), which was mostly explained by water depth, distance to shore, and chlorophyll concentration (Supplementary Figure S3). The paired-niche comparison revealed that the PCA1 was primarily responsible for the niche divergence between the BDPS and ODPS (Figure 2b), which was primarily explained by water depth, ice thickness, and current velocity (Supplementary Figure S3); the PCA1 was responsible for the niche divergence between the BDPS and SDPS (Figure 2c), which was primarily explained by water temperature, chlorophyll concentration, and salinity

(Supplementary Figure S3); and the PCA1 was also responsible for the niche divergence between the BDPS and SDPS (Figure 2d), which was primarily explained by ice thickness, chlorophyll concentration and water temperature (Supplementary Figure S3).

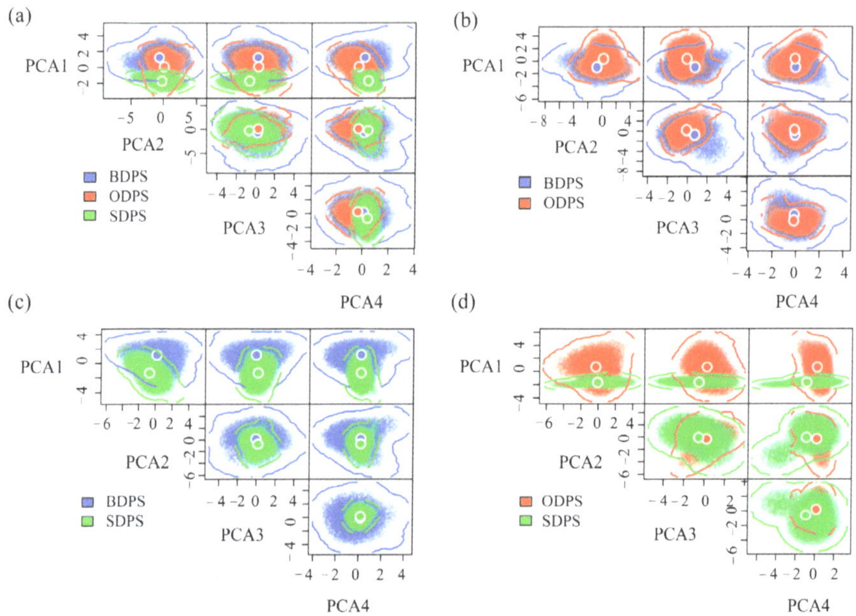

Figure 2. Three spotted seal populations with realized niches quantified by four-dimensional hypervolume. The larger blue, orange and green dots indicate the mean niche position (niche centroid) of the Bering population (BDPS), Okhotsk population (ODPS), and southern population (SDPS), respectively. The overlapping hypervolumes for the three populations (**a**), BDPS and ODPS (**b**), BDPS and SDPS (**c**) and ODPS and SDPS (**d**).

3.2. Current SDMs Projections

Based on the TSS and AUC values in the model results, eight models were selected to build the weighted species-level ensemble model after removing MaxEnt and SRE, and nine models were selected to build the weighted population-level ensemble model after removing SRE from the ten single models. The higher values of AUC and TSS for all four ensemble models indicated high predictive performance (Table 3).

Table 3. Number of models and evaluating indicators for the ensemble models built at the species level and population level. NME, number of models used in ensemble modeling; TSS, the true skill statistics; AUC, the area under the receiver operating characteristic curve.

Ensemble	TSS	AUC	NME
Species model	0.857	0.953	8
BDPS model	0.861	0.949	9
ODPS model	0.932	0.975	9
SDPS model	0.950	0.978	9

The species-level model showed that water temperature and depth considerably contributed to the distribution of spotted seals, while current velocity and chlorophyll concentration contributed little to the model (Figure 3a). According to this model, spotted seals preferred to live in waters with a temperature range of 0 °C to 15 °C and a depth of 0 m to 1000 m (Supplementary Figure S4). The population-level model showed that the main factors influencing the potential distribution of spotted seals differ between populations.

Specifically, the BDPS distribution was most influenced by water temperature and ice thickness; this population preferred to inhabit areas with water temperature ranging from 0 °C to 8 °C (Supplementary Figure S5a) and had a higher probability of occurrence at ice thicknesses of 0–1 m (Supplementary Figure S5b). The ODPS distribution was most influenced by water temperature and depth; this population preferred to inhabit areas with water temperature ranging from 2 °C to 12 °C (Supplementary Figure S5c) and water depth ranging from 0 m to 2000 m (Supplementary Figure S5d). The SDPS distribution was mainly influenced by water temperature and distance to shore (Figure 3b); this population preferred to live in a water temperature range of 2 °C to 15 °C (Supplementary Figure S5e) and had a higher probability of living in near-shore waters not more than 800 meters from the coast (Supplementary Figure S5f).

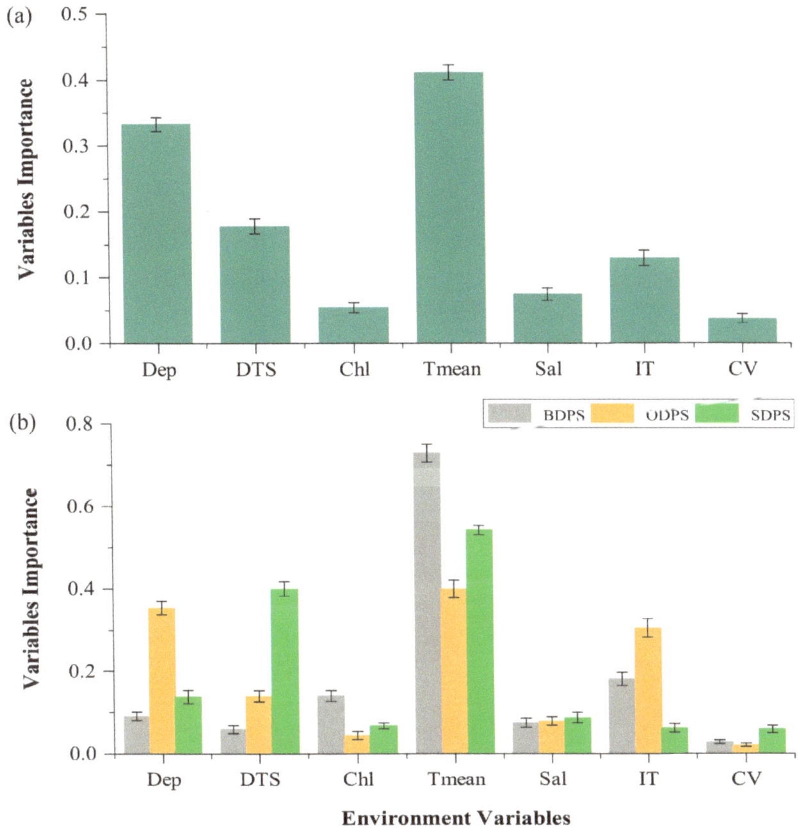

Figure 3. Importance of environmental variables in driving whole-species distribution based on the species-level model (**a**) and population distributions based on the population-level model (**b**). Dep, water depth; DTS, distance to shore; Tmean, water temperature; Sal, salinity; IT, ice thickness; CV, current velocity.

According to the results of the model, the distribution and suitable habitat of the three spotted seal populations under current environmental conditions are as follows: the BDPS is mainly distributed in the Chukchi Sea, the Bering Sea, the coast of Kamchatka Island and the northern part of the Sea of Okhotsk, with the largest area of suitable habitat (6.24×10^6 km^2) (Figure 4c,f). The ODPS is mainly distributed in the Sea of Okhotsk and near Sakhalin Island, extending northward to the Aleutian Peninsula, with the second largest area of suitable habitat (3.54×10^6 km^2) (Figure 4d,g). The SDPS is mainly distributed on Hokkaido

Island, the Sea of Japan and the Yellow Sea of China, reaching as far as the East China Sea, with the smallest area of suitable habitat (1.08×10^6 km^2) (Figure 4e,h).

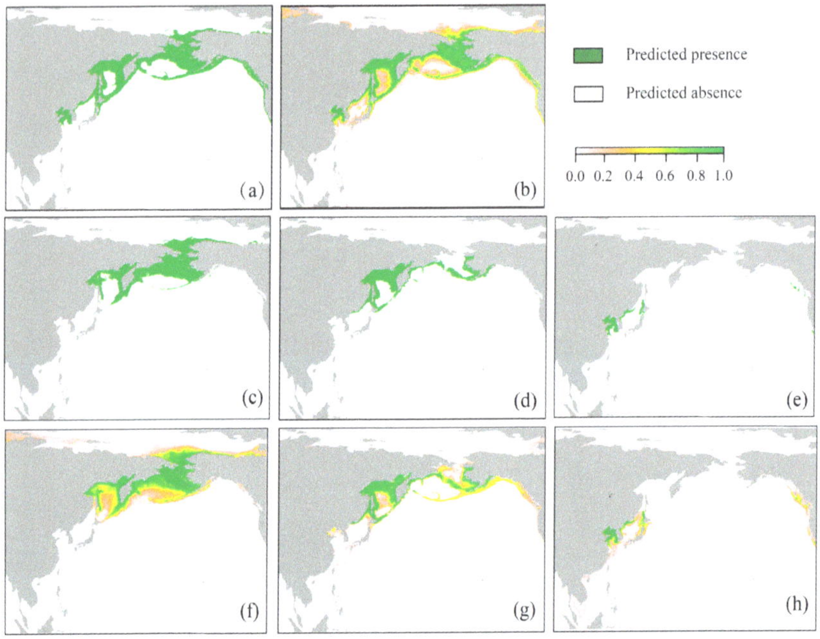

Figure 4. Habitat suitability map of spotted seals projected by the ensemble models under current climate scenarios. (**a**,**b**) Respective binary and continuous plots of the species; (**c**,**f**) respective binary and continuous plots for BDPS; (**d**,**g**) corresponding plots for ODPS; (**e**,**h**) corresponding plots for SDPS.

The species model results are shown in Figure 4a,b. Both the continuous and binary projections showed that the predicted suitable area from the species model is similar to that from the BDPS model in the Chukchi Sea and the Bering Sea, is similar to that from the ODPS model in the Okhotsk Sea and is smaller than that from the SDPS model in the southernmost region.

3.3. Habitat Suitability under Future Climate Scenarios

This study built SDMs at both the species and population levels to predict climate change impacts on potentially suitable habitats for spotted seals. Habitat-scale changes are influenced by climate change scenarios, particularly under the pessimistic scenario of uncontrolled greenhouse gas emissions (RCP 8.5), and the suitable habitat is predicted to vary considerably within the appropriate range (Table 4). For all climate change scenarios, both the BDPS and ODPS show a decreasing trend in the extent of suitable habitat, with the greatest decrease under RCP 8.5 in the 2050s, while the SDPS shows an increasing trend in the extent of suitable habitat. For the BDPS, suitable habitat for spotted seals is mainly stable in the Chukchi Sea and around the Bering Strait, with the northern coasts of the Chukchi and Taymyr Peninsulas also serving as potential habitats (Figure 5c,f). For the ODPS, suitable habitat is mainly stable along the northwestern coast of the Kamchatka Peninsula, while most of the suitable habitat in the Sea of Okhotsk will be lost (Figure 5d,g). For the SDPS, although large areas of suitable habitat are found around the Aleutian Islands and the Gulf of Alaska, the southernmost spotted seal colony in Liaodong Bay will be lost (Figure 5e,h).

Table 4. Predicted size of changes [9] in species range based on the species-level and population-level models under future climate scenarios. RCP 2.6 (8.5), the representative concentration pathway 2.6 (8.5); 2050s (2100s), at the middle (end) of the 21st century.

RCP	BDPS		ODPS		SDPS		Species	
	2050s	2100s	2050s	2100s	2050s	2100s	2050s	2100s
RCP 2.6	−32.48	−38.34	−32.80	−47.61	13.15	17.86	−1.43	−1.73
RCP 8.5	−38.24	−63.94	−44.25	−66.51	9.57	62.91	2.27	36.16

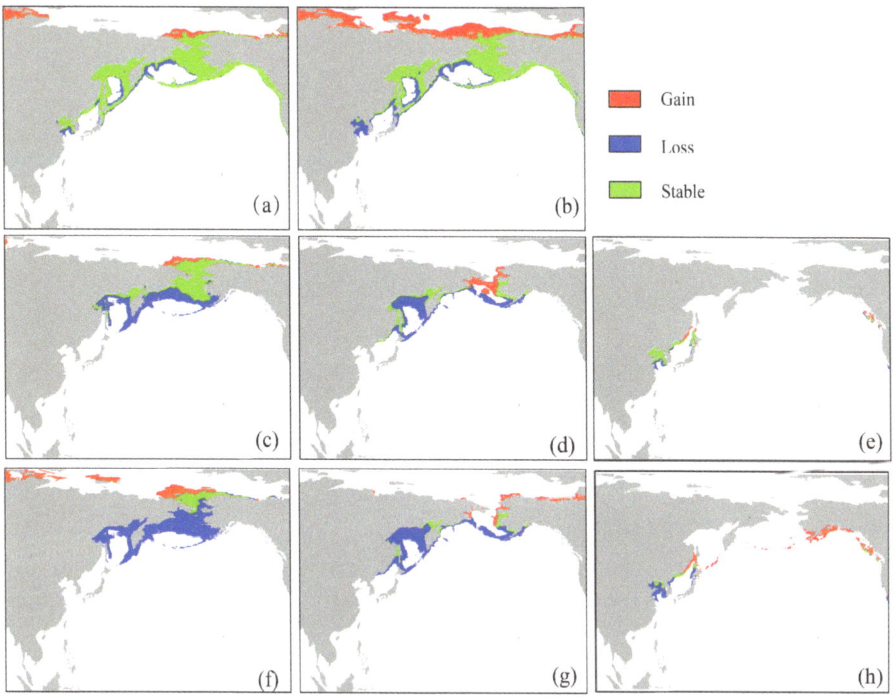

Figure 5. Predicted change in the suitable habitat based on the ensemble models in the 2050s under RCP 8.5 (5085) and in the 2100s under RCP 8.5 (0085) for the species (**a**,**b**), BDPS (**c**,**f**), ODPS (**d**,**g**) and SDPS (**e**,**h**).

At the species level, the southernmost Liaodong Bay spotted seal breeding area is gradually being lost; as the range of the Arctic Ocean seas north of the Chukchi Peninsula expands toward higher latitudes, the distribution of spotted seals may occur at higher latitudes in the Thamel Peninsula (Figure 5a,b). Range changes in the size of the predicted species tended to decrease under RCP 2.6 and increase under RCP 8.5, with the change reaching 36.16% under RCP 8.5 in the 2100s.

3.4. Spotted Seal Conservation Gap Analysis

According to the results of the protection gap analysis, 278,617 km^2 was found to be protected, representing only 5.65% of the spotted seal range, i.e., more than 94% of the area is not covered by protected areas (Figure 6a). In the overlaid map, the Eastern Bering Sea, the Aleutian Islands and the Sea of Okhotsk overlap significantly with the range of spotted seals. Although the number of protected areas around the Yellow and Bohai Seas in China is high, they are very small. Further analysis suggests that outside these existing marine protected areas (MPAs), there are both important nonbreeding areas and breeding habitat for spotted seals, such as the western coast of the Kamchatka Peninsula, the western Bering Sea and Peter the Great Bay, which remain unprotected.

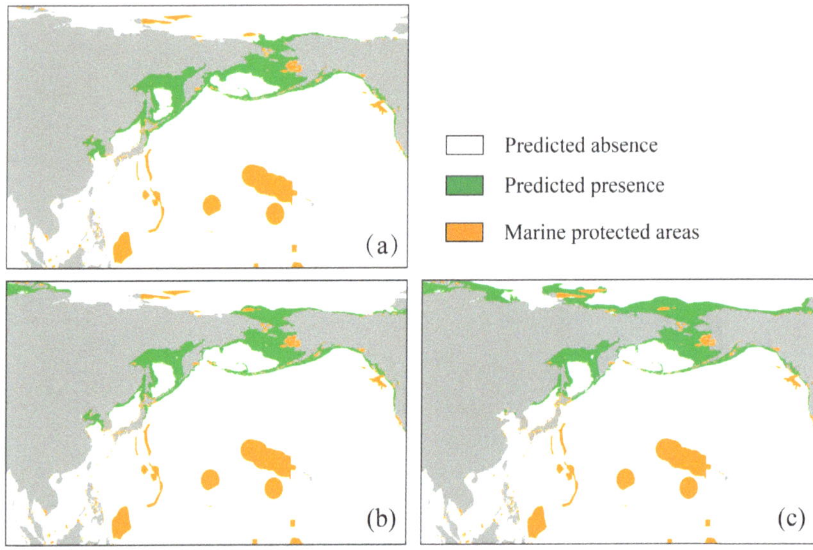

Figure 6. Analysis of the conservation gap for spotted seals under current and future climate scenarios. (**a**) Analysis of the conservation gap under the climate scenario; (**b**) analysis of the conservation gap in the 2050s under RCP 8.5 (5085); (**c**) analysis of the conservation gap in the 2100s under RCP 8.5 (0085).

In this study, the potential range of spotted seals under future climate scenarios was overlaid with existing MPAs, and it was found that spotted seals were protected in only 6.04–6.22% of their range under the RCP 2.6 and RCP 8.5 scenarios in the 2050s and 2100s (Table 5). In summary, the overall change in the protected gap area is insignificant in the four future climate scenarios (Figure 6b,c).

Table 5. Spotted seal areas (km^2) and corresponding percentages protected under the current and future climate scenarios.

Climate Scenario	Area Protected (km^2)	Percentage of Protection [9]
current	278,617	5.65
2050s RCP 2.6	278,062	6.17
2050s RCP 8.5	277,776	6.04
2100s RCP 2.6	279,403	6.22
2100s RCP 8.5	277,449	5.56

4. Discussion

4.1. Consideration of Local Adaptation

The study quantified the realized niches of three spotted seal populations and found that the BDPS exhibited a substantially broader niche space than the ODPS and SDPS, primarily due to niche contraction/expansion. These findings underscore that geographically separated spotted seal populations inhabit dissimilar ecological niches; therefore, niche conservatism does not apply to this species. Then, we constructed SDMs for the BDPS, ODPS, and SDPS individually, which accounted for local adaptation, and these models revealed that the three populations showed differing responses to climate change predictors. In summary, this study emphasizes the importance of considering local adaptation in projecting the potential distribution of species to inform conservation and management decisions in a climate change scenario.

Local adaptation and intraspecific variation were incorporated into SDMs based on the recognition that populations of species inhabiting widely different habitats over significant time scales will often show adaptations to their respective local conditions, resulting in intraspecific niche variation. This local adaptation may be reflected by morphological and genetic differences [43]. For instance, studies have revealed significant differences in the nonmetric features of the skull between spotted seals from the central and eastern Bering Sea [44] and a phylogeographic break between spotted seals breeding in the Yellow Sea–Japan Sea region and those breeding in the Sea of Okhotsk, Bering Sea and Chukchi Sea. The above studies, in conjunction with the observed niche divergence among the BDPS, ODPS, and SDPS established in this study, stress the importance of building SDMs at the population level to account for local adaptation.

In this study, despite the fact that the species and population level SDMs predicted a similar change trend in the species range, with a reduction in suitable habitat for the BDPS and ODPS and an expansion of suitable habitat for the SDPS under future climate change scenarios, the magnitudes of the range change predicted by the two types of models varied. The population level model produced more encouraging findings for the BDPS and ODPS with less loss of appropriate habitat. Due to the inclusion of potential local responses in population-level models, our climate change estimates were in fact less pessimistic. Our results are in line with a number of published studies that suggest adaptive genetic variation within a species can reduce the species' susceptibility to climate change [45–47].

4.2. Impacts of Climate Change on Spotted Seals

The niche divergence of the BDPS, ODPS and SDPS was mainly due to niche contraction/expansion, to which water temperature, ice thickness and chlorophyll concentration contributed most. Spotted seals are cool-temperature marine mammals and water temperature is an important influence on both the physiology and behavior of spotted seals, affecting them indirectly by altering the distribution of prey, predators and disease-causing vectors [48]. Also, spotted seals are often dependent on sea ice for breeding and foraging, and thicker ice may provide better breeding and foraging conditions, while thinner ice may limit these activities. Moreover, chlorophyll concentration variations may reflect the productivity of marine ecosystems and the base of the food chain [49]. Differences in nutrient conditions and chlorophyll concentration in different regions may lead to different availability of food resources, and spotted seal populations may choose to adapt to different food resources according to chlorophyll concentration in different regions, resulting in population differentiation. Climate-induced changes in these factors cause species to alter their current distribution patterns to track ecological niches. In summary, these three climatic factors directly or indirectly affect the physiology, behavior and availability of food resources of spotted seals, thus leading to the divergence of adaptation strategies among different populations.

Different populations of spotted seals may have different levels of adaptiveness and vulnerability to climate change. Our population-level SDMs predicted that the BDPS will colonize the northern coast of the Chukchi, and as far west as the Taymyr Peninsulas, while retaining most of the current suitable areas, indicating the resilience of the population to climate change. In contrast, the ODPS preserves most of the currently suitable areas, and parts of the suitable habitat in the Sea of Okhotsk will be lost, indicating that the population is less resilient to climate change. The SDPS will have large areas of suitable habitat near the Aleutian Islands and the Gulf of Alaska, but spotted seals of the SDPS would not be capable of shifting their range to the northeast to reach this area and the southernmost spotted seal breeding area in Liaodong Bay will be lost, indicating the vulnerability of this population to climate change.

Overall, the thinning and breaking up of ice caused by climate change will expose vast regions of the northern Bering Sea and Chukchi Sea, which is likely to increase suitable habitat for northern spotted seals, thus offsetting the loss of habitat in the south. Similar findings were reported in studies of climate change-induced geographic translocation of species [15,50–52].

4.3. Model Predictive Accuracy

Integrated habitat suitability models built with a weighted integration technique, especially for rare species, can increase model prediction accuracy and avoid overfitting problems without compromising explanatory power [53]. The TSS values of the four ensemble models were all 0.86, their AUC values were all 0.95, and their errors for the environmental importance results were all less than 0.2, showing that the predictions were very accurate representations of the current and future distributions of spotted seals under various climate scenarios. An integrated model was used in this work to estimate the whole distributions of spotted seals, and the results were mostly in line with the known range of spotted seals (temperate and cold temperate coastal and shore).

Although the results were as expected, there were limitations to the method of model species distribution to predict the range of spotted seals. We used future environmental data that did not occur objectively but were predictions based on atmosphere-ocean general circulation models (AOGCMs), as well as the effects of many factors such as changes in food availability, changes in ocean currents that persist between breeding sites, and the ability of spotted seals to swim on their own. When predicting the suitable distribution of spotted seals, the results of the study may have overestimated the range of the spotted seal. Predictions made in the future that take these elements into account in the model will be closer to the species' actual distribution. Moreover, to improve or validate the predictive power of SDMs, independent geographically or temporally separated data should be collected [54]. Considering the difficulty and cost of field surveys, emerging environmental DNA methods could be used to determine the presence of potential distribution areas for spotted seals as predicted in this study, and these data could be used in future work on SDMs.

4.4. Management and Conservation of Spotted Seals

Spotted seals have received little attention from conservationists or managers despite significant human and climate change challenges. Spotted seals and other marine mammals are vulnerable in the face of global warming, and their potential extinction could have far-reaching consequences for the functioning of global marine ecosystems in the future [55]. Water temperature and ice thickness are two important environmental factors that might affect the distribution of spotted seal populations geographically, yet they are also strongly related to climate change. To develop climate-adapted conservation and management methods, it is imperative to assess how the changing climate is affecting the appropriateness of spotted seal habitat. In general, all three populations exhibit a propensity to migrate toward the poles under a warming climate. However, the SDPS are more vulnerable to climate than the BDPS and ODPS, and their population sizes have already been significantly reduced from historical levels and may be at risk of population genetic extinction. Therefore, the SDPS deserves more attention and protection in the face of climate stress.

Marine protected areas (MPAs) have proven to be an effective tool for protecting endangered species and maintaining ecosystem services [56]. The European Union target of '30 by 30' that is 30% of the ocean protected (as MPAs) by 2030. For the conservation of spotted seals and their habitat, many countries and regions have established MPAs [57,58]. These marine protected areas were crucial for the conservation of spotted seals and their habitats. However, when we overlaid the spotted seal range, we discovered that only 5.65–6.22% of the range was protected, meaning that more than 94% of the area was unprotected. Therefore, protected areas (marine reserves, nature reserves and national parks) need to be expanded, and the establishment of protected areas across international borders should be considered to better protect spotted seals and their habitats [59].

The current and potential future distribution of spotted seals is mainly in coastal waters; as a result, protecting the species from anthropogenic environmental contamination and hunting pressure in these seas is essential. Given the different vulnerabilities of the three populations to climate change, we need to develop local adaptive conservation and management measures for the populations in different areas. Since the SDPS is most sensitive to climate change, we recommend that stricter hunting restrictions, such as bans on poaching for genitalia and culling by fishers, be imposed on this population. For ODPS, although there is a degree of adaptability to climate change and little risk of population extinction, direct or indirect commercial fisheries interactions may have a significant cumulative effect. Therefore, we should increase management efforts to control marine development activities such as sand mining, oil and gas exploration and water pipelines to reduce damage to spotted seals (survival environment). Although the BDPS has the ability to adapt to climate change, we should also pay attention to the timing and routes of the breeding migrations of spotted seals, avoid fishing and shipping operations during migrations, and raise awareness of conservation.

Finally, we must emphasize that MPAs and other suggested conservation measures will only guarantee that adequate habitats for spotted seals are safeguarded from human impact now and in the future. However, as greenhouse gas concentrations rise, the SDPS will lose its ideal habitat. This population will progressively decline towards extinction if the issue is not managed for the long-term. Therefore, reducing human-caused greenhouse gas emissions is the ultimate solution for the sustainability of these populations.

5. Conclusions

In conclusion, this study represents the first step in estimating climate impacts on the potential distribution of spotted seals in the North Pacific considering local adaptation. Population-level SDMs are more reliable than species-level SDMs because of the different responses of the three spotted seal populations to environmental predictor variables. Additionally, conservation efforts should be dedicated to the establishment of MPAs, first in the stable spots predicted to remain climatically suitable for the species, and second in the currently suitable areas. The comparison of current and predicted habitat suitability maps presented in our study serves as a crucial tool allowing us to delineate the most promising regions for establishing both types of measures. In future studies, other analytical methods and multiple data sources should be incorporated to improve our ability to predict the potential distribution of spotted seals and deliver more accurate information for related conservation and management.

Supplementary Materials: The following supporting information can be downloaded at: https://www.mdpi.com/article/10.3390/ani13203260/s1, Figure S1: Results of Pearson correlation analysis for 10 environmental variables; Figure S2: The results of principal component analyses based on all the environmental variables involved in the model using the R package FactoMineR; Figure S3: The degree of contribution of environmental variables to the principal component that has the greatest influence on ecological niche differentiation between populations; Figure S4: The response curves of spotted seals probability of occurrence against water temperature and water depth based on the species-level model; Figure S5: The response curves of population probability of occurrence against the two most important driving factors based on the population-level model. Table S1: Reference list of occurrence data for spotted seal (*Phoca largha*) in this research [21,60–66].

Author Contributions: Conceptualization, L.Y. and L.Z.; methodology, L.Y.; software, L.Y.; validation, S.L., B.C. and W.H.; formal analysis, T.L.; investigation, K.L.; resources, L.Z.; data curation, S.L.; writing—original draft preparation, L.Y.; writing—review and editing, H.Z.; visualization, B.C.; supervision, L.Z.; project administration, L.Z.; funding acquisition, L.Z. All authors have read and agreed to the published version of the manuscript.

Funding: This research was funded by the National Key R&D Program of China (2022YFF0802204) and the National Natural Science Foundation of China (grant no. 32201433).

Institutional Review Board Statement: Ethical review and approval were waived for this study due to all data sources for spotted seals were downloaded online.

Informed Consent Statement: Not applicable.

Data Availability Statement: The data that support the findings of this study are available from the corresponding author upon reasonable request.

Acknowledgments: We thank the modelling help from Linjie Li and Bin Sun in Ocean University of China.

Conflicts of Interest: The authors declare no conflict of interest.

References

1. Dawson, T.P.; Jackson, S.T.; House, J.I. Beyond Predictions: Biodiversity Conservation in a Changing Climate. *Science* **2011**, *332*, 53–58. [CrossRef] [PubMed]
2. Sorte, C.J.B.; Williams, S.L.; Carlton, J.T. Marine range shifts and species introductions: Comparative spread rates and community impacts. *Glob. Ecol. Biogeogr.* **2010**, *19*, 303–316. [CrossRef]
3. Cheung, W.W.L.; Sarmiento, J.L.; Dunne, J.; Frölicher, T.L.; Lam, V.W.Y.; Deng Palomares, M.L.; Watson, R.; Pauly, D. Shrinking of fishes exacerbates impacts of global ocean changes on marine ecosystems. *Nat. Clim. Chang.* **2012**, *3*, 254–258. [CrossRef]
4. Hastings, R.A.; Rutterford, L.A.; Freer, J.J.; Collins, R.A.; Simpson, S.D.; Genner, M.J. Climate Change Drives Poleward Increases and Equatorward Declines in Marine Species. *Curr. Biol.* **2020**, *30*, 1572–1579. [CrossRef]
5. Fossheim, M.; Primicerio, R.; Johannesen, E.; Ingvaldsen, R.B.; Aschan, M.M.; Dolgov, A.V. Recent warming leads to a rapid borealization of fish communities in the Arctic. *Nat. Clim. Chang.* **2015**, *5*, 673–677. [CrossRef]
6. Tian, J.; Lu, Z.; Sanganyado, E.; Gan, Z.; Wang, Z.; Kong, Z.; Wu, J.; Liu, W. Tissue distribution and trophic magnification of trace elements in typical marine mammals in Bohai and north Yellow Seas. *Mar. Pollut. Bull.* **2022**, *184*, 114135. [CrossRef] [PubMed]
7. Pacifici, M.; Foden, W.B.; Visconti, P. Assessing species vulnerability to climate change. *Nat. Clim. Chang.* **2015**, *5*, 215–224. [CrossRef]
8. Smith, A.B.; Godsoe, W.; Rodriguez-Sanchez, F.; Wang, H.H.; Warren, D. Niche Estimation Above and Below the Species Level. *Trends. Ecol. Evol.* **2019**, *34*, 260–273. [CrossRef] [PubMed]
9. Pearson, R.G.; Dawson, T.P. Predicting the impacts of climate change on the distribution of species: Are bioclimate envelope models useful? *Glob. Ecol. Biogeogr.* **2003**, *12*, 361–371. [CrossRef]
10. Lee-Yaw, J.A.; Kharouba, H.M.; Bontrager, M.; Mahony, C.; Csergo, A.M.; Noreen, A.M.; Li, Q.; Schuster, R.; Angert, A.L. A synthesis of transplant experiments and ecological niche models suggests that range limits are often niche limits. *Ecol. Lett.* **2016**, *19*, 710–722. [CrossRef]
11. Carvalho, J.C.; Cardoso, P. Decomposing the Causes for Niche Differentiation Between Species Using Hypervolumes. *Front. Ecol. Evol.* **2020**, *8*, 00243. [CrossRef]
12. Peterson, M.L.; Doak, D.F.; Morris, W.F. Incorporating local adaptation into forecasts of species' distribution and abundance under climate change. *Glob. Chang. Biol.* **2019**, *25*, 775–793.
13. Zhao, X.; Li, X.; Zhang, Z.; Garber, P.A.; Yu, M.; Qiao, H.; Li, M. Differential response to climate change and human activities in three lineages of Sichuan snub-nosed monkeys (*Rhinopithecus roxellana*). *Divers. Distrib.* **2022**, *28*, 2416–2428. [CrossRef]
14. Zhang, P.; Grenouillet, G.; Dong, X.; Zheng, Y.; Lek, S.; Chang, J. Capturing response differences of species distribution to climate and human pressures by incorporating local adaptation: Implications for the conservation of a critically endangered species. *J. Environ. Manag.* **2021**, *284*, 111998. [CrossRef] [PubMed]
15. Li, L.; Zhao, L.; Fu, J.; Sun, B.; Liu, C. Predicting the habitat suitability for populations of Pacific cod under different climate change scenarios considering intraspecific genetic variation. *Ecol. Indic.* **2022**, *142*, 109248. [CrossRef]
16. Hu, Z.M.; Zhang, Q.S.; Zhang, J.; Kass, J.M.; Mammola, S.; Fresia, P.; Draisma, S.G.A.; Assis, J.; Jueterbock, A.; Yokota, M.; et al. Intraspecific genetic variation matters when predicting seagrass distribution under climate change. *Mol. Ecol.* **2021**, *30*, 3840–3855. [CrossRef] [PubMed]
17. Boveng, P.L. Phoca Largha. The IUCN Red List of Threatened Species 2016: e.T17023A45229806. 2016. Available online: https://www.iucnredlist.org/species/17023/45229806 (accessed on 6 September 2023).
18. Wang, F.; Yu, Y.; Wu, H.; Wu, W.; Wang, L.; An, L.; Cai, W. Microplastics in spotted seal cubs (*Phoca largha*): Digestion after ingestion? *Sci. Total Environ.* **2021**, *785*, 147426. [CrossRef]
19. Shaughnessy, P.D.; Fay, F.H. A review of the taxonomy and nomenclature of North Pacific Harbour seals. *J. Zool.* **1977**, *182*, 385–419. [CrossRef]
20. Boveng, P.L.; Bengtson, J.L.; Buckley, T.W.; Cameron, M.F.; Dahle, S.P.; Kelly, B.P.; Megrey, B.A.; Overland, J.E.; Williamson, N.J. *Status Review of the Spotted Seal (Phoca largha)*; Department of Commerce, National Oceanic and Atmospheric Administration, National Marine Fisheries Service: Silver Spring, MD, USA, 2009; 153p, NMFS-AFSC-200.
21. Trukhin, A.M. Changes in the Breeding Range of Spotted Seals, *Phoca largha* (Pallas, 1811) (Carnivora, Pinnipedia), in the Western Sea of Japan: Causes and Effects. *Russ. J. Mar. Biol.* **2022**, *48*, 101–107. [CrossRef]
22. Podlesnykh, A.V.; Katin, I.O. Genetic variation in the spotted seal (*Phoca largha* Pallas, 1811) from the Rimsky-Korsakov Archipelago (Peter the Great Bay, western sea of Japan) as inferred from mitochondrial DNA control region sequences. *Zool. Anz.* **2022**, *301*, 174–178. [CrossRef]
23. Hamann, A.; Aitken, S.N. Conservation planning under climate change: Accounting for adaptive potential and migration capacity in species distribution models. *Divers. Distrib.* **2013**, *19*, 268–280. [CrossRef]

24. The Global Biodiversity Information Facility (GBIF): An On-Line, Free and Open Access to Biodiversity Data. Available online: https://www.gbif.org/occurrence/download/0229487-220831081235567 (accessed on 21 December 2022).
25. The Ocean Biogeographic Information System (OBIS): An On-Line, Worldwide Atlas for Accessing, Modeling and Mapping Marine Biological Data in a Multidimensional Geographic Context. Available online: https://datasets.obis.org/downloads/540d223c-5ef1-458e-b271-6f786e8d1d8b.zip (accessed on 4 December 2022).
26. Aiello-Lammens, M.E.; Boria, R.A.; Radosavljevic, A.; Vilela, B.; Anderson, R.P. spThin: An R package for spatial thinning of species occurrence records for use in ecological niche models. *Ecography* **2015**, *38*, 541–545. [CrossRef]
27. Basher, Z.; Bowden, D.; Costello, M.J. Global Marine Environment Datasets (GMED). World Wide Web Electronic Publication. Version 2.0 (Rev.02.2018). Available online: http://gmed.auckland.ac.nz (accessed on 15 April 2023).
28. Assis, J.; Tyberghein, L.; Bosch, S.; Verbruggen, H.; Serrão, E.A.; De Clerck, O.; Tittensor, D. Bio-ORACLE v2.0: Extending Marine Data Layers for Bioclimatic Modelling. Available online: https://bio-oracle.org/ (accessed on 6 April 2023).
29. Naimi, B.; Hamm, N.A.; Groen, T.A.; Skidmore, A.K.; Toxopeus, A.G. Where is positional uncertainty a problem for species distribution modelling? *Ecography* **2014**, *37*, 191–203. [CrossRef]
30. Fournier, A.; Barbet-Massin, M.; Rome, Q.; Courchamp, F. Predicting species distribution combining multi-scale drivers. *Glob. Ecol. Conserv.* **2017**, *12*, 215–226. [CrossRef]
31. Zhang, Z.; Kass, J.M.; Mammola, S. Lineage-level distribution models lead to more realistic climate change. *Divers. Distrib.* **2021**, *27*, 684–695. [CrossRef]
32. Hutchinson, G.E. Concluding Remarks. *Cold Spring Harb. Symp. Quant. Biol.* **1957**, *22*, 415–427. [CrossRef]
33. Blonder, B.; Morrow, C.B.; Maitner, B.; Harris, D.J.; Lamanna, C.; Violle, C.; Enquist, B.J.; Kerkhoff, A.J. New approaches for delineating n-dimensional hypervolumes. *Methods Ecol. Evol.* **2018**, *9*, 305–319. [CrossRef]
34. Cardoso, P.; Rigal, F.; Carvalho, J.C. BAT–Biodiversity Assessment Tools, an R package for the measurement and estimation of alpha and beta taxon, phylogenetic and functional diversity. *Methods Ecol. Evol.* **2015**, *6*, 232–236. [CrossRef]
35. Thuiller, W.; Georges, D.; Engler, R.; Breiner, F.; Georges, M.D.; Thuiller, C.W. biomod2: Ensemble Platform for Species Dis-tribution Modeling. Available online: https://cran.r-project.org/web/packages/biomod2/index.html (accessed on 6 April 2023).
36. Barbet-Massin, M.; Jiguet, F.; Albert, C.H.; Thuiller, W. Selecting pseudo-absences for species distribution models: How, where and how many? *Methods Ecol. Evol.* **2012**, *3*, 327–338. [CrossRef]
37. Allouche, O.; Tsoar, A.; Kadmon, R. Assessing the accuracy of species distribution models: Prevalence, kappa and the true skill statistic (TSS). *J. Appl. Ecol.* **2006**, *43*, 1223–1232. [CrossRef]
38. Swets, J.A. Measuring the Accuracy of Diagnostic Systems. *Science* **1988**, *240*, 1285–1293. [CrossRef]
39. Liu, C.; White, M.; Newell, G.; Pearson, R. Selecting thresholds for the prediction of species occurrence with presence-only data. *J. Biogeogr.* **2013**, *40*, 778–789. [CrossRef]
40. Zanardo, N.; Parra, G.J.; Passadore, C.; Möller, L.M. Ensemble modelling of southern Australian bottlenose dolphin Tursiops sp. distribution reveals important habitats and their potential ecological function. *Mar. Ecol. Prog. Ser.* **2017**, *569*, 253–266. [CrossRef]
41. UNEP-WCMC and IUCN (2023), Protected Planet: The World Database on Protected Areas (WDPA) and World Database on Other Effective Area-Based Conservation Measures (WD-OECM). Available online: https://www.protectedplanet.net (accessed on 4 May 2023).
42. Zhuang, H.; Shao, F.; Zhang, C.; Xia, W.; Wang, S.; Qu, F.; Wang, Z.; Lu, Z.; Zhao, L.; Zhang, Z. Spatial-temporal shifting patterns and in situ conservation of spotted seal (*Phoca largha*) populations in the Yellow Sea ecoregion. *Integr. Zool.* 2023. [CrossRef] [PubMed]
43. Rice, D.W. *A List of the Marine Mammals of the World*; Department of Commerce, National Oceanic and Atmospheric Administration, National Marine Fisheries Service: Silver Spring, MD, USA, 1977; 13p, NMFS-SSRF-711.
44. Fedoseev, G. *Use of Nonmetrical Characters of Skulls of Bering Sea Seals in a Study of the Phenotypic Structure of Their Populations*; Department of Commerce, National Oceanic and Atmospheric Administration, National Marine Fisheries Service: Silver Spring, MD, USA, 1984; pp. 49–54, NMFS 12.
45. Razgour, O.; Forester, B.; Taggart, J.B.; Bekaert, M.; Juste, J.; Ibanez, C.; Puechmaille, S.J.; Novella-Fernandez, R.; Alberdi, A.; Manel, S. Considering adaptive genetic variation in climate change vulnerability assessment reduces species range loss projections. *Proc. Natl. Acad. Sci. USA* **2019**, *116*, 10418–10423. [CrossRef] [PubMed]
46. Ikeda, D.H.; Max, T.L.; Allan, G.J.; Lau, M.K.; Shuster, S.M.; Whitham, T.G. Genetically informed ecological niche models improve climate change predictions. *Glob. Chang. Biol.* **2017**, *23*, 164–176. [CrossRef]
47. Oney, B.; Reineking, B.; O'Neill, G.; Kreyling, J. Intraspecific variation buffers projected climate change impacts on Pinus contorta. *Ecol. Evol.* **2013**, *3*, 437–449. [CrossRef]
48. Grebmeier, J.M.; Overland, J.E.; Moore, S.E.; Farley, E.V.; Carmack, E.C.; Cooper, L.W.; Frey, K.E.; Helle, J.H.; McLaughlin, F.A.; McNutt, S.L. A major ecosystem shift in the northern Bering Sea. *Science* **2006**, *311*, 1461–1464. [CrossRef]
49. Polovina, J.J.; Howell, E.A.; Kobayashi, D.R.; Seki, M.P. The transition zone chlorophyll front updated: Advances from a decade of research. *Prog. Oceanogr.* **2017**, *150*, 79–85. [CrossRef]
50. Yang, T.; Liu, X.; Han, Z. Predicting the Effects of Climate Change on the Suitable Habitat of Japanese Spanish Mackerel (*Scomberomorus niphonius*) Based on the Species Distribution Model. *Front. Mar. Sci.* **2022**, *9*, 927790. [CrossRef]
51. Sun, B.; Zhao, L.; Shao, F.; Lu, Z.; Tian, J.; Liu, C. Estimating the impacts of climate change on the habitat suitability of common minke whales integrating local adaptation. *Front. Mar. Sci.* **2022**, *9*, 923205. [CrossRef]

52. Chen, Y.; Shan, X.; Ovando, D.; Yang, T.; Dai, F.; Jin, X. Predicting current and future global distribution of black rockfish (*Sebastes schlegelii*) under changing climate. *Ecol. Indic.* **2021**, *128*, 107799. [CrossRef]
53. Breiner, F.T.; Guisan, A.; Bergamini, A.; Nobis, M.P.; Anderson, B. Overcoming limitations of modelling rare species by using ensembles of small models. *Methods Ecol. Evol.* **2015**, *6*, 1210–1218. [CrossRef]
54. Bahn, V.; McGill, B.J. Testing the predictive performance of distribution models. *Oikos* **2013**, *122*, 321–331. [CrossRef]
55. Albouy, C.; Delattre, V.; Donati, G.; Frölicher, T.L.; Albouy-Boyer, S.; Rufino, M.; Pellissier, L.; Mouillot, D.; Leprieur, F. Global vulnerability of marine mammals to global warming. *Sci. Rep.* **2020**, *10*, 548. [CrossRef] [PubMed]
56. Wang, W.; Li, J. conservation of biodiversity in China: Advances and prospects. *Biodivers. Sci.* **2021**, *29*, 133–149. [CrossRef]
57. Ma, Z.; Chen, Y.; Melville, D.S.; Fan, J.; Liu, J.; Dong, J.; Tan, K.; Cheng, X.; Fuller, R.A.; Xiao, X.; et al. Changes in area and number of nature reserves in China. *Conserv. Biol.* **2019**, *33*, 1066–1075. [CrossRef] [PubMed]
58. Witherell, D.; Woodby, D. Application of Marine Protected Areas for Sustainable Production and Marine. *Mar. Fish. Rev.* **2005**, *67*, 1–28.
59. Hamilton, C.D.; Lydersen, C.; Aars, J.; Acquarone, M.; Atwood, T.; Baylis, A.; Biuw, M.; Boltunov, A.N.; Born, E.W.; Boveng, P.; et al. Marine mammal hotspots across the circumpolar Arctic. *Divers. Distrib.* **2022**, *28*, 2729–2753. [CrossRef]
60. Trukhin, A.M.; Permyakov, P.A.; Ryazanov, S.D.; Lobanov, V.B.; Kim, H.W.; Choi, Y.M.; Sohn, H. Migrations of young spotted seals (*Phoca largha*) from Peter the Great Bay, Sea of Japan/East Sea, and the pattern of their use of seasonal habitats. *PLoS ONE* **2021**, *16*, e0244232.
61. Solovyeva, M.A.; Kuznetsova, D.M.; Glazov, D.M.; Burkanov, V.N.; Rozhnov, V.V. The First Data on the Use of the Pacific Ocean Area by the Okhotsk Population of the Spotted Seal *Phoca largha*. *Biol. Bull.* **2021**, *48* (Suppl. S1), S219–S224. [CrossRef]
62. Solov'eva, M.A.; Glazov, D.M.; Kuznetsova, D.M.; Rozhnov, V.V. Movements of spotted seals (*Phoca largha*) in the Sea of Okhotsk according to satellite tagging data. *Russ. J. Ecol.* **2016**, *47*, 405–411. [CrossRef]
63. Han, J.; Lu, Z.; Tian, J.; Ma, Z.; Wang, Z.; Yang, Y.; Wang, Q.; Song, X.; Peng, Z. Release studies on spotted seals (*Phoca largha*) using satellite telemetry tracking technique. *Acta Theriol. Sin.* **2013**, *33*, 300.
64. Mizuno, A.W.; Wada, A.; Ishinazaka, T.; Hattori, K.; Watanabe, Y.; Ohtaishi, N. Distribution and abundance of spotted seals *Phoca largha* and ribbon seals *Phoca fasciata* in the southern Sea of Okhotsk. *Ecol. Res.* **2002**, *17*, 79–96. [CrossRef]
65. Lowry, L.F.; Burkanov, V.N.; Frost, K.J.; Simpkins, M.A.; Davis, R.; DeMaster, D.P.; Suydam, R.; Springer, A. Habitat use and habitat selection by spotted seals (*Phoca largha*) in the Bering Sea. *Can. J. Zool.* **2000**, *78*, 1959–1971. [CrossRef]
66. Lowry, L.F.; Frost, K.J.; Davis, R.; DeMaster, D.P.; Suydam, R.S. Movements and behavior of satellite-tagged spotted seals (*Phoca largha*) in the Bering and Chukchi Seas. *Polar Biol.* **1998**, *19*, 221–230. [CrossRef]

animals

MDPI

Article

Initial Post-Release Performance of Cultured *Cyprinus chilia* Juveniles in a Shallow Lake in Southwestern China

Tingbing Zhu [1], Deguo Yang [1], Jinling Gong [1], Chunyong Wang [2], Xiangjun Miao [3], Yongben Liang [4] and Xuemei Li [1,*]

[1] Key Laboratory of Freshwater Biodiversity Conservation, Ministry of Agriculture and Rural Affairs, Yangtze River Fisheries Research Institute, Chinese Academy of Fisheries Science, Wuhan 430223, China; zhutb@yfi.ac.cn (T.Z.); yangdg@yfi.ac.cn (D.Y.); jlgong123@163.com (J.G.)
[2] Fishery Workstation of Tonghai County, Yuxi 652799, China; wang62018@126.com
[3] Yunnan Institute of Fishery Sciences Research, Kunming 650051, China; kunmingmiao@126.com
[4] Fishery Workstation of Yuxi, Yuxi 653101, China; lyblyben@163.com
* Correspondence: xmli@yfi.ac.cn; Tel.: +86-027-81780112

Simple Summary: The post-release performance of cultured fish is crucial for evaluating the effectiveness of stock enhancement programs. Through a mark–release–recapture experiment, we studied the initial post-release performance characteristics of cultured *Cyprinus chilia* juveniles after being released into a shallow lake in southwestern China. Most of the released fish preferred to inhabit the lake shore zone with water depth less than 300 cm. The majority of long-distance movements (greater than 100 m) occurred within the first 4 days after release. The gut fullness of the released fish showed a trend of initially decreasing and then increasing, and the gut microbial community structure was significantly different from before release. This study suggests that cultured *Cyprinus chilia* juveniles can primarily adapt to the wild environment after 4–5 days of release.

Abstract: The post-release performance of cultured fish is crucial for understanding the viability of cultured fish and assessing the effects of stock enhancement programs. This study aimed to investigate the initial post-release performance of cultured *Cyprinus chilia* juveniles by examining their movement, spatial distribution, gut fullness, and gut microbiota in nature. In July 2022, a total of 20,000 *C. chilia* juveniles, tagged with visible implant fluorescence (VIE), were released into Qilu Lake, a shallow lake in southwestern China. Subsequently, continuous recapture was conducted at fixed recapture sites using trap nets during the first 7 days, one month and three months after release. Out of the released fish, 512 were recaptured, resulting in a recapture rate of 2.56%. The recaptured fish had a 100% tag retention rate. The majority (98.05%) of the recaptured fish were found in the recapture sites located on the eastern or western lakeshore, while only 10 fish were recaptured from the recapture sites in the northern lake area. The water depth range where the recaptured fish were found ranged from 190 to 350 cm, with most fish preferring depths less than 300 cm. The majority of the released fish migrated towards the eastern and western lakeshore, with long-distance movement (greater than 100 m) primarily occurring within the first four days after release. The level of gut fullness in the released fish initially decreased and then increased over time following release. Regarding gut microbiota, the dominant phyla observed in most samples were Firmicutes, Proteobacteria, Cyanobacteria, and Fusobacteria. Furthermore, significant variations in the dominant genera were observed across different samples. Principal coordinates analysis (PCoA) revealed clear separation between the microbial communities of pre-release and post-release *C. chilia* juveniles. This study demonstrated that VIE tagging was a suitable method for short-term marking of *C. chilia* juveniles. Lakeshores with water depths less than 300 cm were identified as preferred habitats for *C. chilia* juveniles. The primary adaptation period for cultured *C. chilia* juveniles released into the natural environment was found to be approximately 4–5 days. These findings contribute to our understanding of the post-release performance of cultured fish and may provide guidance for the management and evaluation of relevant stock enhancement programs.

Citation: Zhu, T.; Yang, D.; Gong, J.; Wang, C.; Miao, X.; Liang, Y.; Li, X. Initial Post-Release Performance of Cultured *Cyprinus chilia* Juveniles in a Shallow Lake in Southwestern China. *Animals* **2023**, *13*, 3196. https://doi.org/10.3390/ani13203196

Academic Editor: Alessia Giannetto

Received: 29 August 2023
Revised: 28 September 2023
Accepted: 9 October 2023
Published: 13 October 2023

Keywords: mark–release–recapture; *Cyprinus chilia*; post-release performance; movement; gut microbiota

1. Introduction

The effectiveness of stock enhancement programs is significantly influenced by the post-release performance of cultured fish. While large-scale stock enhancements are commonly conducted worldwide, successful cases have been limited [1]. One of the reasons for this failure is the insufficient understanding of the biology and ecology of the stocks targeted for enhancement [2]. Additionally, studies on the performance and behavior of fish stocks have been scarce and more challenging compared to research on their general biological characteristics such as growth, food habits, and reproduction. Previous reports on the post-release performance of fish have primarily focused on adult or large-sized fish species, such as *Acipenser sturio*, *Argyrosomus japonicus*, *Anguilla japonica*, *Etelis coruscans*, and *Myxocyprinus asiaticus* [3–7]. In China, numerous stock enhancement programs have been implemented in recent decades, with their scale expanding each year. However, most of these programs in China have targeted small-sized juvenile cultured fish, highlighting the need for further research on the post-release performance of small-sized cultured fish.

There is currently no recognized standard for what indicators to use and how long the research period should be in order to accurately evaluate the performance of released fishes. There are various indicators utilized to assess the post-release performance of cultured fishes, although they lack a unified approach. Previous research has employed indicators such as growth [8,9], survival [10], dispersal or behavior [5,6], and physiology [11] to evaluate post-release performance. In addition, conducting a long-term recapture study is challenging and labor-intensive. Typically, the initial stage of release is of utmost importance in determining the viability of cultured fish, as the majority of mortalities occur within a short timeframe following their release [12]. Furthermore, monitoring and assessing their initial performance is considerably more feasible.

Mark–release–recapture is the fundamental method used to study the post-release performance of artificially released populations. Common fish tagging methods include listing, visible implant fluorescence (VIE), coded wire tags (CWT), and acoustic telemetry. The selection of appropriate tagging methods should be based on factors such as fish species, fish size, tagging cost, and characteristics of the study area. Due to its advantages of individual recognition and no need to recapture, acoustic telemetry was widely used in studies about the post-release movement of fishes, such as applications on *Acipenser sturio* [3], *Argyrosomus japonicus* [4], *Anguilla japonica* [5], *Etelis coruscans* [6], and *Myxocyprinus asiaticus* [7]. However, the acoustic telemetry was expensive and had large tags, which made it unsuitable for small-sized juvenile fish. Alternatively, VIE and CWT were recommended to the mark–release–recapture studies of small-sized fish as they were low cost and did minimal harm. For example, Leber et al. [13] reported on the recruitment patterns of juvenile, cultured Pacific threadfin (*Polydactylus sexfilis*) from the Polynemidae family. The study released the fish along sandy marine shores in Hawaii using CWT. Hervas et al. [9] also revealed the growth, survival, and distribution of released *Atractoscion nobilis* using the CWT method. The initial post-release movement of cultured *Cyprinus pellegrini* juveniles was evaluated using the VIE method [14].

Cyprinus chilia is endemic to the plateau lakes in Yunnan Province, China [15]. *Cyprinus chilia* is omnivorous and mainly feeds on large benthic invertebrates. Under natural conditions, *C. chilia* can grow up to 4 kg and is an important economic fish [15]. However, the natural resource of *C. chilia* has declined rapidly over the last decades due to overfishing, deterioration of the water environment, habitat loss, invasion of alien species, and other factors. At present, it is difficult to find wild populations of *C. chilia* in its naturally distributed lakes. To protect this endangered species, artificial breeding [16] and stock enhancement of *C. chilia* have been successfully implemented. However, there is a lack of information regarding the post-release performance and evaluation of these stock enhancement efforts.

Also, a suitable tagging method for *C. chilia* juveniles has not been evaluated yet. Therefore, the management and planning of *C. chilia* stock enhancement programs remain unclear.

VIE was one of the most popular tagging methods used to mark released fish in China, but there are few reports about the effects of VIE tagging on the assessment of stock enhancement. Therefore, the objectives of this study are (1) to assess the effectiveness of VIE as a short-term tagging method for *C. chilia* juveniles and (2) to investigate the initial post-release performance of cultured *C. chilia* juveniles by monitoring their movement, spatial distribution, gut fullness, and gut microbiota. Correspondingly, two hypotheses are proposed: (1) the VIE should be suitable for the short-term tagging of *C. chilia* juveniles, and (2) cultured *C. chilia* juveniles should go through an obvious transitional period in the new environment. The findings of this study will contribute to a better understanding of the post-release performance of cultured Cyprinus fishes and will be valuable for the management and evaluation of stock enhancement programs related to *C. chilia*.

2. Materials and Methods

2.1. Fish

The fish were artificially hatched *C. chilia* juveniles produced at the Raofu fish farm, Tonghai County. Before the mark–release–recapture trial, fish were about 4 months old and had a mean total length (L) of 10.8 ± 1.2 cm (mean ± standard deviation, n = 46) and mean body weight (W) of 17.54 ± 5.19 g.

2.2. Mark–Release–Recapture Trial

A mark–release–recapture trial was conducted to examine the movement, spatial distribution, gut fullness, and gut microbiota of *C. chilia* juveniles after release. VIE tags were used in the study as its low cost and minimal harm to small-sized fishes.

A total of 20,000 *C. chilia* juveniles were selected for the mark–release–recapture trial. All of the selected fish were tagged with VIE tags to help the identification of fish samples in the subsequent recapture study. The tagging work was carried out on 25–26 July 2022. Prior to the tagging operation, fish were anesthetized using 30 mg/L of MS-222 for 1~2 min. VIE tagging procedures were as follows: red VIEs were injected into the epidermis of the head skin using hand-pressurized syringes (Northwest Marine Technology, Anacortes, WA, USA). The presence of the tags was visually confirmed. Following the tagging process, the fish were transferred to clean water to facilitate their recovery (recovery rate 100%).

The designated water area for release is the Qilu Lake in Yunnan Province, China (Figure 1), as it is one of the main distribution lakes of *C. chilia*. The Qilu Lake has a maximum length of 10.4 km, maximum width of 4.4 km, water area of 38.86 km^2, a maximum water depth of 540 cm, and an average water depth of 403 cm. There were 11 species of fish historically distributed in Qilu Lake, including *C. chilia*, *Cyprinus pellegrini*, *Cyprinus ilishaestomus*, *Cyprinus yunnanensis*, *Anabarilius qiluensis*, *Carassius auratus*, *Misgurnus anguillicaudatus*, *Oryzias latipes*, *Channa argus*, *Monopterus albus*, *Silurus grahami* [17]. There has been no record of *C. chilia* in Qilu Lake since 2000s. According to our investigation of the entire Qilu Lake on 12–14 July 2022, the present fish species include *Cyprinus carpio*, *Carassius auratus*, *Abbottina rivularis*, *Hemicculter Leuciclus*, *Pseudorasbora parva*, *Cultrichthys erythropterus*, *Pelteobagrus fulvidraco*, *Misgurnus anguillicaudatus*, *Paramisgurnus dabryanus*, *Rhinogobius giurinus*, *Hyporhamphus intermedius*. The release site was situated on the south shore of Qilu Lake, with a water depth of 220 cm, transparency of 37 cm, water temperature of 27.4 °C, dissolved oxygen level of 10.46 mg/L, a pH of 8.83, and conductivity of 891 μs/cm. On 27 July 2022, the tagged fish were packed into double-layered nylon fish bags (80 cm × 40 cm) with a density of 150~200 fish per bag. The volume ratio of oxygen to water in the bag was approximately 3:1. The fish bags were then transported (approximately 0.5 h) to the release site for release. The entire release process was conducted manually. The fish bags were initially submerged in the water at the lakeshore to balance the temperature for 0.5 h. Afterward, the fish bags were untied, and the tagged fish were carefully released into the lake.

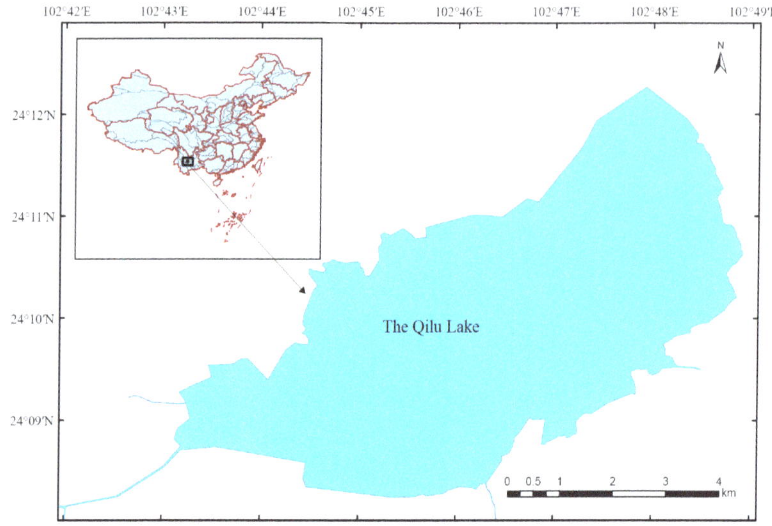

Figure 1. Geographic location diagram of Qilu Lake.

The recapture work was divided into two stages, including a 7-day continuous recapture and a long-term recapture.

The 7-day continuous recapture was conducted immediately following the release of fish. The lake was divided into key recapture area and regular recapture area. The key recapture area was around the release site (Figure 2a). A total of 13 fixed recapture sites were set up in the key recapture area, including 1 site (S1) at the release site, 4 sites (E1, E2, E3, E4) on the eastern lakeshore, 4 sites (W1, W2, W3, W4) on the western lakeshore, and 4 sites (N1, N2, N3, N4) in the northern lake area (Figure 2a). This created a 4-layer arc recapture monitoring network, with the release point as the origin. The distances from the recapture sites on the 1st, 2nd, 3rd, and 4th layers to the release site were set as 150, 300, 500, and 800 m, respectively. However, due to limitations in terrain, interference from wind and waves on the positioning ship, and other factors, there was some variation between the actual distance between adjacent recapture sites and the preset values (Tables 1 and 2). The regular recapture area included lake areas outside of key recapture area (Figure 2b). A total of 12 fixed recapture sites were set up in the regular recapture area, including sites D1, D2, D3, D4, D5, D6, D7, D8, D9, D10, D11, and D12 (Figure 2b).

Table 1. Water depth of the recapture sites and their distance to the release site in the mark–release–recapture trial of cultured *Cyprinus chilia* juveniles released into Qilu Lake.

Recapture Site	Water Depth (cm)	Distance to the Release Site (m)
S1	220	0
E1	200	157
E2	230	310
E3	200	524
E4	190	813
N1	350	104
N2	430	220
N3	490	454
N4	300	676
W1	300	145
W2	250	345

Table 1. *Cont.*

Recapture Site	Water Depth (cm)	Distance to the Release Site (m)
W3	250	551
W4	230	960
D1	160	2932
D2	200	5082
D3	120	4625
D4	180	4832
D5	250	5048
D6	160	7019
D7	100	6251
D8	150	3792
D9	190	1586
D10	395	3143
D11	540	3072
D12	403	5213

Table 2. Distances between the adjacent sites in the key recapture area of the mark–release–recapture trial of cultured *Cyprinus chilia* juveniles released into Qilu Lake.

Adjacent Recapture Site	Distance (m)
E1–E2	165
E2–E3	237
E3–E4	297
N1–N2	147
N2–N3	234
N3–N4	224
W1–W2	202
W2–W3	212
W3–W4	409

The long term recapture was conducted one month and three months after release. The recapture sites for the long term recapture was the same as the recapture sites in the regular recapture area of the 7-day continuous recapture, plus the release sites (S1). For each long-term recapture, 3-day continuous recapture was conducted.

Recapture gears used were trap nets with dimensions of 10 m in length, 45 cm in width, 33 cm in height, and a mesh size of 7 mm. The maximum volume of the net was 1.485 m^3. A density of one net per site was set, and the nets were deployed 2 h prior to fish release. After the fish were released, the catch in the nets was collected every 24 h, and the nets were then repositioned during the recapture periods. It should be noted that the net at S1 had a water entry time of only 6 h on the fifth day after release, as it was affected by an incorrect removal of the net.

2.3. Processing of the Recaptured Fish Samples

All recaptured fish samples were identified by species. For *C. chilia* juveniles, the tag presence of fish was determined through visual examination. The recaptured *C. chilia* juveniles were counted separately based on the date and recapture site. When fewer than 30 released fish samples were recaptured at a site per day, the total length and weight of all fish samples were measured. If more than 30 released fish were recaptured at a site per day, a random sample of 30 fish was measured for total length (accurate to 0.1 cm) and weight (accurate to 0.1 g).

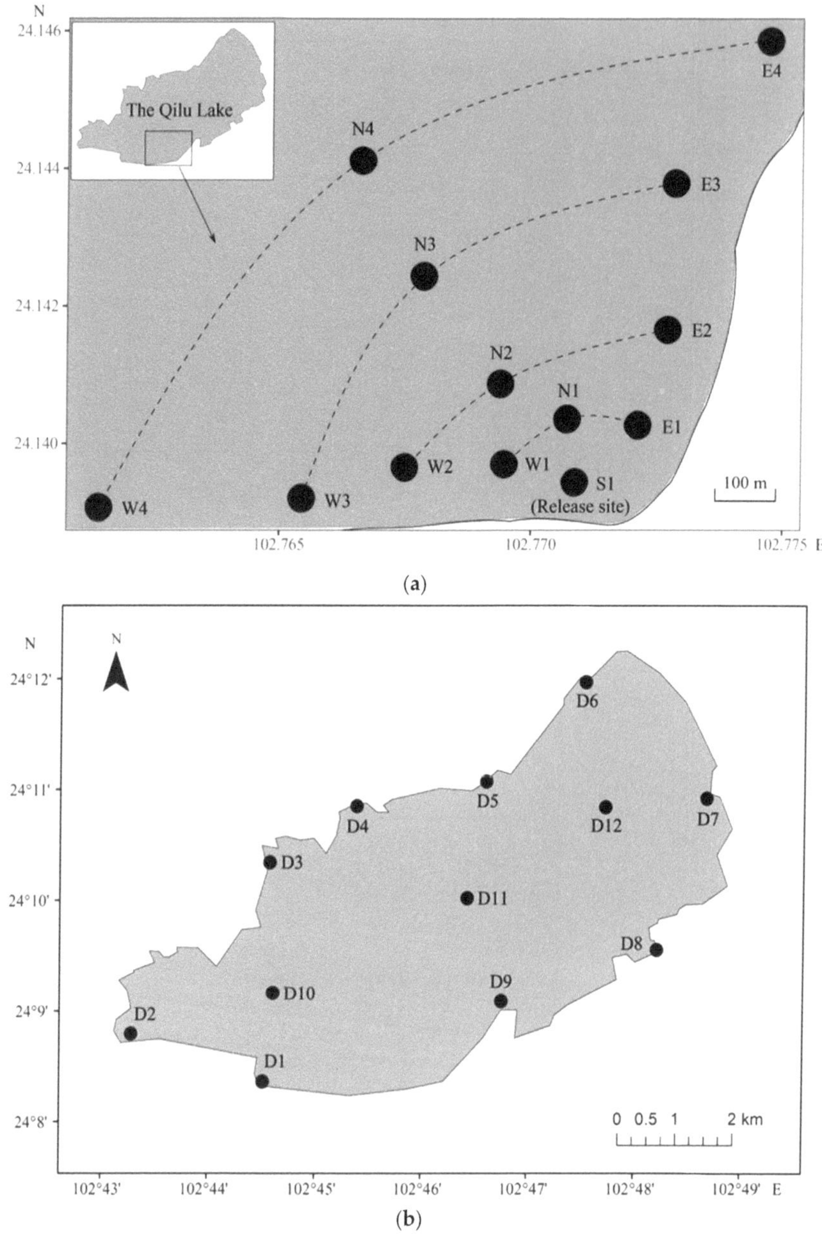

Figure 2. Recapture sites of cultured *Cyprinus chilia* juveniles released into Qilu Lake. (**a**) Sites in key recapture area; (**b**) sites in regular recapture area.

According to the number of fish samples, 3–20 fish were randomly selected for dissection every day to assess the feeding intensity of the released fish by examining their gut fullness levels [18]. The level of gut fullness ranged from 0 to 5, representing feed intake from low to high (Table 3). Additionally, freshly dissected guts of three fish samples were collected every day for analysis of gut microbiota structure. According to the post-release days and sampling order, the samples were named as D0–1, D0–2, D0–3, D1–1, D1–2, D1–3, D2–1, D2–2, D2–3, D3–1, D3–2, D3–3, D4–1, D4–2, D4–3, D5–1, D5–2, D5–3, D6–1, D6–2,

D6–3, D7–1, D7–2, and D7–3, respectively. After sampling, any remaining surviving fish were transferred to indoor breeding tanks for temporary cultivation.

Table 3. The categorization of fish gut fullness level.

Gut Fullness Level	Classification Criteria
0	No food or very little food in fish gut
1	The food volume accounts for 1/4 of fish gut
2	The food volume accounts for 1/2 of fish gut
3	The food volume accounts for 3/4 of fish gut
4	Food fills the whole fish gut
5	Food fills the whole fish gut and makes the gut swell

2.4. Gut Microbiota Structure Analysis

Total microbiota genomic DNA was extracted from approximately 0.3 g of the freshly dissected intestine of each fish using the FastDNA spin kit for soil (MP, Solon, OH, USA) following the manufacturer's protocols. The 515F (5′-GTGYCAGCMGCCGCGGTA-3′) and 907R (5′-CCGTCAATTCMTTTRAGT-3′) primers [19,20] were used to amplify the bacterial 16S rRNA gene V4–V5 fragments. PCR integration and protocols were carried out as follows: 94 °C for 3 min followed by 30 cycles of 94 °C for 40 s, 56 °C for 60 s, 72 °C for 60 s, and a final extension at 72 °C for 10 min until the reaction was halted by the user. The amplicons were purified and sequenced using the Illumina Miseq platform at Guangdong Meilikang Bio-Science Ltd. (Foshan, China).

Raw reads were merged using FLASH 1.2.8 and subsequently processed using QIIME 1.9.0, as described [21]. Briefly, all the merged sequences were assigned to each sample based on their barcode sequences, and the trimmed barcodes and primer sequences were removed using QIIME 1.9.0 software. Low-quality and chimeric sequences were removed using QIIME 1.9.0 and UCHIME 4.2.40, respectively. Subsequently, the remaining high-quality sequences were clustered into operational taxonomic units (OTUs) with 97% identity using UPARSE 7.1 [22]. The taxonomy of each OTU was assigned using the RDP classifier [23] in the gg_13_8 database. Alpha diversity indices were calculated using QIIME version 1.9.0. The Illumina sequencing raw data obtained from this study were deposited in the NCBI Sequence Read Archive with accession No. PRJNA999371.

2.5. Data Analysis

The condition factor, recapture rate, tag retention rate, fish distribution density, water depth selectivity, and daily movement velocity were calculated.

Condition factor:

$$CF = 100\% \times W/L^3$$

where *CF* was condition factor, *W* was body weight (g), and *L* was total length (cm).

Recapture rate:

$$C = 100\% \times n/N$$

where C was recapture rate, *n* was the number of recaptured releasing fish, and *N* was the number of releasing fish.

Tag retention rate:

$$R = 100\% \times n_p/n$$

where *R* was the tag retention rate, n_p was the number of recaptured *C. chilia* with VIE tag, and *n* was the total number of recaptured *C. chilia*.

Fish density:

$$D_i = N_i/V_i$$

where D_i was the density (ind./m^3) of fish at recapture site *i*, N_i was the number of recaptured releasing fish per day at recapture site *i*, and V_i was the fully extended volume of the trap net used at recapture site *i*.

ever, there was no significant size difference between fishes moving towards the western lakeshore and those that stayed at the release site (Table 5).

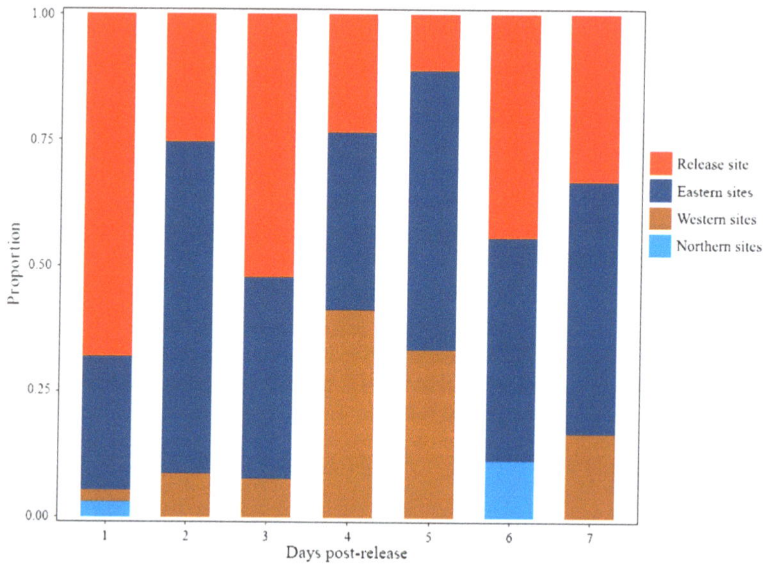

Figure 5. Relative proportion of recaptured cultured *Cyprinus chilia* juveniles from different direction sites post-release into Qilu Lake.

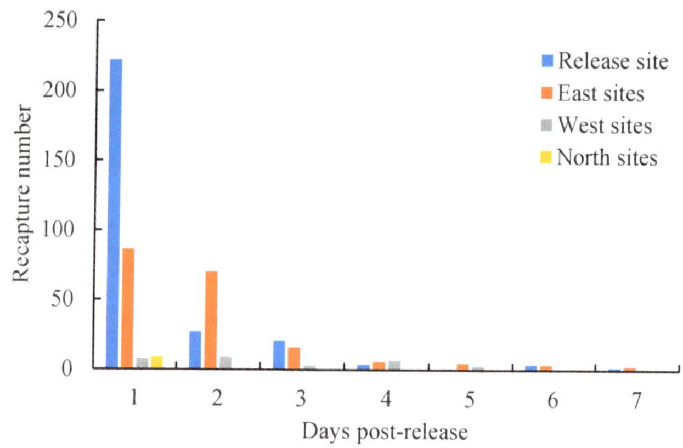

Figure 6. Number of recaptured cultured *Cyprinus chilia* juveniles from different direction sites post-release into Qilu Lake.

Table 5. Size of cultured *Cyprinus chilia* juveniles that moved by direction post-release into Qilu Lake combined for the entire 7 day period.

	Stay in the Release Site	Eastern Lakeshore	Western Lakeshore	Northern Lake Area
Mean total length (cm)	9.5 ± 1.7^{b}	10.3 ± 1.6^{a}	9.7 ± 1.2^{ab}	10.8 ± 2.0^{a}
Mean body weight (g)	11.39 ± 7.02^{b}	14.28 ± 7.50^{a}	11.72 ± 5.04^{ab}	16.39 ± 8.34^{a}
Sample size	100	135	31	10

Note: Means followed by standard deviation. Different superscripts indicate significant difference.

3.4. Spatial Distribution Patterns

Upon release, fish initially formed large schools at the release site and in the vicinity. Subsequently, they gradually dispersed towards adjacent lakeshore regions (Figure 7).

3.5. Water Depth Selectivity

The recapture sites, where fish were successfully retrieved, exhibited a water depth range of 190–350 cm. It was observed that the majority of fish displayed a preference for water depths less than 300 cm (Table 6).

Table 6. Water depth selectivity of cultured *Cyprinus chilia* juveniles released into Qilu Lake.

Water Depth (cm)	Selectivity						
	Day 1	Day 2	Day 3	Day 4	Day 5	Day 6	Day 7
≤200	0.40	0.78	0.58	0.50	0.54	0.53	0.30
201~300	0.48	0.22	0.42	0.50	0.46	0.47	0.70
301~400	0.12	0	0	0	0	0	0
401~500	0	0	0	0	0	0	0

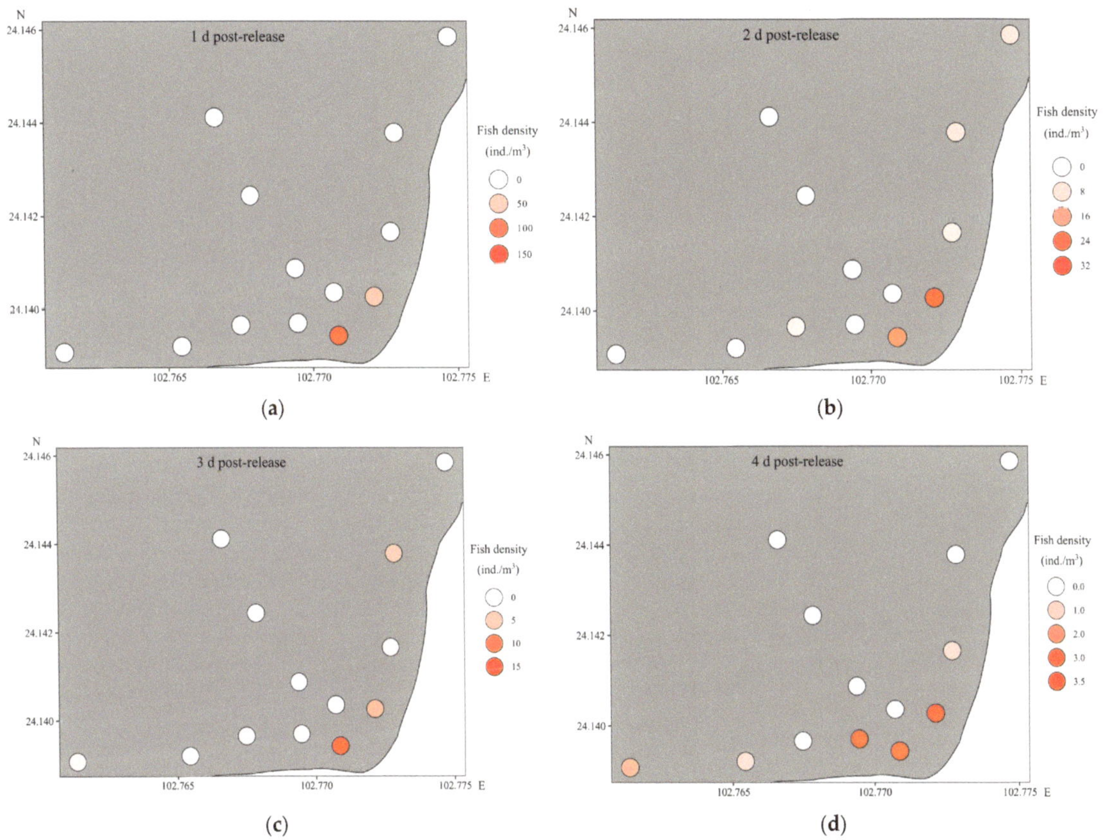

Figure 7. *Cont.*

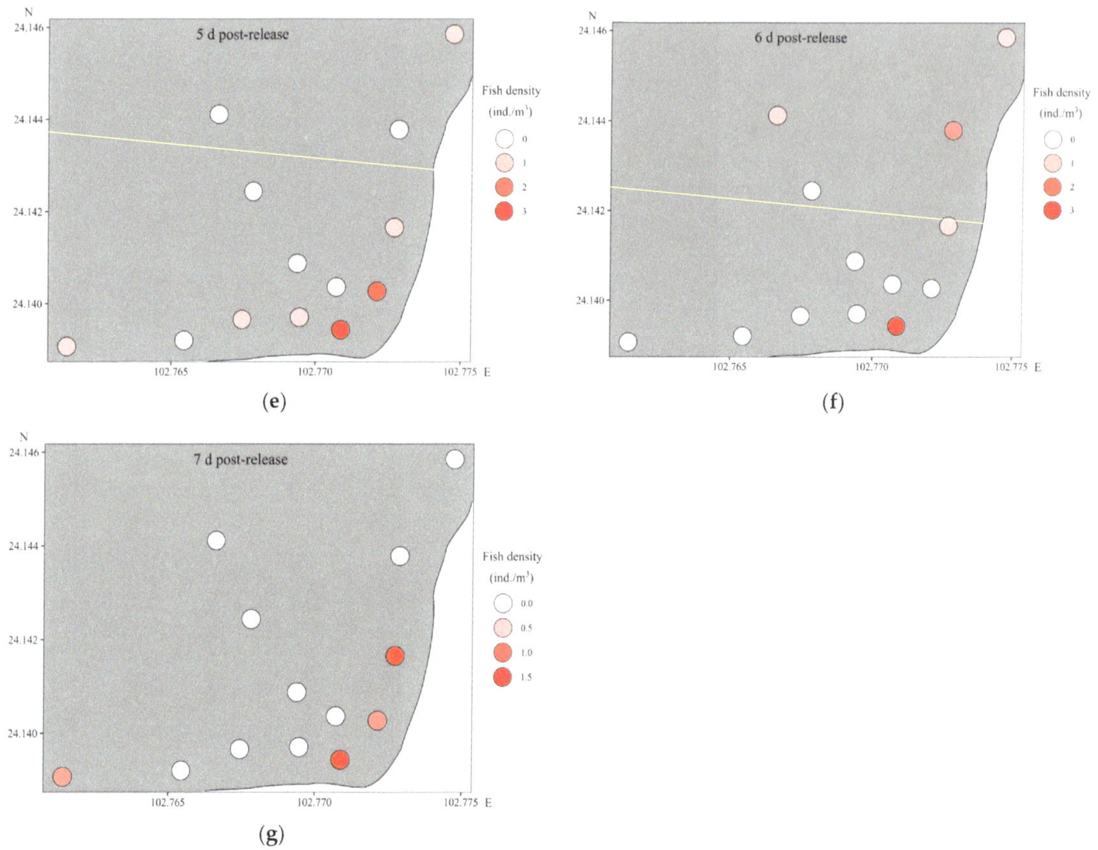

Figure 7. Spatial density dynamics of cultured *Cyprinus chilia* juveniles after being released into Qilu Lake: (**a**) 1 d post-release; (**b**) 2 d post-release; (**c**) 3 d post-release; (**d**) 4 d post-release; (**e**) 5 d post-release; (**f**) 6 d post-release; (**g**) 7 d post-release.

3.6. Movement Velocity

The movement velocity of the fish in the eastern and western lakeshore initially decreased and then increased during the 7-day recapture period, with changes occurring on the 4th day (Figure 8). Comparatively, the movement velocity of fish was higher on the eastern lakeshore than on the western lakeshore. An individual fish with the highest velocity (813 m/d) was recaptured at site E4 on the first day after being released. This fish had a total length of 11.6 cm and a body weight of 19.6 g. However, it was difficult to assess the movement velocity of the fish in the northern lake area due to the limited number of fish samples.

3.7. Gut Fullness

There was no significant difference in the levels of gut fullness among samples collected on different recapture days. However, released fish exhibited a pattern in the levels of gut fullness, when the average values were considered. Specifically, the level of gut fullness in the released fish decreased from day 1 to day 4 after release, and subsequently increased from day 4 to day 7 after release (Figure 9).

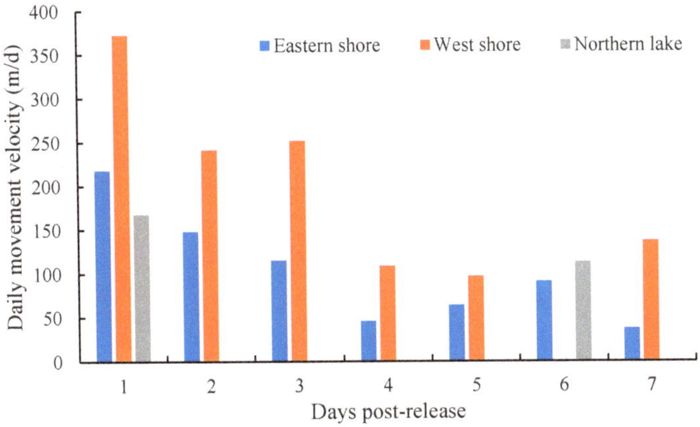

Figure 8. Daily movement velocity of cultured *Cyprinus chilia* juveniles in different directions post-release into Qilu Lake.

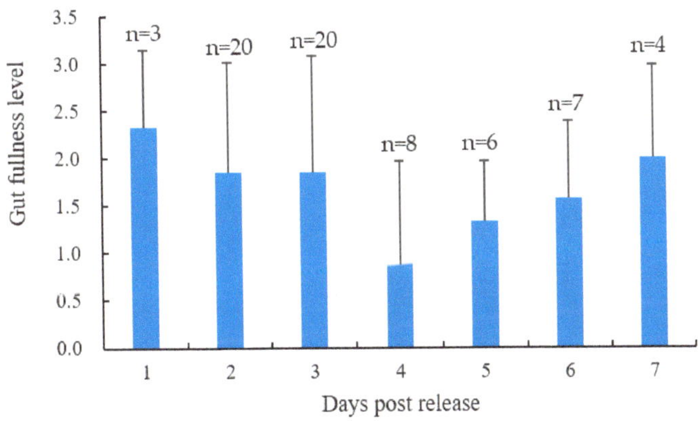

Figure 9. Gut fullness level dynamics of cultured *Cyprinus chilia* juveniles post-release into Qilu Lake. Error bars represent the standard deviation.

3.8. Gut Microbiota

After eliminating low-quality sequences, a total of 1,974,160 high-quality sequences were obtained from 24 samples of gut microbiota collected from *C. chilia* on different release days. To minimize the potential influence of varying sample sequencing depths on the subsequent analysis results, we randomly selected 30,485 sequences from each sample for further analysis. Among the detected sequences, a total of 56 phyla were identified. The dominant phyla in the gut microbiota were Firmicutes, Tenericutes, Proteobacteria, Cyanobacteria, Fusobacteria, Actinobacteria, Planctomycetes, Bacteroidetes, Chloroflexi, and Euryarchaeota. These dominant phyla accounted for 98.18 ± 0.05% of the high-quality sequences analyzed in this study (Figure 10).

A total of 90,575 operational taxonomic units (OTUs) were identified. However, an average of 3773.96 ± 580.96 OTUs were detected in each sample, with a significant difference between the pre-release group and the group 4 days post-release (Figure 11a). The Shannon and Simpson indices of the gut microbiota showed no significant difference between the different groups. However, the Chao1 index of the 5-day post-release group was significantly different from the other groups (Figure 11b–d).

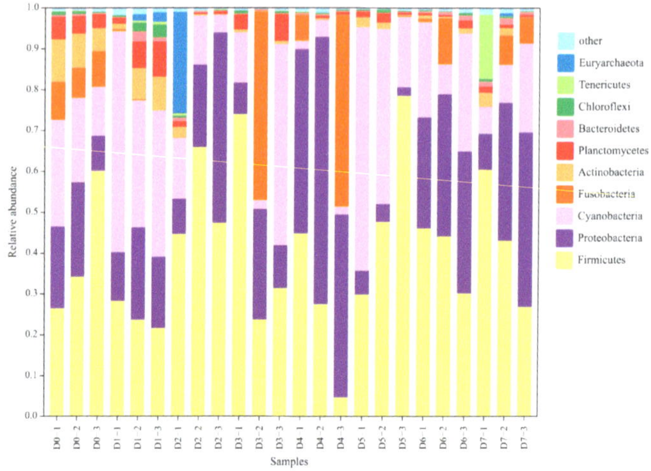

Figure 10. Relative abundance dynamics of dominant gut microbial phyla of cultured *Cyprinus chilia* juveniles post-release into Qilu Lake. D0–1 indicates the first sample of fish before release, D1–2 indicates the second sample of fish post-release for 1 day, etc.

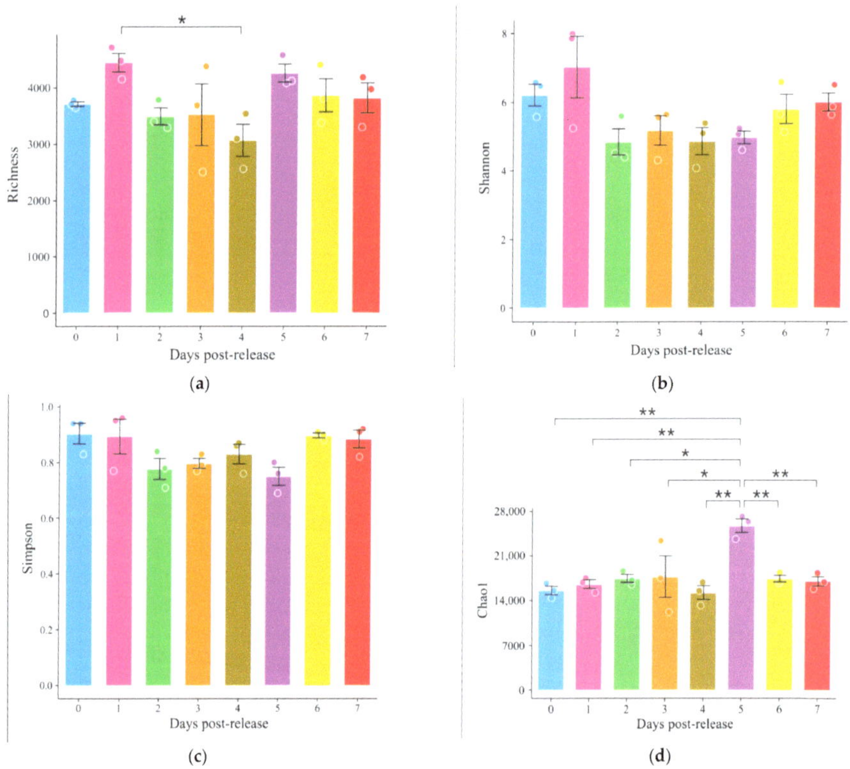

Figure 11. α-diversity index dynamics of the gut microbiota of cultured *Cyprinus chilia* juveniles post-release into Qilu Lake. (**a**) Richness index; (**b**) Shannon index; (**c**) Simpson index; (**d**) Chao1 index. Error bars represented the standard deviation. "*" indicate significant differences between groups ($p < 0.05$). Different colors represent different days post-release. "**" indicate extremely significant difference between groups ($p < 0.01$).

At the genus level, a total of 2819 genera were detected in the gut microbiota, out of which 46 were found to be dominant (Figure 12). The dominant genera before the release were Firmicutes other, Microcystis PCC-7914, *Cetobacterium*, PeM15_unclassified genus, and Planktothrix NIVA-CYA 15. After the release, the dominant genera were Firmicutes other, Microcystis PCC-7914, *Aeromonas*, *Romboutsia*, and *Cetobacterium*.

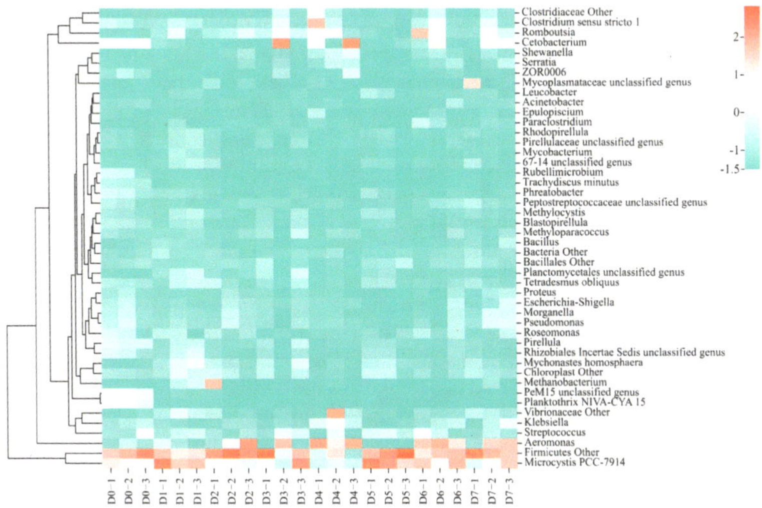

Figure 12. Heatmap profile of dominant gut microbial genera of cultured *Cyprinus chilia* juveniles post-release into Qilu Lake. D0–1 indicates the first sample of fish before releasing, D1–2 indicates the second sample of fish post-release for 1 day, etc.

The results of PCoA showed that the microbial communities in the D0 group were clearly separated from the other groups (Figure 13).

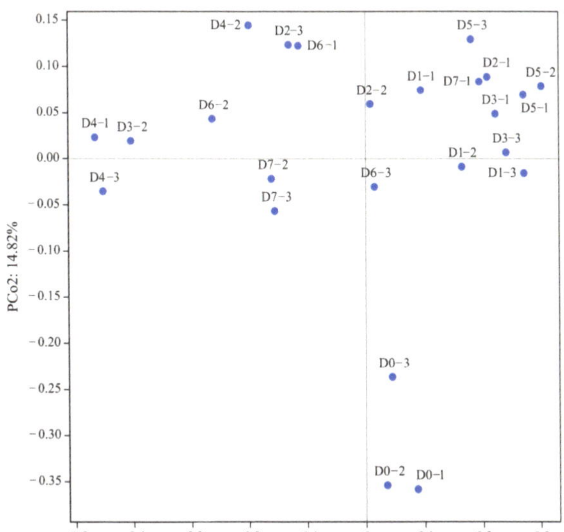

Figure 13. Principal coordinate analysis at the OTU levels of gut microbiota of cultured *Cyprinus chilia* juveniles post-release into Qilu Lake. D0–1 indicates the first sample of fish before releasing, D1–2 indicates the second sample of fish post-release for 1 day, etc.

4. Discussion

Marking methods suitable for juvenile fish are crucial to improving the efficiency of behavioral studies on released fish and evaluating the effects of numerous ongoing stock enhancement programs. Previous studies on the post-release behaviors of fish have primarily relied on acoustic telemetry [4–7,24] and satellite archival tags [25]. These methods enable the monitoring of individual fish movement and habitat selection without the need for physical capture. However, these marking methods had large tags and were therefore unsuitable for smaller-sized juvenile fish. Furthermore, the cost associated with acoustic telemetry and satellite archival tags makes it impractical to tag a large number of fish. It is worth noting that electronic tags may also affect the short-term behavior of fish [26]. Consequently, marking methods such as VIE and CWT are preferred for mark–release–recapture studies of large quantities of small-sized fishes due to their affordability, minimal damage, ease of operation, and ease of identification. The findings of the present study demonstrate high VIE tag retention and minimal damage in all recaptured fish samples, indicating that VIE is a suitable marking method for short-term mark–release–recapture studies of *C. chilia*.

Many studies have confirmed that the hatchery environment can affect the post-release behavior of fishes [27]. On the first day after release, the majority of individuals remained near the release point, which is referred to as the "residency period" [4]. The localized behavior exhibited by the cultured *C. chilia* juveniles on the first day after release might be due to unfamiliarity in a new environment, or it could be an extension of the settlement behavior developed in the artificial rearing environment. Similarly, hatchery-reared Mulloway (*Argyrosomus japonicus*) [4] and *Micropterus cataractae* [28] also showed a period of minimal movement when released into new environments with varying durations. From the second day after release, the *C. chilia* juveniles began to increase their movement and exploration of the environment. This may reflect the tradeoff between avoiding predation and resource utilization in the new environment. As time passes, factors such as hunger may motivate *C. chilia* juveniles to leave the release point and spread to more unfamiliar waters in order to increase their opportunity to obtain resources.

The released fish showed significant habitat selectivity. As an important parameter of fish habitat selectivity, water depth preference were reported in many fish species. For example, *Acanthopagrus schlegelii* prefers water depths of 8–10 m, while *Lutjanus argentimaculatus* prefers a water depth of 1–3.5 m [29]. The movements of land-locked Atlantic salmon (*Salmo salar*) in a large lake were nearshore (<2 km from shore) from spring to summer at ~20 m bathymetric depths [30]. The present study found the released fish mainly remained in shallow waters near the release point. Selectivity of water depth showed that released fish preferred water depths of up to 300 cm. Furthermore, the movement of released fish towards the eastern lakeshore was greater in scale than that towards the western lakeshore. It is speculated that the difference in complexity of the lake shorelines in the east and west may be related. The shoreline near the western side of the lake was relatively straight, while the shoreline near the eastern side of the lake was winding and complex. This complex shoreline provides more diverse habitat conditions, which can be beneficial for fish by providing shelter and feeding opportunities. Other species, like *Acanthopagrus schlegelii*, also prefer to stay in areas where food is more abundant and easy to hide in the initial stages after release [31].

The post-release movement velocity of *C. chilia* juveniles fluctuated over the 7-day trial. The movement of *C. chilia* juveniles decreased over the first three days after release and remained low for the following four days. This suggests that the majority of movement by *C. chilia* juveniles occurs within the first three days after release. The dispersal or movement of different fish often varies after release. For example, the horizontal movement velocity of released deep-water longtail red snapper (*Etelis coruscans*) was 2.2 km/d [6], while the movement velocity of artificially bred Japanese eels (*Anguilla japonica*) after release into the sea can reach 2.31 km/h [5]. Lake trout (*Salvelinus namaycush*) released into Lake Ontario had an average swimming speed of 1.64 km/h over the first day [32]. Cyprinus fishes are generally sedentary. For example, the mean daily movements of grass carp (*Ctenopharyngodon idella*) in Lake Erie ranged from <0.01 to 2.49 km/d, and only 25% of studied grass carp had mean daily movements greater than 0.88 km/d [33]. Therefore, the movement of *C. chilia* juveniles in our study was also relatively slow. The maximum recorded movement velocity in the present study was 813 m/d, and the overall movement gradually decreased over time, which is similar to that of some other fish species. For example, the movement velocity of mature Chinese sucker (*Myxocyprinus asiaticus*) released into the Yangtze River also showed a declining trend over time [7]. The gradual adaptation to the new environment and the acquisition and retention of an ideal habitat space may be reasons for the declining daily movement distance of *C. chilia* juveniles in this study.

The levels of gut fullness in released *C. chilia* juveniles showed a trend of initially decreasing and then increasing, although no statistical differences were found among recapture days. Limited sample sizes may explain these results. However, the mean levels of gut fullness in the released fish decreased from day 1 to day 4 after release, and then increased from day 4 to day 7 after release. This suggests that the released fish may begin feeding on the 5th day after release. Based on changes in the movements of released fish, it may be inferred that the first four days post-release are a transitional period for *C. chilia* juveniles in adapting to the new environment. During this period, they need to find suitable habitats and adapt to consuming natural prey. By the fifth day after release, released fish have entered an adaptation period during which they have relatively stable habitats and feeding conditions. Similarly, the gut fullness of the hatchery-reared honmasu salmon *Oncorhynchus rhodurus* × *masou* parr, released a week prior, was found to be comparable to that of wild parr [11].

This study indicates that the structure of the gut microbial community in *C. chilia* juveniles undergoes significant changes after they were released. Our study provided the first report on the gut microbiota of *C. chilia*. From the perspective of the composition of gut microbial phyla, the variations within 7 days after release were not significant. However, there is a fluctuation in the dominant genus composition of gut microbes in *C. chilia* juveniles after release. The dominant genera of gut microbiota on the first day after release mainly included Firmicutes other, Microcystis PCC-7914, *Cetobacterium*, PeM15_unclassified genus, and Planktothrix NIVA-CYA 15. The members of Firmicutes and *Cetobacterium* perform digestion functions. For example, PeM15_unclassified genus is a member of the Actinobacteria phylum, which has been verified as an important group of PAOs (phosphorus-accumulating organisms) in enhanced biological phosphorus removal systems. Additionally, some members of this genus may contain nitrite reductase genes that are involved in denitrification [34]. Microcystis PCC-7914 and Planktothrix NIVA-CYA 15 are indicators of water body eutrophication, which can be used to assess the eutrophication levels in ponds used for fish rearing. Starting from the second day after release, the dominant genera continued to change, indicating that the *C. chilia* juveniles were undergoing a transitional period of adaptation to the new environment. Generally, the dominant genera after release were *Aeromonas* and *Romboutsia*. *Aeromonas* can secrete enzymes related to pathogenicity and environmental adaptation, such as hemolytic enterotoxin, lipase, protease, and amylase [35]. These enzymes improve the digestibility of food and play an important role in the digestion process. *Romboutsia* can ferment and metabolize macromolecular carbohydrates that are difficult for a host fish to digest into short-chain fatty acids, such as butyric acid. This process helps reduce intestinal pH, improve the host's immune regulation ability, and maintain the balance of intestinal microecology [36]. The increase in these two genera may be an adaptive strategy of *C. chilia* juveniles to cope with the increased risk of food scarcity and stress infection after release.

Only the 7-day continuous recapture collected the released fish successfully, and the reasons for the failure to obtain samples over a longer period (1 month and 3 months after release) are unknown. We speculate that the dispersal of *C. chilia* after release may lead to a continuous decrease in density, thereby reducing the probability of catching released fish, or perhaps they experienced high post-release mortality. However, the present results may be sufficient to reveal the initial post-release performance of cultured *C. chilia* juveniles.

5. Conclusions

Our study highlights the initial post-release performance of cultured *C. chilia* juveniles and examines the applicability of VIE tag, water depth selectivity, feeding adaptation to natural diet, and gut microbiota of cultured *C. chilia* juveniles after release. Overall, the VIE tag was a suitable method for short-term marking of *C. chilia* juveniles. Both the results of behavioral and feeding indicated that an obvious transitional period of cultured *C. chilia* juveniles was 4–5 days post-release. However, there are still many remaining questions that need to be studied in future, such as the performance of released *C. chilia* in a longer term, reasons for the failure of recapture after *C. chilia* being released for more than 7 days, and the habitat selection mechanism of the released *C. chilia*.

Based on the present results, we propose the following suggestions: Firstly, stocking of *C. chilia* juveniles should be conducted in waters with diverse habitats and a depth not exceeding 300 cm. Secondly, due to their limited movement velocity, the release sites of *C. chilia* juveniles should be multiple (see example [37]) and dispersed to avoid intense local resource competition. Thirdly, natural bait should be acclimated prior to release to shorten the time it takes for *C. chilia* juveniles to resume feeding after release.

Author Contributions: Conceptualization, T.Z., X.L. and D.Y.; methodology, T.Z., J.G. and D.Y.; software, T.Z. and X.L.; validation, T.Z. and X.L.; formal analysis, T.Z.; investigation, T.Z., J.G. and C.W.; data curation, J.G.; writing—original draft preparation, T.Z.; writing—review and editing, X.L. and D.Y.; supervision, C.W., X.M. and Y.L.; funding acquisition, T.Z. and D.Y. All authors have read and agreed to the published version of the manuscript.

Funding: This research was funded by National key research and development program of China (grant number 2020YFD09005), China Agriculture Research System of MOF and MARA (grant number CARS-46), and the Central Public Interest Scientific Institution Basal Research Fund, the Chinese Academy of Fishery Sciences (CAFS) (grant number 2020TD57).

Institutional Review Board Statement: All animal experiments in the present study were approved by the Institutional Animal Care and Use Committee of the Yangtze River Fisheries Research Institute, Chinese Academy of Fishery Sciences (approval no. 2022YFI-ZTB-01, 6 July 2022), and they were performed following the institutional ethical guidelines for experimental animals.

Informed Consent Statement: Not applicable.

Data Availability Statement: All DNA sequences were deposited in the NCBI Sequence Read Archive database with the accession number PRJNA999371.

Acknowledgments: The authors are grateful to Feifei HU, Kairun Liu, Lili Cui, Cuiliang Wang, Yi Sun, Yuanchao Zhang, and Zhongyuan Wang for their help in tagging and recapturing the experimental fish and sample collection.

Conflicts of Interest: The authors declare no conflict of interest.

References

1. Zhang, C.; Xu, B.; Xue, Y.; Ren, Y. Fisheries stock enhancement assessment, Progress and prospect. *J. Fish. China* **2022**, *46*, 1509–1524.
2. Molony, B.W.; Lenanton, R.; Jackson, G.; Norriss, J. Stock enhancement as a fisheries management tool. *Rev. Fish Biol. Fish.* **2003**, *13*, 409–432. [CrossRef]
3. Carrera-García, E.; Rochard, E.; Acolas, M. Effects of rearing practice on post-release young-of-the-year behavior: *Acipenser sturio* early life in freshwater. *Endanger. Species Res.* **2017**, *34*, 269–281. [CrossRef]
4. Taylor, M.D.; Laffan, S.W.; Fairfax, A.V.; Payne, N.L. Finding their way in the world, Using acoustic telemetry to evaluate relative movement patterns of hatchery-reared fish in the period following release. *Fish. Res.* **2017**, *186*, 538–543. [CrossRef]
5. Noda, T.; Wada, T.; Iwasaki, T.; Sato, T.; Narita, K.; Matsumoto, I.; Hori, T.; Mitamura, H.; Arai, N. Post-release behaviors and movements of cultured and wild Japanese eels (*Anguilla japonica*) in a shallow brackish water lagoon in northeastern Japan. *Environ. Biol. Fish.* **2019**, *102*, 1435–1456. [CrossRef]

6. Okuyama, J.; Shishidou, H.; Hayashibara, T. Post-release horizontal and vertical behavior and philopatry of deepwater longtail red snapper *Etelis coruscans* around a bank. *Fish. Sci.* **2019**, *85*, 361–368. [CrossRef]
7. Wu, J.; Li, F.; Du, H.; Zhang, H.; Wang, C.; Li, J.; Wei, Q. Movement of the matured Chinese sucker (*Myxocyprinus asiaticus*) after releasing in the Yangtze River. *J. Fish. Sci. China* **2019**, *26*, 1239–1246.
8. Costas, N.; Álvarez, M.; Pardo, I. Stocking efficiency and the effects of diet preconditioning on the post-release adaptation of hatchery-reared juveniles of Atlantic salmon (*Salmo salar* L.) in an Atlantic temperate stream. *Environ. Biol. Fish.* **2013**, *96*, 33–44. [CrossRef]
9. Hervas, S.; Kai, L.; Shane, M.A.; Drawbridge, M.A. Quantitative assessment of a white seabass (*Atractoscion nobilis*) stock enhancement program in California, Post-release dispersal, growth and survival. *Fish. Res.* **2010**, *105*, 237–243. [CrossRef]
10. Whitney, N.M.; White, C.F.; Gleiss, A.C.; Schwieterman, G.D.; Anderson, P.; Hueter, R.E.; Skomal, G.B. A novel method for determining post-release mortality, behavior, and recovery period using acceleration data loggers. *Fish. Res.* **2016**, *183*, 210–221. [CrossRef]
11. Munakata, A.; BjÖrnsson, B.T.; Jönsson, E.; Amano, M.; Ikuta, K.; Kitamura, S.; Kurokawa, T.; Aida, K. Post-release adaptation processes of hatchery-reared honmasu salmon parr. *J. Fish Biol.* **2000**, *56*, 163–172. [CrossRef]
12. Howell, B.R. Fitness of hatchery-reared fish for survival in the sea. *Aquac. Fish. Manag.* **1994**, *25*, 3–17.
13. Leber, K.M.; Brennan, N.P.; Arce, S.M. Recruitment patterns of juvenile, cultured pacific threadfin, *Polydactylus sexfilis* (Polynemidae), Released along sandy marine shores in Hawaii. *Bull. Mar. Sci.* **1998**, *62*, 389–408.
14. Zhu, T.; Hu, F.; Gong, J.; Du, H.; Yang, D.; Wu, X.; Li, X.; Sun, Y.; Liu, K.; Liang, Y.; et al. Initial post-release migration of hatchery-reared *Cyprinus pellegrini* juveniles in the Qilu Lake. *J. Fish. Sci. China* **2022**, *29*, 1358–1365.
15. Chu, X.; Chen, Y. *The Fishes of Yunnan, China (Part II Cprinidae)*; Science Press: Beijing, China, 1989.
16. Zhang, S.; Zhang, Y.; Yang, C.; Peng, J. Study on artificial domestication and breeding of *Cyprinus chilia* in Fuxian Lake. *Mod. Agric. Sci. Technol.* **2014**, 293–294. Available online: https://kns.cnki.net/KCMS/detail/detail.aspx?dbname=cjfd2014&filename=anhe201413176&dbcode=cjfq (accessed on 28 August 2023).
17. Yin, M. *Fish Ecology*; China Agricultural Press: Beijing, China, 1993.
18. Yang, J.; Chen, Y.; He, Y. Studies on fish diversity in palteau lakes of the central Yunnan. *Chin. Biodivers.* **1994**, *2*, 204–209.
19. Alcon-Giner, C.; Caim, S.; Mitra, S.; Ketskemety, J.; Wegmann, U.; Wain, J.; Belteki, G.; Clarke, P.; Hall, L.J. Optimisation of 16S rRNA gut microbiota profiling of extremely low birth weight infants. *BMC Genom.* **2017**, *18*, 841. [CrossRef]
20. Wear, E.K.; Wilbanks, E.G.; Nelson, C.E.; Carlson, C.A. Primer selection impacts specific population abundances but not community dynamics in a monthly time-series 16S rRNA gene amplicon analysis of coastal marine bacterioplankton. *Environ. Microbiol.* **2018**, *20*, 2709–2726. [CrossRef]
21. Ni, J.; Fu, C.; Huang, R.; Li, Z.; Li, S.; Cao, P.; Zhong, K.; Ge, M.; Gao, Y. Metabolic syndrome cannot mask the changes of faecal microbiota compositions caused by primary hepatocellular carcinoma. *Lett. Appl. Microbiol.* **2021**, *73*, 73–80. [CrossRef]
22. Edgar, R.C. UPARSE: Highly accurate OTU sequences from microbial amplicon reads. *Nat. Methods* **2013**, *10*, 996–998. [CrossRef]
23. Wang, Q.; Garrity, G.M.; Tiedje, J.M.; Cole, J.R. Naive Bayesian classifier for rapid assignment of rRNA sequences into the new bacterial taxonomy. *Appl. Environ. Microb.* **2007**, *73*, 5261–5267. [CrossRef] [PubMed]
24. Ferter, K.; Hartmann, K.; Kleiven, A.R.; Moland, E.; Olsen, E.M. Catch-and-release of Atlantic cod (*Gadus morhua*), post-release behaviour of acoustically pretagged fish in a natural marine environment. *Can. J. Fish. Aquat. Sci.* **2015**, *72*, 252–261. [CrossRef]
25. Hoolihan, J.P.; Luo, J.; Abascal, F.J.; Campana, S.E.; Metrio, G.D.; Dewar, H.; Domeier, M.L.; Howey, L.A.; Lutcavage, M.E.; Musyl, M.K.; et al. Evaluating post-release behaviour modification in large pelagic fish deployed with pop-up satellite archival tags. *J. Mar. Sci.* **2011**, *68*, 880–889. [CrossRef]
26. Wilson, A.D.M.; Hayden, T.A.; Vandergoot, C.S.; Kraus, R.T.; Dettmers, J.M.; Cooke, S.J.; Krueger, C.C. Do intracoelomic telemetry transmitters alter the post-release behaviour of migratory fish? *Ecol. Freshw. Fish* **2017**, *26*, 292–300. [CrossRef]
27. Watz, J. Structural complexity in the hatchery rearing environment affects activity, resting metabolic rate and post-release behaviour in brown trout *Salmo trutta*. *J. Fish Biol.* **2019**, *95*, 638–641. [CrossRef]
28. Ingram, T.R.; Tannehill, J.E.; Young, S.P. Post-release survival and behavior of adult shoal bass in the flint river, Georgia. *N. Am. J. Fish. Manag.* **2013**, *33*, 717–722. [CrossRef]
29. Wang, Z.; Chen, G.; Zeng, L. Study of fish behavior using acoustic fish tags and wireless tracker. *South China Fish. Sci.* **2018**, *14*, 51–59.
30. Larocque, S.M.; Lake, C.; Johnson, T.B.; Fisk, A.T. Patterns in spatial use of land-locked Atlantic salmon (*Salmo salar*) in a large lake. *J. Great Lakes Res.* **2022**, *48*, 381–391. [CrossRef]
31. Liu, Y.; Yang, C.; Shan, B.; Sun, D.; Liu, S.; Li, T.; Liu, M.; Xie, Q. Investigation of a mark-recapture method of black porgy, *Acanthopagrus schlegelii*, in Daya Bay using plastic oval tags. *J. Fish. Sci. China* **2019**, *26*, 63–70. [CrossRef]
32. Alexander, J.G.; Furgal, S.L.; Gorsky, D.; Marsden, J.E.; Biesinger, Z.F.; Lantry, B.F. Evaluation of post-stocking dispersal and mortality of juvenile lake trout Salvelinus namaycush in Lake Ontario using acoustic telemetry. *J. Great Lakes Res.* **2022**, *48*, 572–580.
33. Harris, C.; Brenden, T.O.; Vandergoot, C.S.; Faust, M.D.; Herbst, S.J.; Krueger, C.C. Tributary use and large-scale movements of grass carp in Lake Erie. *J. Great Lakes Res.* **2021**, *47*, 48–58. [CrossRef]
34. Rodríguez, E.; García-Encina, P.A.; Stams, A.J.M.; Maphosa, F.; Sousa, D.Z. Metaomics approaches to understand and improve wastewater treatment systems. *Rev. Environ. Sci. Biotechnol.* **2015**, *14*, 385–406. [CrossRef]

35. Pemberton, J.M.; Kidd, S.P.; Schmidt, R. Secreted enzymes of Aeromonas. *FEMS Microbiol. Lett.* **1997**, *152*, 1–10. [CrossRef] [PubMed]
36. Qin, R.; Wang, J.; Chao, C.; Yu, J.; Wang, S. RS5 produced more butyric acid through regulating the microbial community of human gut microbiota. *J. Agr. Food Chem.* **2021**, *69*, 3209–3218. [CrossRef] [PubMed]
37. Molony, B.W.; Church, A.R.; Cabassi, T. Tag loss and movement of stocked yearling rainbow trout (*Oncorhynchus mykiss*) in southwestern Australia. *N. Z. J. Mar. Freshw. Res.* **2014**, *48*, 60–69. [CrossRef]

animals

Article

The Food Niche Overlap and Interspecific Relationship between the Sympatric Tibetan Macaque and Grey Snub-Nosed Monkey

Li Yue [1,†], Cheng Wang [2,†], Bingshun Meng [1], Bo Xie [1], Heqin Cao [1], Haijun Su [1] and Mingming Zhang [1,*]

[1] College of Forestry, Research Center for Biodiversity and Nature Conservation, Guizhou University, Guiyang 550025, China
[2] Zhangjiajie Giant Salamander National Nature Reserve Affairs Center, Zhangjiajie 427400, China
* Correspondence: mmzhang@gzu.edu.cn; Tel.: +86-182-1200-7787
† These authors contributed equally to this work.

Simple Summary: The dietary habits of animals that inhabit the same region can reveal valuable information about their food composition, nutritional strategies, and competition for resources. Analysis of their dietary habits can provide insight into differences in food consumption, thus identifying the potential overlap and competition for resources. In this study, we used DNA metabarcoding to investigate the winter dietary habits of Tibetan macaques and grey snub-nosed monkeys through an analysis of 40 fecal samples. The results showed that Tibetan macaques consumed plants from 117 families and 184 genera, while grey snub-nosed monkeys consumed plants from 109 families and 165 genera. The aim of the research was to assess the winter dietary habits of both monkey species living in the same area, to increase our knowledge of their food preferences and its composition, and to reveal the possible relationship between the overlap of their food niche and interspecific relations, providing useful information for the conservation of the resources in their natural habitat.

Abstract: Assessing the trophic niche and interspecific relationships between related species and determining how the species maintain differences in nutritional niches while coexisting in the same area are important topics in ecological research. Therefore, exploring the mechanism of food resource utilization, competition and coexistence among species distributed in the same region is important. In this study, we used fecal samples and metagenome sequencing technology to study the plant feeding habits and coexistence mechanisms of Tibetan macaques (*Macaca thibetana*) and grey snub-nosed monkeys (*Rhinopithecus brelichi*) within the same area. In the winter of 2020, we collected a total of 40 fecal samples from Tibetan macaques and grey snub-nosed monkeys; of those, 29 samples were considered valid and were analyzed using DNA metabarcoding. The results showed that in winter, Tibetan macaques consumed plants from 117 families and 184 genera, whereas grey snub-nosed monkeys consumed plants from 109 families and 165 genera. Diversity analysis revealed that there was a significant difference in the food composition of Tibetan macaques and grey snub-nosed monkeys. Tibetan macaques had a broader food niche width than grey snub-nosed monkeys at the family and genus levels. In winter, the food niches of Tibetan macaques and grey snub-nosed monkeys almost entirely overlapped (0.99). Our research provides detailed dietary data for Tibetan macaques and grey snub-nosed monkeys and valuable information that can aid in conservation efforts targeting these species.

Keywords: diet composition; grey snub-nosed monkeys; DNA metabarcoding; food niche overlap; Tibetan macaques

Citation: Yue, L.; Wang, C.; Meng, B.; Xie, B.; Cao, H.; Su, H.; Zhang, M. The Food Niche Overlap and Interspecific Relationship between the Sympatric Tibetan Macaque and Grey Snub-Nosed Monkey. *Animals* **2023**, *13*, 2536. https://doi.org/10.3390/ani13152536

Academic Editors: Chunwang Li and Anna Nekaris

Received: 5 July 2023
Revised: 26 July 2023
Accepted: 4 August 2023
Published: 6 August 2023

1. Introduction

Competition for resources among sympatric species is a central research topic in community ecology [1]. Since they share habitat resources, sympatric species may only utilize a subset of all resources available in the area [2]. The fundamental concept of resource utilization posits that two species occupying the same n-dimensional niche are likely to experience intense competition [3]. The competitive exclusion principle further indicates that given limited resources, two completely competing species cannot coexist in the long term [4]. Therefore, to minimize overlap with other species, each sympatric species develops a unique niche during the long-term evolutionary process; however, the interplay of ecological requirements and conspecifics leads to complex interspecies relationships, where habitat resource differentiation is a crucial mechanism for the coexistence of sympatric species [3,5,6]. Niche differentiation and interspecific relationships have always intrigued researchers and attracted scholarly attention. To unravel interspecific relationships, some scholars have studied resource allocation related to the ecological niches of fish [7], carnivores [8], ungulates [9], and primates [10,11], among other taxa.

Diet plays a crucial role in shaping the nutritional ecological niche of animals, which devote considerable efforts to foraging for food that provides the energy necessary for their survival [1,12]. However, in winter, food resources become scarce, leading to competitive interactions among coexisting species [13]. Differences in dietary composition among sympatric species arise from the inevitable reduction in competition mechanisms [13]. At the same time, body size also represents a key competitive factor relevant to coexisting species when foraging for food [14]. Furthermore, animal foraging behavior is also influenced by nutrient richness and food palatability.

Traditional diet analysis is commonly conducted using a microscope to inspect the solid remains of food in the feces or stomach contents of animals or through direct observation documenting the food consumed [15]. Microscopic examination heavily relies on the skills of the observer to identify incompletely digested flora and fauna in animal feces or stomach contents. However, it may be challenging to implement direct observation for animals living in densely vegetated areas and those feeding on smaller organisms. The fecal metagenome technique utilized in this study has several advantages over the traditional morphological analysis of fecal contents. First, it is based on the gene level and is therefore unaffected by the external morphology, developmental stage or environmental conditions of the individual [16]. Second, it is efficient and has a high level of accuracy, enabling the classification of mixed samples from various individuals and species simultaneously, thus facilitating the identification of known or new species [17]. Third, the sequencing fragments utilized are usually short and of low quality, making it possible to analyze degraded DNA samples [18]. Last, deep sequencing enhances the detection rate of species with scant representation in mixed or environmental samples [18]. Due to its many advantages, the fecal metagenome has been utilized in dietary studies of various animals, including *Ficedula hypoleuca* [19], European catfish *Silurus glanis* [20], *Chlorocebus pygerythrus* [21], and *Discoglossus scovazzi* [22].

As a vital component of forest ecosystems, primates play a crucial role in maintaining the balance of the ecosystem [23]. To gain a comprehensive understanding of primates, studies are conducted on individual species to determine phylogenetics, migration, feeding patterns, population densities, and less frequently interspecific interactions [23]. However, the most commonly used method by researchers to obtain dietary data for primates is to collect their fecal samples in the wild and then analyze them using DNA metabarcoding. Previous studies have employed DNA metabarcoding technology to obtain comprehensive dietary data for species such as *Macaca fascicularis* [24], *Presbytis femoralis* [25], and lemur species [26]. By analyzing the diet data, we are able to enhance our understanding of the dietary diversity and variations among primate species and investigate their adaptation to environmental changes and threats in greater depth [27].

Fanjing Mountain is a critical area for the conservation of biodiversity and a crucial habitat for protected subtropical forest species in the upper reaches of the Yangtze River [28,29]. The area is inhabited by five species of nonhuman primates, including Tibetan macaques (*Macaca thibetana*) and grey snub-nosed monkeys (*Rhinopithecus brelichi*) ([30], pp. 444–445). Tibetan macaques, also known as red-faced monkeys, are classified as second-class protected animals in China and are widely distributed in central and southeastern China, including 49 counties and cities in Guizhou Province, China, such as Songtao, Yinjiang, Jiangkou, and Chishui [31]. Grey snub-nosed monkeys, also known as white shouldered snub-nosed monkeys or grey golden monkeys, are endemic to Guizhou, China [32], are critically endangered according to The International Union for Conservation of Nature Red List (IUCN Red List), and are listed as a first-class protected animal in China [33]. Fanjing Mountain is the sole global habitat of grey snub-nosed monkeys [34]. Current population data indicate that there are only approximately 700 grey snub-nosed monkeys remaining in the wild [35]. Scholars studying the population size, ecology, behavior, distribution, and diet of grey snub-nosed monkeys have provided qualitative information that can support the more efficient protection and management of this critically endangered species [36–38]. In the Fanjing Mountain Nature Reserve, Tibetan macaques primarily reside in evergreen broad-leaved forests and in evergreen and deciduous broad-leaved mixed forests at elevations between 400 and 1900 m. On the other hand, grey snub-nosed monkeys inhabit the evergreen broad-leaved mixed forests and the deciduous broad-leaved forests at elevations between 1500 and 2200 m, and their home range overlaps with that of Tibetan macaques [31]. Grey snub-nosed monkeys have a narrow distribution and a small population, whereas Tibetan macaques exhibit a more consistent population over time and space [39]. The home range of grey snub-nosed monkeys, which have a small population, is confined; however, Tibetan macaques have a wider home range than their grey snub-nosed counterparts, which allows them to acquire more food resources. The highly complex topography of Fanjingshan Mountain gives rise to a variable regional microclimate that results in a diverse ecological environment, with distinct forest community types present even in small, localized areas ([30], p. 3). Based on these facts, our research aims to address two main questions: (1) Are there any variations in the dietary composition between Tibetan macaques and grey snub-nosed monkeys? (2) Is there a distinct separation of the nutritional niche between Tibetan macaques and grey snub-nosed monkeys? Ultimately, we hypothesized the following: (1) there are differences in the food composition of Tibetan macaques and grey snub-nosed monkeys, and (2) there is a divergence in the food niches of Tibetan macaques and grey snub-nosed monkeys.

Using a DNA metagenomic approach, we studied the resource utilization and diets of Tibetan macaques and grey snub-nosed monkeys in the ecologically unique region of the Fanjing Mountains. Our aims, through the analysis of fecal samples of sympatric Tibetan macaques and grey snub-nosed monkeys, were (i) to understand their feeding preferences, (ii) to reveal feeding differences between them, and (iii) to identify food niche overlap and interspecific relationships.

2. Materials and Methods

2.1. Study Area and Fecal Collection

The Fanjing Mountain National Nature Reserve (FJMR) is located at 27°49′50″ to 28°1′130″ N, 108°45′55″ to 108°48′30″ E, covering an area of 419 km^2. Fanjing Mountain is the central peak of the Wuling Mountain range in the transition zone from the Yunnan Guizhou Plateau to the hills of western Hunan. The FJMR is one of the areas of highest protection priority in the forest biodiversity protection priority area in the upper reaches of the Yangtze River [29]. The complex topography and landform cause a changeable regional microclimate, leading to the emergence of different types of forest communities in some small areas [40]. The reserve is rich in animal and plant resources, with more than 210 species of rare and globally threatened wild plants, such as *Davidia involucrata* and *Abies fanjingshanensis*. There are more than 110 species of rare and globally threatened wild animals, such as *Viverricula indica*, *Syrmaticus ellioti* and *Capricornis milneedwardsii*.

We conducted a field study in January 2020, collecting fresh fecal samples from populations of Tibetan macaques and grey snub-nosed monkeys located in the northern FJMR (Figure 1). During sampling, disposable medical gloves were worn to place the samples into 100 mL sterile centrifuge tubes, which were then labeled and placed inside a self-sealing bag, and the GPS location, habitat type and other information were recorded.

To avoid the cross-contamination of samples, the gloves were changed every time a sample was taken. The fecal samples were soaked in 95% ethanol on the day of sampling, the ethanol was then poured out after soaking for 24 h, and the samples were transferred to silica gel for drying and preservation. The samples were returned to the laboratory and stored in a special low-temperature freezer at −80 °C.

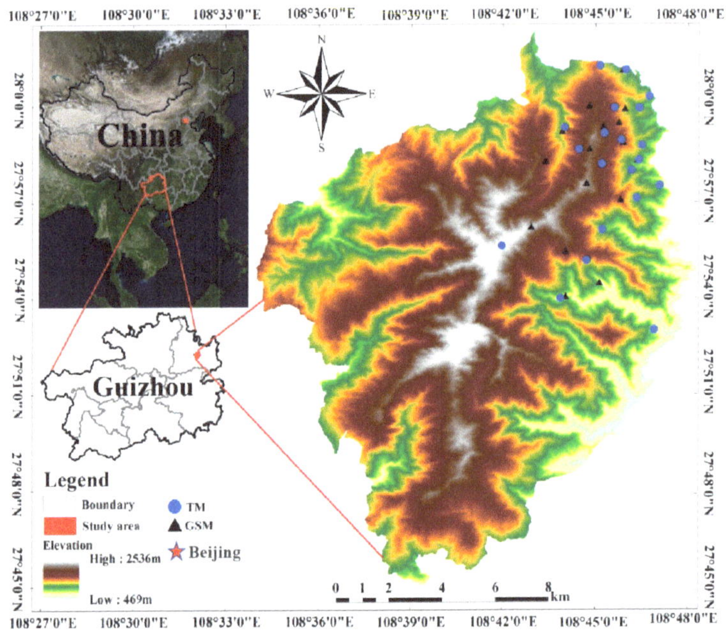

Figure 1. Feces sampling points for the study of grey snub-nosed monkeys and Tibetan macaques in Fanjing Mountain National Nature Reserve (GSM is an abbreviation for grey snub-nosed monkeys, TM is an abbreviation for Tibetan macaques).

2.2. DNA Extraction

The total DNA of the fecal samples was extracted using the 2CTAB/PCI method [41]. Each DNA extraction included approximately 100 mg of the outer surface for the molecular identification of the species. The confirmed fecal samples from Tibetan macaques and grey snub-nosed monkeys were homogenized, and 100 mg of homogenous feces was used for DNA extraction and molecular dietary analysis [42].

2.3. PCR Amplification for Fecal Species Identification

Fecal mitochondrial DNA 16S rRNA fragments were amplified using the following primers Z1aF: 5′-ATGTCACCACCAACAGAGACTAAAGC-3′; hp2R:5′-CGTCCTTTGTAA CGATCAAG-3′ [43]; COIintF: 5′-GGWACWGGWTGAACWGTWTAYCCYCC-3′ [44]; COIjg HCO2198: 5′-TANACYTCNGGRTGNCCRAARAAYCA-3′ [45]. The PCR amplification procedure was as follows: predenaturation at 95 °C for 5 min, denaturation at 95 °C for 30 s, annealing at 60 °C for 30 s, 30 cycles of extension at 72 °C for 40 s, and final extension at 72 °C for 7 min. After the reaction, 3 μL PCR products were used for 1% agarose gel electrophoresis to confirm the PCR-amplified fragments. The spliced sequence file was compared with the NCBI nucleic acid database data using the NCBI Blast program (https://blast.ncbi.nlm.nih.gov/Blast.cgi, accessed on 1 July 2023), and the species producing the fecal sample was identified when the similarity of the sequences was over 98%.

During the field survey at the FJMR, a total of 40 fecal samples were collected from Tibetan macaques and grey snub-nosed monkeys. However, after DNA extraction and

PCR amplification, it was found that 11 samples did not yield any food data. Therefore, we considered these 11 samples to be invalid. We obtained 29 samples that provided food data, including 18 from Tibetan macaques and 11 from grey snub-nosed monkeys (Table 1).

Table 1. Fecal sampling information.

Group Information	Sample Number
Tibetan macaques	DH1, DH2, DH3, DH4, LJB1, LJB2, EB2, KZB2, KZB4, DH_a, HTP_a, KZB_a, KZB_c, EB_b, LJB_a, LJB_b, LJB_c, LJB_d
Grey snub-nosed monkeys	BP1, BP2, EB1, KZB1, KZB3, BP_c, BP_e, KZB_b, KZB_d, KZB_e, EB_a

2.4. Amplicon Sequencing

Double-ended sequencing was performed using the Illumina HiSeq X Ten system with a read length of 150 bp at each end. The fecal DNA was purified using the AxyPrep DNA Gel Recovery Kit and Shanghai Personal Biotechnology Co. Ltd. sequenced the purified PCR products. After the splicing, quality control, deduplication, filtering, and chimera removal steps, the reads obtained via sequencing were divided into operational taxonomic units (OTUs) according to a classification level of 98% similarity, and OTU cluster analysis and species taxonomic analysis were performed according to the sample information [46]. Sequence species identification principles included the following: (1) when the sequence comparison results were consistently $\geq$ 98% and corresponded to only a single species, the sequence was considered to come from that species; (2) when the sequence comparison results were consistently $\geq$ 98% and corresponded to multiple species, the local undivided species were excluded, and if there were still multiple species, the identification results were recorded as the lowest taxonomic unit covering these species; (3) when the highest recorded consistency of the sequence comparison results was $\geq$95%, the identification results were recorded as the lowest taxonomic unit covering these species; (4) when the consistency of the sequence comparison result was <95%, the sequence could not be determined and was recorded as "unknown"; and (5) sequences with differences $\leq$2 were merged [42,46,47].

2.5. Food Niche Overlap Analysis

Relative read abundance (*RRA*) is the abundance of sequences in a food group as a percentage of the total food sequence in a valid sample and is used to reflect the relative biomass [48]. Therefore, we used *RRA* to measure the amount of food eaten by Tibetan macaques and grey snub-nosed monkeys [49]. The calculation formula is as follows:

$$RRA_i = \frac{1}{S}\sum_{j=1}^{S} \frac{N_{ij}}{\sum_{i=1}^{T} N_{ij}} \times 100\%$$

where S represents the total number of valid samples, T represents the number of food groups eaten, and $N_{i,j}$ represents the number of sequences of food group i in sample j.

The food overlap between Tibetan macaques and grey snub-nosed monkeys was calculated using the food niche width index and the Schoener food overlap index [47,50]; the calculation formulas are as follows:

Food niche width index:

$$B = 1/\sum P_i^2$$

where P_i represents the proportion of plants in a certain family/genus to the total number of all plant families/genera;

Schoener food overlap index:

$$D_{ij} = 1 - 0.5\left(\sum \left| P_{ik} - P_{jk} \right| \right)$$

where i and j denote endemic plants eaten by Tibetan macaques and grey snub-nosed monkeys, respectively, and k denotes plants eaten by Tibetan macaques and grey snub-nosed monkeys. P_{ik} and P_{jk} represent the proportion of endemic plants eaten by Tibetan macaques and grey snub-nosed monkeys, respectively. D_{ij} values are between 0 (no overlap) and 1 (complete overlap).

2.6. Statistical Analysis

Data were analyzed using IBM SPSS Software v.23. Alpha diversity (Chao1 index, Shannon index, Simpson index and richness index) was calculated using R (4.2.1) software to measure the species abundance and diversity of samples. The nonparametric Kruskal-Wallis rank-sum test for independent samples was used to analyze significant differences in alpha diversity between Tibetan macaques and grey snub-nosed monkeys. Based on the unweighted UniFrac distance, a similarity analysis was performed to test the significance of differences between groups, and a master coordinate analysis plot was generated.

3. Results

A total of 3,757,544 original sequences of the target fragment, with a band size of 420 bp, were obtained in 29 fecal samples from Tibetan macaques and grey snub-nosed monkeys. A total of 2,743,662 valid sequences were obtained through primer removal, splicing, mass filtration, deduplication, chimera removal, and clustering of the reads, of which 1,769,343 were from Tibetan macaques and 974,279 were from grey snub-nosed monkeys. The number of sequences obtained in the Tibetan macaques and grey snub-nosed monkeys samples ranged from 63,680 to 119,768, and from 74,853 to 108,268, respectively. The average number of effective sequences obtained from the Tibetan macaques and grey snub-nosed monkeys samples was 98,297 and 88,571, respectively, with an average length of 202 bp. There were 1067 and 1060 OTUs identified in the Tibetan macaques and grey snub-nosed monkeys samples, respectively, clustered with a similarity of 98%. Sorted from 29 samples, a total of 1075 OTUs accounted for 45.9% of the total, 882 OTUs from Tibetan macaques accounted for 37.7% of the total, and 385 OTUs from grey snub-nosed monkeys accounted for 16.4% of the total (Figure 2).

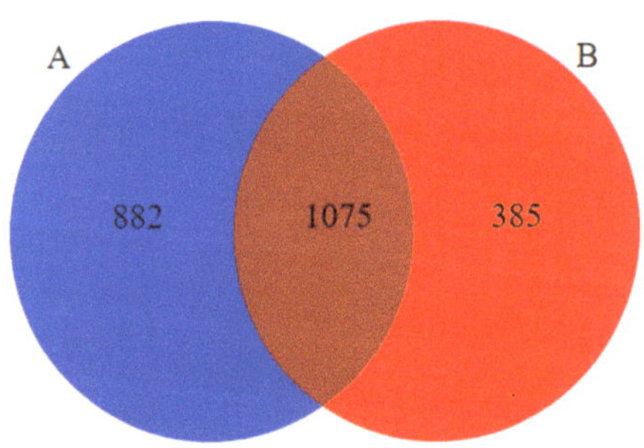

Figure 2. Venn diagram of OTUs between Tibetan macaques (A) and grey snub-nosed monkeys (B) for composition of food contained in the feces sample; the intersection between (A) and (B) represents the number of OTUs shared by Tibetan macaques and grey snub-nosed monkeys. (OTU, which stands for Operational Taxonomic Unit, is a method used in microbial ecology and biodiversity research. It involves grouping similar genetic data sequences together to represent a taxonomic unit. This method is commonly employed when the exact species or taxonomic identity is unknown or difficult to determine).

3.1. Food Composition of Tibetan Macaques and Grey Snub-Nosed Monkeys

A total of 63 orders were identified in 29 fecal samples. A total of 117 families and 184 genera were identified in the fecal samples of Tibetan macaques. At the family level, the preponderant food composition, accounting for more than 1% of the total, was

Pentaphylacaceae (3.26%), Rubiaceae (2.99%), Brassicaceae (2.64%) and Poaceae (1.28%) (Figure 3A), and at the genera level, it was *Eurya* (3.25%) and *Morinda* (1.27%) (Figure 3B); 165 genera of 109 families were identified in samples from grey snub-nosed monkeys (Attached Table 1), of which the families that were fed on and accounted for more than 1% of the total were Brassicaceae (2.60%), Lauraceae (1.59%) and Rubiaceae (1.25%) (Figure 3C), and the genera was *Morinda* (1.19%) (Figure 3D).

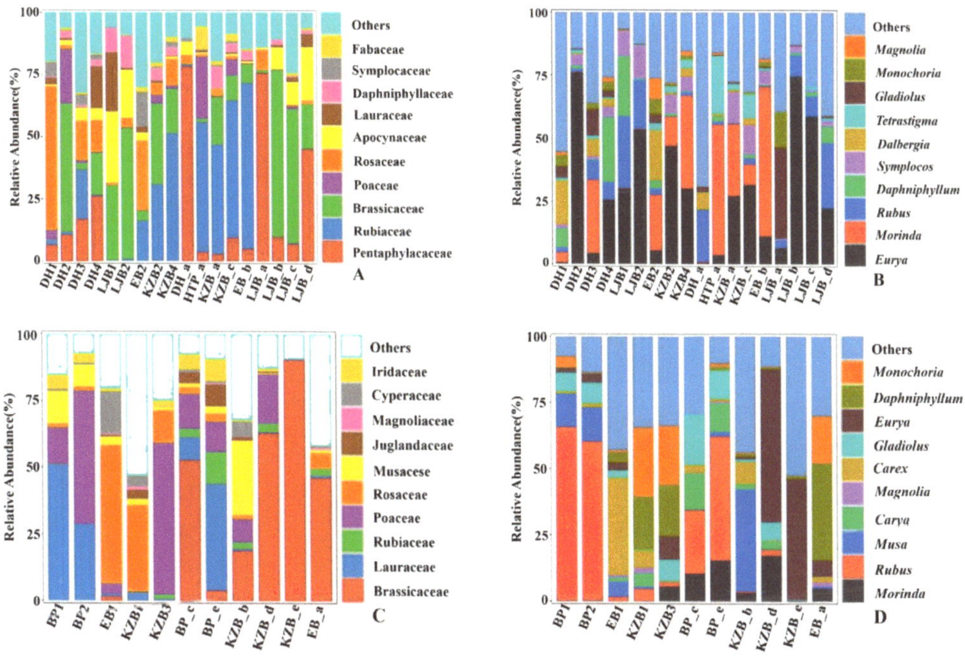

Figure 3. The plant food sources of Tibetan macaques and grey snub-nosed monkeys are among the top 10 in terms of abundance within their respective plant families and genera. (**A**,**B**) are at the family and genus levels of Tibetan macaques, and (**C**,**D**) are the families and genera foraged by grey snub-nosed monkeys, respectively.

3.2. Analysis of Food Diversity in Tibetan Macaques and Grey Snub-Nosed Monkeys

The α diversity index showed that there was no significant difference in the Chao1, Shannon, Simpson and richness indices between Tibetan macaques and grey snub-nosed monkeys ($p > 0.05$) (Figure 4). PERMANOVA-based β diversity analysis showed that there were significant differences in the diets of Tibetan macaques and grey snub-nosed monkeys ($p < 0.05$) (Figure 5).

3.3. Difference Analysis of Food in Tibetan Macaques and Grey Snub-Nosed Monkeys

The linear discriminant analysis effect size (LEfSe) showed that 33 plant compositions in the diets of Tibetan macaques and grey snub-nosed monkeys had discriminant characteristics. In the food composition of grey snub-nosed monkeys, there were five families and three genera of plants that differed from Tibetan macaques, with the family levels Musaceae, Papaveraceae, Hydrocharitaceaey, Linderniaceae and Pontederiaceae, and the genera levels *Musa*, *Hydrilla* and *Lindernia*. Among the dietary compositions of Tibetan macaques, there were 6 families and 8 genera that differed from grey snub-nosed monkeys, including the families Caryophyllaceae, Betulaceae, Platypodaceae, Lauraceae and Magnoliaceae, and the genera *Prunus*, *Spinosaurus*, *Ficus*, *Prunus*, *Astragalus*, *Cardamom*, *Viburnum* and *Pods* (Figure 6A); there were significant differences in nine genera, including

Tephroseris, *Cardamine* and *Stachyurus*, while the grey snub-nosed monkeys samples had significant differences in three genera, namely *Lindernia*, *Hydrilla* and *Musa* (Figure 6B).

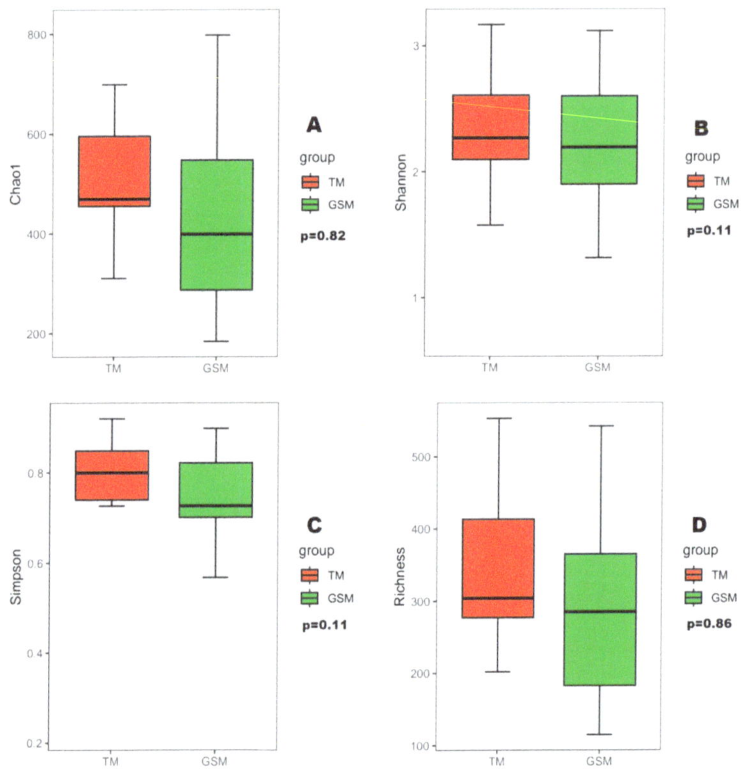

Figure 4. Alpha diversity, including Chao1 (**A**), Shannon (**B**), Simpson (**C**) and Richness (**D**) indices, between Tibetan macaques and grey snub-nosed monkeys. (GSM is an abbreviation for grey snub-nosed monkey, TM is an abbreviation for Tibetan macaque).

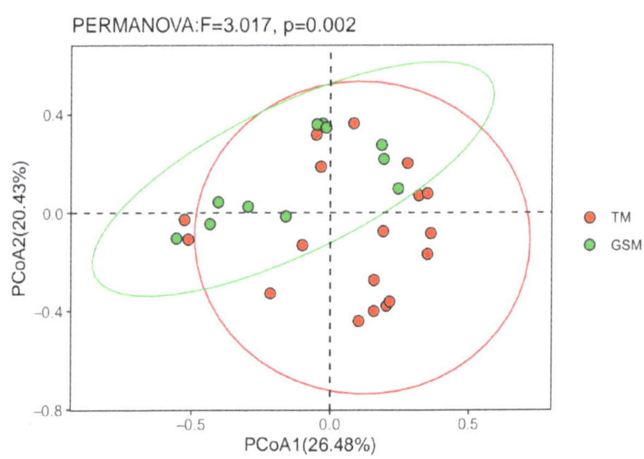

Figure 5. Beta diversity of Tibetan macaques and grey snub-nosed monkeys. (GSM is an abbreviation for grey snub-nosed monkey, TM is an abbreviation for Tibetan macaque).

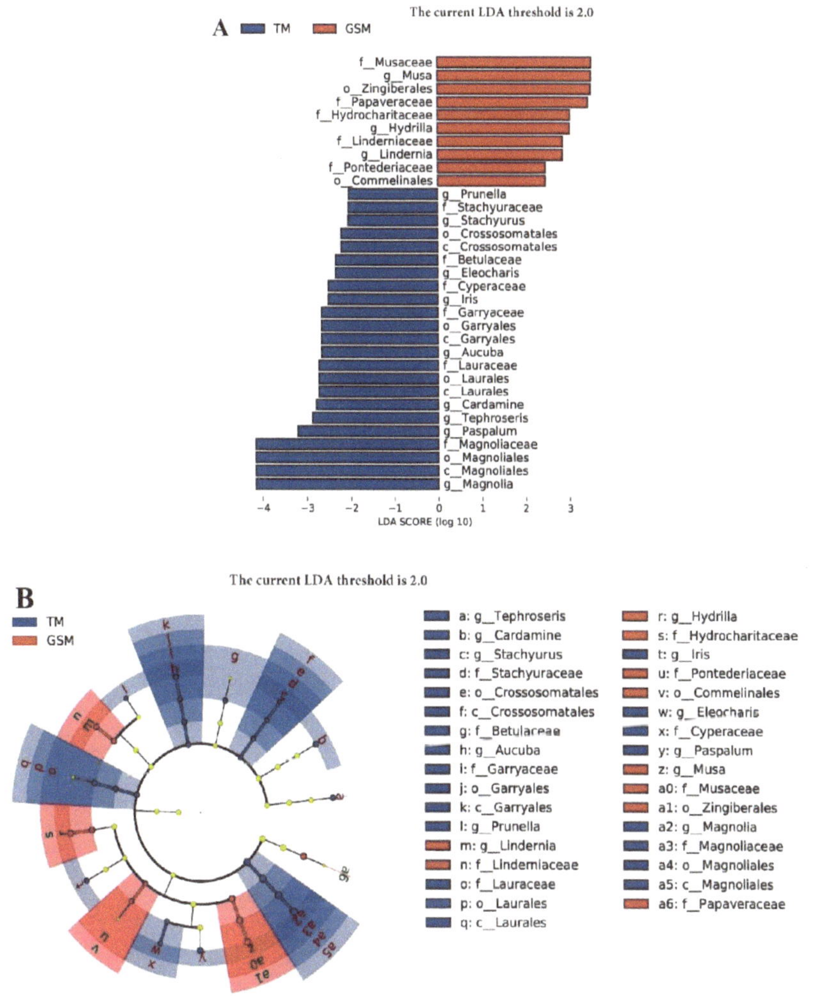

Figure 6. Linear discriminant analysis (LDA) integrated with effect size (LEfSe). (**A**) The differences in abundance between Tibetan macaque and the grey snub-nosed monkey groups. (**B**) Cladogram indicating the phylogenetic distribution of plants correlated with the Tibetan macaque and grey snub-nosed monkey groups (GSM is an abbreviation for grey snub-nosed monkey, TM is an abbreviation for Tibetan macaque).

3.4. Food Niche Overlap of Tibetan Macaques and Grey Snub-Nosed Monkeys

There were 8 families and 26 genera of endemic plants in the diet of Tibetan macaques, while there were 8 genera of endemic plants in the diet of grey snub-nosed monkeys. There were jointly 158 genera and 109 families in the diets of Tibetan macaques and grey snub-nosed monkeys (Figure 7).

At the family and genus levels, the food niche width of Tibetan macaques was wider than that of grey snub-nosed monkeys. At the family and genus levels, the food overlap of Tibetan macaques and grey snub-nosed monkeys reached almost 1 (Table 2).

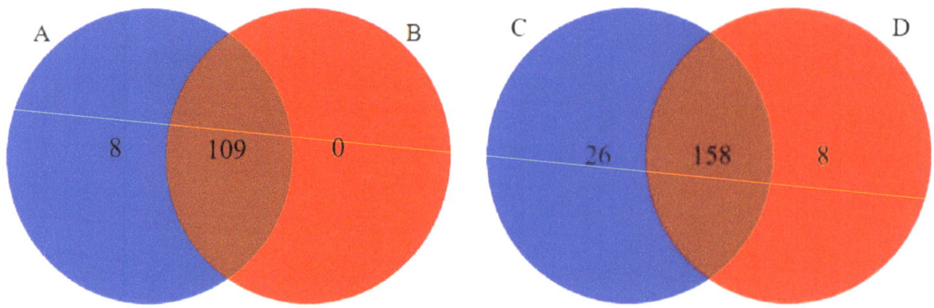

Figure 7. Tibetan macaque and grey snub-nosed monkey unique food with shared food Venn diagram. (A) is a family specific to Tibetan macaques, (B) is a family specific to grey snub-nosed monkeys, (C) is a genus specific to Tibetan macaques, (D) is a genus specific to grey snub-nosed monkeys.

Table 2. Comparison of food diversity index and niche width between Tibetan macaques and grey snub-nosed monkeys.

	B		D_{ij}	
	Family	**Genus**	**Family**	**Genus**
Tibetan macaques	9.3865	10.2150	0.9998	0.9930
Grey snub-nosed monkeys	8.1951	9.2386		

4. Discussion

4.1. Diet of the Two Primates

In this study, the winter diets of Tibetan macaques and grey snub-nosed monkeys, two primates with overlapping distributions at Fanjingshan, were investigated. The dietary preferences of the two primates can be inferred from their respective diets. We found that Tibetan macaques at Fanjingshan had a more varied diet than those found in the Anhui province of China [12]. Our study showed that the dietary composition of grey snub-nosed monkeys differed from the results of previous research on this topic [35,38,51]. Nonetheless, given the disparity in research methods, it is not possible to make a direct comparison of the plant species consumed in each primate study. The differences between these results and those of our study are likely related to differences in research methods. Previous research has shown that fecal samples can be effectively employed for metagenomic sequencing, resulting in the rapid acquisition of dietary data in animals with complex diets [25]. Consequently, the metagenomic analysis of fecal samples has emerged as a popular method in animal dietary research, particularly for species with complex dietary patterns such as fish [7], primates [52,53], carnivores [8], and ungulates [9].

Liu et al. [12] used quick scanning sampling to study Tibetan macaques in Huangshan city, Anhui Province, and found that Tibetan macaques fed on the plants of 23 families and 31 genera in winter. Nie et al. [51] reported that grey snub-nosed monkeys feed on 49 genera of plants in 25 families; Guo et al. [54] reported that grey snub-nosed monkeys feed on 31 families and 51 genera; Xiang et al. [50] reported that the diet of grey snub-nosed monkeys contains 107 plants from 28 families and 58 genera, and includes mainly mature leaves and buds in winter. However, through the use of the DNA metabolism barcoding analysis method, we found that the diet of Tibetan macaques consists of food items from 117 families and 184 genera, while the diet of grey snub-nosed monkeys includes food items from 109 families and 165 genera. This is slightly different from the results of this paper. The feeding behaviors of Tibetan macaques and grey snub-nosed monkeys have been examined by researchers using the instantaneous scanning sampling method. Nonetheless, this method is impacted by both internal and external factors at the study site, such as the geographical location of observation points and vegetation distribution, leading researchers to record solely the tall trees and shrubs the Tibetan macaques and grey snub-nosed monkeys consume, thus limiting observations in areas with dense vegetation and low herb coverage. DNA metabarcoding analysis provides a more comprehensive approach than direct observation in food studies [55].

By employing DNA metagenomic sequencing, this study benefited from efficient and accurate deep sequencing, resulting in the acquisition of more comprehensive dietary data than in previous studies. However, identifying specific edible plant materials consumed by Tibetan macaques and grey snub-nosed monkeys remains challenging, despite the effectiveness of DNA metagenomic sequencing for accurately deep-sequencing fecal contents. To comprehensively study animal diets, the optimal approach in future research is to combine direct observation with DNA microbiome sequencing. This will enable the comprehensive acquisition of diet data, which is essential to improving wildlife protection.

4.2. Food Diversity in the Two Primates

The alpha diversity analysis of the diets of Tibetan macaques and grey snub-nosed monkeys indicated no significant difference between their diets. However, beta diversity analysis and linear discriminant analysis effect size (LEfSe) suggested that Tibetan macaques consume a wider variety of food than grey snub-nosed monkeys.This is consistent with our hypothesis (1). The availability of plant species that animals consume is primarily limited by the distribution of vegetation in their habitat [56]. This view is supported by studies on *Indri indri* [57], Macaca [58] and *Cercopithecus mitis kandti* [59], which also adapt their diets to available food resources in their environment. Tibetan macaques are distributed in forest vegetation types at elevations ranging from 700 to 2400 m, while grey snub-nosed monkeys prefer forest vegetation at altitudes ranging from 1600 to 1900 m [60]. Tibetan macaques have a broader habitat range and thus more access to a wider range of food sources, enabling them to eat a greater variety of vegetation and food types than grey snub-nosed monkeys.

In primates, the availability of food resources within the home range, as well as their body size, limits their diet. The body size of primates not only restricts the types of food they can consume, but also influences their preference for concentrated or dispersed food resources [61,62]. Larger primates tend to have a lower quality and variety of food [63]. Tibetan macaques and grey snub-nosed monkeys are two species of primates with comparable body lengths of 51.0 to 71.0 cm, and 63.7 to 69.0 cm [31,64], respectively. However, Tibetan macaques weigh between 12.5 to 20 kg, while grey snub-nosed monkeys weigh only 13.25 to 15.75 kg [31,65]. Together, they consumed 158 genera of plants belonging to 109 families, with only a few being exclusively consumed by one species. Tibetan macaques fed on 8 families and 26 genera, while grey snub-nosed monkeys fed on 8 genera only. Although both species had similar food preferences, their specific diet compositions differed. Nevertheless, since there is no significant difference in the body shape of Tibetan macaques and grey snub-nosed monkeys, the diet compositions of the two species are also not significantly different, except for the exclusive foods particular to each species.

Previous research has shown that primates possess selectivity toward specific types of food based on their nutritional composition [66]. The quality, quantity, composition, and water content of food also affect the food choices of primates [67]. When high-nutrition value food is in abundance, predators choose it over other low-nutrition food items. Even when multiple food items are available, high-nutrition-value food remains the first choice of predators, with little consideration for low-nutrition food items [68]. Primates have nutritional needs that must be met for life activities, such as metabolism; they primarily consume plant-based foods rich in protein, fat, carbohydrates, water, vitamins, and trace elements. A study by Li et al. [69] found that the nutritional composition of most plants varies throughout the year and *Rhinopithecus roxellana* preferred plants with a high protein content and a low cellulose content. In winter, plants with abundances greater than 1 may also be higher in nutritional value; yet, these plants contribute little nutrition to the diet of Tibetan macaques and grey snub-nosed monkeys as these primates feed primarily on a small subset of staple foods. As proposed by the theory of nutritional balance, animals balance their nutrition by mixing different foods and nutrients rather than maximizing the availability of a single nutrient [70]. In winter, food resources are scarce, and to obtain sufficient energy, Tibetan macaques and grey snub-nosed monkeys consume a significant amount of foods with a low nutritional value, leading to a relatively high proportion of food items with an occurrence rate of less than 1 in their respective diets. Since the nutritional components of the primates' diets were not analyzed, inferences about their foraging strategies were based solely on the type and ratio of plant species they consumed. Therefore, studying the nutritional composition of their diet will be the primary focus of our future research.

4.3. Coexistence Mechanism of the Two Primates

The niche width index, which reflects the proportion of species to spatial resources, spatial distribution range, and uniformity [71], was used in this study. The results of this study indicated that Tibetan macaques have a greater advantage in resource competition than grey snub-nosed monkeys since the niche width index of Tibetan macaques was higher. Wang et al. [72] discovered that winter Tibetan macaques and grey snub-nosed monkeys had different temporal and spatial niches, which could minimize the competition for space. Nevertheless, in our study, we found that Tibetan macaques and grey snub-nosed monkeys had almost identical diets, with a food overlap index close to 1. This finding contradicts our original hypothesis (2), which suggests that the coexistence mechanism between the two primates occurs via the differentiation of their temporal and spatial niches rather than via specialization in trophic niches. The results of our study regarding trophic niche overlap contrasted with those from studies focusing on coexisting lowland gorillas and chimpanzees in Gabon and studies on *Saguinus mystax* and *Saguinus fuscicollis* in the Amazon. The different species demonstrate distinct foraging strategies, leading to niche separation [73,74]. Scholars have utilized DNA metagenomics to investigate the dietary overlap between sika deer and wild buffalo within the Northeast China Tiger and Leopard National Park, and the findings indicated a significant degree of dietary convergence in both species [75]. Notably, food resources vary across different study areas. Additionally, each species has a unique foraging preference, which means that different research results are possible.

To a significant extent, the coexistence of species relies on the resource differentiation among them. According to the competitive exclusion theory, coexisting species in the same habitat require exclusive temporal, spatial, and food niches in order to obtain sufficient resources for survival and reproduction [4,76]. While the winter trophic niches of Tibetan macaques and grey snub-nosed monkeys distributed in the FJMR were almost entirely overlapping, the niche breadth of both species throughout other seasons remains unknown. Therefore, it is necessary to conduct additional studies to determine whether feeding niches diverge in spring, summer, and autumn.

5. Conclusions

In this study, we utilized DNA metabarcoding to explore the diets of Tibetan macaques and grey snub-nosed monkeys. The results indicated differences in their food consumption. Tibetan macaques had a higher food niche width index than grey snub-nosed monkeys; however, both had a high trophic niche overlap. Therefore, trophic niche differentiation is not the mechanism by which Tibetan macaques and grey snub-nosed monkeys coexist.

Author Contributions: Conceptualization, H.S. and M.Z.; methodology, L.Y. and C.W.; software, C.W.; validation, L.Y.; formal analysis, B.M.; investigation, C.W., B.M. and B.X.; resources, C.W.; data curation, L.Y.; writing—original draft preparation, L.Y.; writing—review and editing M.Z. and H.C.; visualization, L.Y.; supervision, M.Z.; project administration, M.Z.; funding acquisition, M.Z. All authors have read and agreed to the published version of the manuscript.

Funding: This research was funded by the Guizhou Provincial Science and Technology Foundation (Grant No. [2019]2842) and the National Natural Science Foundation of China (Grant No. 32260331).

Institutional Review Board Statement: Ethical review and approval were waived for this study because the sampling did not involve the hunting of experimental animals.

Informed Consent Statement: Not applicable.

Data Availability Statement: The data presented in this study are available upon request from the corresponding author.

Acknowledgments: We sincerely thank Lei Shi, Haibo Li, Wei Ran, Yaqiang Wang, and Xiangwen Hou from the Fanjing Mountain National Nature Reserve Administration of Guizhou Province for helping us to collect experimental samples in this study.

Conflicts of Interest: The authors declare no conflict of interest.

References

1. Wang, G.L.; Yin, H.B. Advances on foraging behavioral ecology in nonhuman primates. *J. Biol.* **2008**, *5*, 10–12.
2. Hutchinson, G.E. Concluding Remarks. *Cold Spring Harb. Lab. Press* **1957**, *22*, 415–427. [CrossRef]
3. Chillo, V.; Rodríguez, D.; Ojeda, R.A. Niche partitioning and coexistence between two mammalian herbivores in the Dry Chaco of Argentina. *Acta Oecologica* **2010**, *36*, 611–616. [CrossRef]
4. Hardin, G. The Competitive Exclusion Principle. *Science* **1960**, *131*, 1292–1297. [CrossRef]
5. Jonathan, M.; Levine, C.M.D. Elton revisited: A review of evidence linking diversity and invasibility. *Oikos* **1999**, *87*, 15–26. [CrossRef]
6. Chu, C.J.; Wang, Y.S.; Liu, Y.; Jiang, L.; He, F.L. Advances in species coexistence theory. *Biodivers. Sci.* **2017**, *25*, 345–354. [CrossRef]
7. Eloranta, A.P.; Finstad, A.G.; Sandlund, O.T.; Knudsen, R.; Kuparinen, A.; Amundsen, P.A. Species interactions, environmental gradients and body size shape population niche width. *J. Anim. Ecol.* **2022**, *91*, 154–169. [CrossRef]
8. Lanszki, Z.; Horváth, G.F.; Bende, Z.; Lanszki, J. Differences in the diet and trophic niche of three sympatric carnivores in a marshland. *Mammal Res.* **2020**, *65*, 93–104. [CrossRef]
9. Namgail, T.; Mishra, C.; de Jong, C.B.; van Wieren, S.E.; Prins, H.H.T. Effects of herbivore species richness on the niche dynamics and distribution of blue sheep in the Trans-Himalaya. *Divers. Distrib.* **2009**, *15*, 940–947. [CrossRef]
10. Hadi, S.; Ziegler, T.; Waltert, M.; Syamsuri, F.; Mühlenberg, M.; Hodges, J.K. Habitat Use and Trophic Niche Overlap of Two Sympatric Colobines, Presbytis potenziani and Simias concolor, on Siberut Island, Indonesia. *Int. J. Primatol.* **2012**, *33*, 218–232. [CrossRef]
11. Zhou, Q.; Huang, Z.; Wei, H.; Huang, C.; Zhou, Q.; Huang, Z.; Wei, H.; Huang, C. Variations in diet composition of sympatric Trachypithecus francoisi and Macaca assamensis in the limestone habitats of Nonggang, China. *Zool. Res.* **2018**, *39*, 284–290. [CrossRef] [PubMed]
12. Liu, C.; Li, W.B.; Li, B.W.; Yang, P.P.; Xia, D.P.; Li, J.H. Food composition and nutrients of wild Tibetan macaques (*Macaca thibetana*) in winter. *Acta Theriol. Sin.* **2021**, *41*, 649–657.
13. Ribeiro, L.B.; Freire, E.M.X. Trophic ecology and foraging behavior of Tropidurus hispidus and Tropidurus semitaeniatus (Squamata, Tropiduridae) in a caatinga area of northeastern Brazil. *Série Zool.* **2011**, *101*, 225–232. [CrossRef]
14. Juri, G.L.; Naretto, S.; Mateos, A.C.; Chiaraviglio, M.; Cardozo, G. Influence of Life History Traits on Trophic Niche Segregation between Two Similar Sympatric Tupinambis Lizards. *S. Am. J. Herpetol.* **2015**, *10*, 132–142. [CrossRef]
15. Pompanon, F.; Deagle, B.E.; Symondson, W.O.C.; Brown, D.S.; Jarman, S.N.; Taberlet, P. Who is eating what: Diet assessment using next generation sequencing. *Mol. Ecol.* **2012**, *21*, 1931–1950. [CrossRef] [PubMed]
16. Guo, Y.P.; Zhang, H.; Zhao, X.G.; Luo, H.L.; Zhang, Y.J. Applications of DNA Metabarcoding in Diet Identification of Herbivores. *Biotechnol. Bull.* **2021**, *37*, 252–260. [CrossRef]
17. Liu, G.; Ning, Y.; Xia, X.F.; Gong, M.H. The application of high-throughput sequencing technologies to wildlife diet analysis. *Acta Ecol. Sin.* **2018**, *38*, 3347–3356.
18. Tillmar, A.O.; Dell'Amico, B.; Welander, J.; Holmlund, G. A Universal Method for Species Identification of Mammals Utilizing Next Generation Sequencing for the Analysis of DNA Mixtures. *PLoS ONE* **2013**, *8*, e83761. [CrossRef]
19. Verkuil, Y.I.; Nicolaus, M.; Ubels, R.; Dietz, M.W.; Samplonius, J.M.; Galema, A.; Kiekebos, K.; de Knijff, P.; Both, C. DNA metabarcoding quantifies the relative biomass of arthropod taxa in songbird diets: Validation with camera-recorded diets. *Ecol. Evol.* **2022**, *5*, 12.
20. Guillerault, N.; Bouletreau, S.; Iribar, A.; Valentini, A.; Santoul, F. Application of DNA metabarcoding on faeces to identify European catfishSilurus glanis diet. *J. Fish Biol.* **2017**, *90*, 2214–2219. [CrossRef]
21. Brun, L.; Schneider, J.; Carrió, E.M.; Dongre, P.; Taberlet, P.; Waal, V.D.; Fumagalli, L. Focal vs. fecal: Seasonal variation in the diet of wild vervet monkeys from observational and DNA metabarcoding data. *Ecol. Evol.* **2022**, *10*, 12. [CrossRef] [PubMed]
22. Pereira, A.; Samlali, M.A.; Khifa, A.S.; Slimani, T.; Harris, D.J. A pilot study on the use of DNA metabarcoding for diet analysis in a montane amphibian population from North Africa. *Afr. J. Herpetol.* **2021**, *70*, 68–74. [CrossRef]

23. Norconk, M.A.; Boinski, S.; Forget, P. Primates in 21st century ecosystems: Does primate conservation promote ecosystem conservation? *Am. J. Primatol.* **2011**, *73*, 3–8. [CrossRef]
24. Osman, N.A.; Abdul-Latiff, M.A.B.; Mohd-Ridwan, A.R.; Yaakop, S.; Karuppannan, K.V.; Md-Zain, B.M.; Yaakop, K.V.K.B. Metabarcoding data analysis revealed the plant dietary variation of long-tailed macaque Macaca fascicularis (Cercopithecidae, Cercopithecinae) living in disturbed habitats in Peninsular Malaysia. *Biodivers. Data J.* **2022**, *10*, e89617. [CrossRef]
25. Srivathsan, A.; Ang, A.; Vogler, A.P.; Meier, R. Fecal metagenomics for the simultaneous assessment of diet, parasites, and population genetics of an understudied primate. *Front. Zool.* **2016**, *13*, 17. [CrossRef]
26. Rowe, A.K.; Donohue, M.E.; Clare, E.L.; Drinkwater, R.; Koenig, A.; Ridgway, Z.M.; Martin, L.D.; Nomenjanahary, E.S.; Zakamanana, F.; Randriamanandaza, L.J.; et al. Exploratory analysis reveals arthropod consumption in 10 lemur species using DNA metabarcoding. *Am. J. Primatol.* **2021**, *83*, e23256. [CrossRef] [PubMed]
27. Machado, F.F.; Jardim, L.; Dinnage, R.; Brito, D.; Cardillo, M. Diet disparity and diversity predict extinction risk in primates. *Anim. Conserv.* **2023**, *26*, 331–339. [CrossRef]
28. Xia, W.P.; Zhang, Y.Z. *Primae Research and Conservation*; China Forestry Publishing House: Beijing, China, 1995; pp. 150–151.
29. Zhu, C.Q.; Li, D.Q.; Dong, K.; Wang, X.L.; Shi, P.L. Setting biodiversity conservation priorities in the Forests of the Upper Yangtze Ecoregion based on ecoregion conservation methodology. *Biodivers. Sci.* **2006**, *2*, 87–97.
30. Zhou, X.Z.; Yang, Y.Q. *Fanjingshan Research*; Guizhou People's Publishing House: Guiyang, China, 1990; pp. 3, 444–445.
31. Luo, R. *The Mammalian Fauna of Guizhou*; Guizhou Science and Technology Publishing House: Guiyang, China, 1993; pp. 155–162.
32. Yang, Y.Q.; Lei, X.P.; Yang, C.D. *Fanjingshan Research*; Guizhou Science and Technology Publishing House: Guiyang, China, 2002; p. 6.
33. IUCN Red List of Threatened Species. Available online: https://www.iucnredlist.org/en (accessed on 12 February 2023).
34. Cui, D.Y.; Niu, K.F.; Tan, C.L.; Yang, M.; Shi, L.; Zhang, W.Y.; Zhang, J.G.; Yang, Y.Q. Golden snub-nosed monkey distributed in Fanjing Mountain, Guizhou, China. *Chin. J. Zool.* **2014**, *49*, 792. [CrossRef]
35. Guo, Y.Q.; Zhou, J.; Song, X.H.; Deng, H.Q.; Qiu, Y.; Shi, L.; Zhou, J. The population of *Rhinopithecus brelichi* in Fanjingshan National Nature Reserve, Guizhou, China. *Acta Theriol. Sin.* **2017**, *37*, 104–108. [CrossRef]
36. Bleisch, W.; Cheng, A.S.; Ren, X.D.; Xie, J.H. Preliminary results from a field study of wild Guizhou snub-nosed monkeys (Rhinopithecus brelichi). *Folia Primatol.* **1993**, *60*, 72–82. [CrossRef] [PubMed]
37. Bleisch, W.; Xie, J.H. Ecology and behavior of the Guizhou snub-nosed langur (*Rhinopithecus brelichi*). In *The Natural History of the Doucs and Snub-Nosed Monkeys*; World Scientific: Singapore, 1998; pp. 217–239. [CrossRef]
38. Xiang, Z.; Liang, W.; Nie, S.; Li, M. Diet and Feeding Behavior of *Rhinopithecus brelichi* at Yangaoping, Guizhou. *Am. J. Primatol.* **2012**, *74*, 551–560. [CrossRef] [PubMed]
39. Chang, Y.; Zhang, M.M.; Liu, Z.S.; Hu, T.H.; Li, Z.G. Summer diets of Sympatric Blue Sheep (*Pseudois nayaur*) and Red Deer (*Cervus elaphus alxaicus*) in the Helan Mountains, China. *Acta Ecol. Sin.* **2010**, *30*, 1486–1493.
40. Zhou, Z.X.; Yang, Y.Q. *Research on Fanjing Mountain*; Guizhou People's Publishing House: Guiyang, China, 1990; pp. 95–96.
41. Vallet, D.; Petit, E.J.; Gatti, S.; Levréro, F.; Ménard, N. A new 2CTAB/PCI method improves DNA amplification success from faeces of Mediterranean (*Barbary macaques*) and tropical (lowland gorillas) primates. *Conserv. Genet.* **2008**, *9*, 677–680. [CrossRef]
42. Xiong, M.Y.; Wang, D.J.; Bu, H.L.; Shao, X.N.; Zhang, D.; Li, S.; Wang, R.J.; Yao, M. Molecular dietary analysis of two sympatric felids in the Mountains of Southwest China biodiversity hotspot and conservation implications. *Sci. Rep.* **2017**, *7*, 41909. [CrossRef]
43. Hofreiter, M.; Poinar, H.N.; Spaulding, W.G.; Bauer, K.; Martin, P.S.; Possnert, G.; Paabo, S. A molecular analysis of ground sloth diet through the last glaciation. *Mol. Ecol.* **2000**, *9*, 1975–1984. [CrossRef]
44. Geller, J.; Meyer, C.; Parker, M.; Hawk, H. Redesign of PCR primers for mitochondrial cytochrome c oxidase subunit I for marine invertebrates and application inall-taxabioticsurveys. *Wiley Online Libr.* **2013**, *13*, 851–861. [CrossRef]
45. Leray, M.; Yang, J.Y.; Meyer, C.P.; Mills, S.C.; Agudelo, N.; Ranwez, V.; Boehm, J.T.; Machida, R.J. A new versatile primer set targeting a short fragment of the mitochondrial COI region for metabarcoding metazoan diversity: Application for characterizing coral reef fish gut contents. *Front. Zool.* **2013**, *10*, 34. [CrossRef]
46. Edgar, R.C. UPARSE: Highly accurate OTU sequences from microbial amplicon reads. *Nat. Methods* **2013**, *10*, 996–998. [CrossRef]
47. Shao, X.N.; Song, D.Z.; Huang, Q.W.; Li, S.; Yao, M. Fast surveys and molecular diet analysis of carnivores based on fecal DNA and Fast surveys and molecular diet analysis of carnivores based on fecal DNA and metabarcoding. *Biodivers. Sci.* **2019**, *27*, 543–556.
48. Cao, H.; Yang, X.; Peng, C.; Wang, Y.; Guo, Q.; Su, H.J. Winter Dietary Analysis Reveals the Foraging Differences of Wild Boar (*Sus scrofa*) in Different Regions of a Karst Mountainous Area. *Animals* **2023**, *13*, 727. [CrossRef] [PubMed]
49. Deagle, B.E.; Thomas, A.C.; McInnes, J.C.; Clarke, L.J.; Vesterinen, E.J.; Clare, E.L.; Kartzinel, T.R.; Eveson, J.P. Counting with DNA in metabarcoding studies: How should we convert sequence reads to dietary data? *Mol. Ecol.* **2019**, *28*, 391–406. [CrossRef] [PubMed]
50. Cui, D.Y.; Liu, Z.S.; Xiao, M.; Zhai, H.; Hu, T.H.; Li, Z.G. Winter Food-habits of Red Deer (*Cervus elaphus alxaicus*) in Helan Mountains, China. *Zool. Res.* **2007**, *04*, 383–388.
51. Nie, S.G.; Xiang, Z.F.; Li, M. Preliminary report on the diet and social structure of gray snub-nosed monkeys (*Rhinopithecus brelichi*) at Yangaoping, Guizhou, China. *Acta Theriol. Sin.* **2009**, *29*, 326–331.
52. Tilson, R.L.; Tenaza, R.R. Interspecific Spacing Between Gibbons (*Hylobates klossii*) and Langurs (*Presbytis potenziani*) on Siberut Island, Indonesia. *Am. J. Primatol.* **1982**, *2*, 355–361. [CrossRef]

53. Osman, N.A.; Abdul-Latiff, M.A.B.; Mohd-Ridwan, A.R.; Yaakop, S.; Nor, S.M.; Md-Zain, B.M. Diet Composition of the Wild Stump-Tailed Macaque (*Macaca arctoides*) in Perlis State Park, Peninsular Malaysia, Using a Chloroplast tRNL DNA Metabarcoding Approach: A Preliminary Study. *Animals* **2020**, *10*, 2215. [CrossRef]
54. Guo, Y.; Zhou, J.; Xie, J.; Garber, P.A.; Bruford, M.; Ren, B.; Li, D.; Zhou, J. Altitudinal ranging of the Guizhou golden monkey (*Rhinopithecus brelichi*): Patterns of habitat selection and habitat use. *Glob. Ecol. Conserv.* **2018**, *16*, e473. [CrossRef]
55. Mallott, E.K.; Garber, P.A.; Malhi, R.S. trnL outperforms rbcL as a DNA metabarcoding marker when compared with the observed plant component of the diet of wild white-faced capuchins (*Cebus capucinus*, Primates). *PLoS ONE* **2018**, *13*, e199556. [CrossRef]
56. Sengupta, A.; Radhakrishna, S. Influence of Fruit Availability on Fruit Consumption in a Generalist Primate, the Rhesus Macaque *Macaca mulatta*. *Int. J. Primatol.* **2016**, *37*, 703–717. [CrossRef]
57. Britt, A.; Randriamandratonirina, N.J.; Glasscock, K.D.; Iambana, B.R. Diet and Feeding Behaviour of Indri indri in a Low-Altitude Rain Forest. *Folia Primatol.* **2002**, *73*, 225–239. [CrossRef]
58. Tsuji, Y.; Hanya, G.; Grueter, C.C. Feeding strategies of primates in temperate and alpine forests: Comparison of Asian macaques and colobines. *Primates* **2013**, *54*, 201–215. [CrossRef] [PubMed]
59. Tuyisingize, D.; Eckardt, W.; Caillaud, D.; Kaplin, B.A. High flexibility in diet and ranging patterns in two golden monkey (*Cercopithecus mitis kandti*) populations in Rwanda. *Am. J. Primatol.* **2022**, *84*, e23347. [CrossRef]
60. Wang, C.; Su, H.J.; Hu, C.S.; Shi, L.; Zhang, M.M. Spatial distribution of sympatric *Rhinopithecus brelichi* and *Macaca thibetana* in Fanjingshan National Nature Reserve, Guizhou, China. *Acta Theriol. Sin.* **2020**, *40*, 193–200. [CrossRef]
61. Fleagle, J.G.; Meldrum, D.J. Locomotor behavior and skeletal morphology of two sympatric pitheciine monkeys, *Pithecia pithecia* and *Chiropotes satanas*. *Am. J. Primatol.* **1988**, *16*, 227–249. [CrossRef] [PubMed]
62. Dagosto, M. Primate Origins and Evolution. A phylogenetic reconstruction. *Int. J. Primatol.* **1991**, *12*, 77–79. [CrossRef]
63. Ford, S.M.; Davis, L.C. Systematics and body size: Implications for feeding adaptations in new world monkeys. *Am. J. Phys. Anthropol.* **1992**, *88*, 415–468. [CrossRef]
64. Li, X.T. *The World's Endangered and Extinct Mammals*; Knowledge Publishing House: Beijing, China, 2002; p. 4.
65. Quan, G.Q.; Xie, J.H. Data on the Guizhou subspecies *Rhinopithecus roxellanae* brelichi Thomas. *Acta Theriol. Sin.* **1981**, *2*, 113–116. [CrossRef]
66. Zhou, Q.H.; Huang, C.M. Advances in ecological research on the limestone langurs in China. *Acta Theriol. Sin.* **2021**, *41*, 59–70. [CrossRef]
67. Hu, G. Distribution and conservation status of black langur resources in China. In *Proceedings of the 9th Member Congress and 10th Academic Symposium of Sichuan Zoological Society*; Sichuan Zoological Society: Chengdu, China, 2011.
68. Anthony, R.E.; Sinclair, A.R. *Wildlife Ecology, Conservation and Management*; Wiley-Blackwel: New York, NY, USA, 2006; pp. 60–77.
69. Li, J.Y. Abundance of Diet Plants and Seasonal Changes of Diet Nutrients of *Rhinopithecus roxellana* in Shennongjia Natural Reserve, China. Ph.D. Thesis, Beijing Forestry University, Beijing, China, 2015.
70. Raubenheimer, D.; Simpson, S.J. Organismal Stoichiometry: Quantifying Non-Independence among Food Components. *Ecology* **2004**, *85*, 1203–1216. [CrossRef]
71. Jiang, W.X.; Fu, X.C.; Tang, T.; Cai, Q.H. Community Structure and Niche of Macroinvertebrates in the Xiangxi River in Hubei, China. *Chin. J. Appl. Environ. Biol.* **2009**, *03*, 314–337. [CrossRef]
72. Wang, C. Spatiotemporal Coexistence Pattern of *Rhinopithecus brelichi* and *Macaca thibetana* in Fanjingshan Nature Reserve based on Infrared Camera. Master's Thesis, Gui Zhou University, Guizhou, China, 2021.
73. Tutin, C.E.G.; Fernandez, M.; Rogers, M.E.; Williamson, E.A.; McGrew, W.C.; Altmann, S.A.; Southgate, D.A.T.; Crowe, I.; Tutin, C.E.G.; Whiten, A.; et al. Foraging profiles of sympatric lowland gorillas and chimpanzees in the Lopé Reserve, Gabon. *Philos. Trans. Biol. Sci.* **1991**, *334*, 179–186. [CrossRef]
74. Heymann, E.W.; Knogge, C.; Tirado, H.E. Vertebrate predation by sympatric tamarins, *Saguinus mystax* and *Saguinus fuscicollis*. *Am. J. Primatol.* **2000**, *51*, 153–158. [CrossRef]
75. Wu, F.; Zhu, D.; Wen, P.; Tang, Z.; Bao, L.; Guan, Y.; Ge, J.; Wang, H. Domestic Cattle in a National Park Restricting the Sika Deer Due to Diet Overlap. *Animals* **2023**, *13*, 561. [CrossRef] [PubMed]
76. Schoener, T.W. Resource Partitioning in Ecological Communities. *Science* **1974**, *185*, 27–39. [CrossRef] [PubMed]

animals

MDPI

Article

Time to Step Up Conservation: Climate Change Will Further Reduce the Suitable Habitats for the Vulnerable Species Marbled Polecat (*Vormela peregusna*)

Xiaotian Cheng [1], Yamin Han [1], Jun Lin [2], Fan Jiang [3], Qi Cai [4], Yong Shi [3], Dongyang Cui [3] and Xuanye Wen [3,*]

[1] The Station of Forest Seedling Quarantine and Pest Management, Changji 831100, China; chengxiaotian8862@163.com (X.C.); hym0727@126.com (Y.H.)
[2] Locust and Rodent Control Headquarters of Xinjiang Uygur Autonomous Region, Urumqi 830000, China; jun_lin2013@163.com
[3] Center for Biological Disaster Prevention and Control, National Forestry and Grassland Administration, Shenyang 110031, China; jiangfan930430@163.com (F.J.); yshisyau@163.com (Y.S.); cuidy@bdpc.org.cn (D.C.)
[4] Institute of Ecological Conservation and Restoration, Chinese Academy of Forestry, Beijing 100091, China; caiqilinda1010@126.com
* Correspondence: wenxuanye_1116@163.com; Tel.: +86-186-0440-0387

Simple Summary: Assessing suitable habitats for species has great potential to guide the management and conservation of threatened species, especially rare species that are poorly studied and remote from human populations. *Vormela peregusna*, a mustelid mammal endemic to Eurasia, was assessed as vulnerable by the International Union for Conservation of Nature in 2015 due to its low population size and increasing human threats. We used the ensemble model to assess the distribution of suitable areas for *V. peregusna* under current and future climate change conditions to contribute to the better protection of endangered animals.

Abstract: Habitat loss and human threats are putting the marbled polecat (*Vormela peregusna)* on the brink of extinction. Numerous recent studies have found that climate change will further deteriorate the living environment of endangered species, leading to their eventual extinction. In this study, we used the results of infrared camera surveys in China and worldwide distribution data to construct an ensemble model consisting of 10 commonly used ecological niche models to specify potential suitable habitat areas for *V. peregusna* under current conditions with similar environments to the sighting record sites. Changes in the suitable habitat for *V. peregusna* under future climate change scenarios were simulated using mid-century (2050s) and the end of the century (2090s) climate scenarios provided by the Coupled Model Intercomparison Project Phase 6 (CMIP6). We evaluated the accuracy of the model to obtain the environmental probability values (cutoff) of the *V. peregusna* distribution, the current distribution of suitable habitats, and future changes in moderately and highly suitable habitat areas. The results showed that the general linear model (GLM) was the best single model for predicting suitable habitats for *V. peregusna*, and the kappa coefficient, area under the curve (AUC), and true skill statistic (TSS) of the ensemble model all exceeded 0.9, reflecting greater accuracy and stability than single models. Under the current conditions, the area of suitable habitat for *V. peregusna* reached 3935.92×10^4 km^2, suggesting a wide distribution range. In the future, climate change is predicted to severely affect the distribution of *V. peregusna* and substantially reduce the area of suitable habitats for the species, with 11.91 to 33.55% of moderately and highly suitable habitat areas no longer suitable for the survival of *V. peregusna*. This shift poses an extremely serious challenge to the conservation of this species. We suggest that attention be given to this problem in Europe, especially the countries surrounding the Black Sea, Asia, China, and Mongolia, and that measures be taken, such as regular monitoring and designating protected areas for the conservation of vulnerable animals.

Keywords: *Vormela peregusna*; suitable area; red list; species distribution models; Biomod2

Citation: Cheng, X.; Han, Y.; Lin, J.; Jiang, F.; Cai, Q.; Shi, Y.; Cui, D.; Wen, X. Time to Step Up Conservation: Climate Change Will Further Reduce the Suitable Habitats for the Vulnerable Species Marbled Polecat (*Vormela peregusna*). *Animals* **2023**, *13*, 2341. https://doi.org/10.3390/ani13142341

Academic Editor: Chunwang Li

Received: 8 June 2023
Revised: 12 July 2023
Accepted: 14 July 2023
Published: 18 July 2023

1. Introduction

Determining how to protect endangered species is one of the main problems faced in conservation biology [1,2]. The most recent evaluation showed that over 42,100 species are at risk of extinction to varying degrees, a number that amounts to 28% of all species assessed [3]. Habitat loss and fragmentation are among the most important issues currently observed. Understanding habitat suitability and the factors that influence species habitats is the basis for the protection of threatened species [4,5]. In general, the risk of warming due to climate change will exacerbate the loss of species habitat [6]. Additionally, the loss of fauna may be imminent if scientifically sound conservation measures are not developed, especially for animals that are already scarce.

Habitat analysis is an important prerequisite for conservation planning and managerial decision making [7] and is urgently needed to enhance the conservation of endangered species. Species distribution models (SDMs) use environmental variables associated with species distribution sites to predict the ecological needs and potential distribution of species and are widely used in invasion biology, conservation biology, global change biology, and risk of disease transmission [8]. Similar to other ecological models, ecological niche models have uncertainty in their predictions, which are closely related to species distribution sites, environmental variables, and model algorithms and parameters, among which the species distribution sites and model algorithms have the greatest effects [9]. There are more than 30 SDMs available [10], each with different advantages; however, it is difficult to choose the optimal assessment model for species that have rarely been studied [11]. Biomod2 is an ensemble model based on a variety of single models [12] and provides better prediction accuracy and spatial sensitivity to small sample sets than do single models, which are prone to overfitting and uncertainty [13–15].

Vormela peregusna is the only small mammal in the genus *Vormela* of the order Carnivora of the class Mammalia [16]. *V. peregusna* is valuable for ecosystem stability and medical research. On the one hand, it is a major predator of desert rodents [17] and can effectively protect desert vegetation from rodent damage; on the other hand, *V. peregusna* is associated with many tick-borne pathogens [18], and research on this species has helped constrain the transmission mechanisms of zoonotic diseases. Unfortunately, despite the value of this species, the risk to the survival of *V. peregusna* is very concerning. As early as 2008, *V. peregusna* was included on the Red List of Threatened Species by the International Union for Conservation of Nature (IUCN), and a subsequent reassessment found that the *V. peregusna* population declined by 30% in a decade [19], reaching a classification of vulnerable. Although China, Bulgaria [20], and other countries have classified *V. peregusna* as a protected species, thus far, no countries have established targeted conservation measures to save *V. peregusna* from extinction.

We hypothesize that there is a direct relationship between the species distribution of *V. peregusna* and its food sources, and since there are no efficient methods to monitor *V. peregusna* populations at a broad scale, currently, most records come from opportunistic sightings [21]. Observations from southern Europe show that *V. peregusna* prefers to appear in farmland–grassland interlaced areas with good vegetation conditions [22]. However, this situation is completely different in Asia and the Middle East, with records of occurrences in countries such as Iraq, China, and Mongolia coming from sparsely vegetated hilly and desert areas [23,24]. Another strong piece of evidence is that, in infrared camera monitoring of *V. peregusna*'s main food *Rhombomys opimus*, it was found that the activity rhythm of *V. peregusna* was positively correlated with that of *R. opimus* [25]. When the activity of *R. opimus* increased or decreased with seasonal changes, the activity pattern of *V. peregusna* also changed accordingly. Although the effect of food on the survival of *V. peregusna* has not yet been clearly determined, it will be of great help to protect *V. peregusna* if protected areas can be delineated by considering the suitable range and occurrence of rodents.

Habitat loss is the main reason for the endangerment of *V. peregusna*. In Europe, the conversion of much grassland to cropland has reduced the suitable habitat for *V. peregusna*, while in Asia, land desertification is the main threat to *V. peregusna* [26]. Several past studies have found that climate change will have a dramatic impact on the stability of desert and grassland ecosystems [27,28]. Studies from Central Asian grasslands have shown that changes in precipitation are the main factor causing changes in grassland vegetation [29], and studies on the survival of desert fauna have also shown a significant decline in desert bird populations over the past century due to temperature and precipitation [30]. It is also important to address whether the habitat of *V. peregusna*, as a desert and grassland habitat species, will be affected by climate change.

In this study, we collected monitoring data obtained with infrared cameras in the desert areas of northwest China and reports of *V. peregusna* distribution worldwide to improve the accuracy of model-based distribution predictions. Ten single models were established based on Biomod2, and the ability of different models to predict suitable habitat areas for *V. peregusna* was compared. An ensemble model was used to predict the distribution of *V. peregusna* and changes in suitable habitat areas in current and future situations. Specifically, our aims were to (a) provide new strategies for conducting suitable habitat studies of species with small populations, (b) enhance the conservation of vulnerable animals, and (c) recommend conservation priority areas for effective conservation in the future.

2. Materials and Methods

2.1. Occurrence Data

In previous studies, we used infrared cameras to monitor rodent pests in desert areas of China, and 12 *V. peregusna* distribution points were identified [25]. Additionally, a total of 491 distribution points was obtained by searching the database of the Global Biodiversity Information Facility (http://www.gbif.org, accessed on 18 October 2022) and research articles related to the distribution of *V. peregusna* [31–34]. Due to the duplication of many point co-ordinates, we removed redundant data to reduce the error caused by the clustering effect so that only one distribution point was retained in each grid (100 km^2). Ultimately, 101 valid points were obtained (Figure 1), and the latitude and longitude co-ordinates of each point are shown in Table S1.

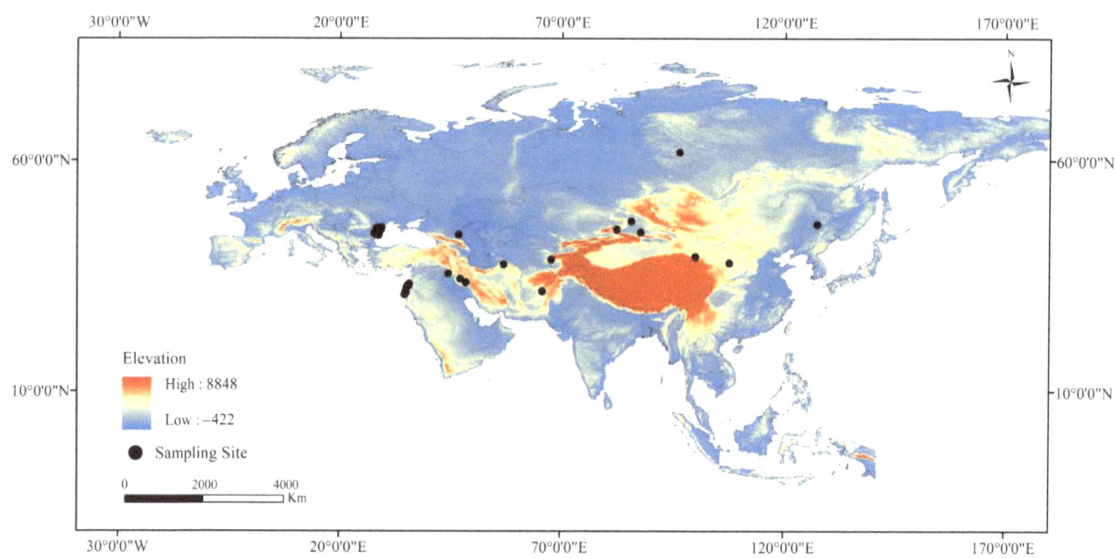

Figure 1. Distribution of records of *V. peregusna*.

2.2. Selection and Processing of Environmental Variables

We used 31 environmental variables for modeling, and these environmental variables were shown to be directly related to *R. opimus* distribution in our earlier modeling [35]:

19 bioclimatic factors, 9 soil factors, and 3 topographic factors. The climate data were downloaded from the WorldClim database (http://worldclim.org, accessed on 19 October 2022), and the future climate scenarios were downloaded from the BCC-CSM (Beijing Climate Center, China Meteorological Administration, Beijing, China). These scenarios included those with low (SSP126), moderate (SSP245), and high (SSP585) emissions of greenhouse gases [36]. The soil and topographic factor data were obtained from the Harmonized World Soil Database (HWSD) of the Food and Agriculture Organization of the United Nations (http://www.fao.org/faostat/en/#data, accessed on 19 October 2022), with the spatial resolution of each factor set to 10 arc min [37].

The interactions between environmental factors can lead to collinearity issues during modeling and subsequent overfitting [38]. First, variance inflation factor (VIF) analysis was used to select the most important environmental variables. Second, R was used to conduct a Pearson correlation analysis, reduce the complexity of the model, and improve its prediction accuracy. Factors with a correlation less than 0.8 were preliminarily selected, and, from those, factors with a VIF less than 10 were retained. Then, based on the Pearson correlation test results, factors with a correlation coefficient less than 0.8 were retained, and factors with correlation coefficients greater than 0.8 were omitted. A total of 18 environmental variables were selected (Table 1).

Table 1. Environmental variables with their contributions and suitable value ranges.

Code	Environmental Variable	Variable Importance
bio16	Precipitation in wettest quarter	23.34
t_caco3	Topsoil calcium carbonate content	14.46
bio18	Precipitation in warmest quarter	13.94
t_teb	Topsoil teb.	12.36
bio3	Isothermality	10.65
elev	Elevation	7.27
t_cec_clay	Topsoil CEC (CLAY)	4.77
bio19	Precipitation in coldest quarter	4.26
bio17	Precipitation in driest quarter	2.18
bio15	Precipitation seasonality	2.04
t_caco4	Topsoil gypsum content	1.00
slope	Slope	0.99
bio5	Max. temperature	0.66
t_ece	Topsoil salinity (Elco)	0.56
t_gravel	Topsoil gravel content	0.43
t_oc	Topsoil organic carbon	0.41
t_esp	Topsoil sodicity (ESP)	0.37
t-sand	Topsoil sand fraction	0.30

2.3. Model Construction

To reduce the modeling bias caused by the uncertainty inherent in models, we first performed fitting with 10 single methods included in Biomod2: a generalized additive model (GAM), a generalized boosted regression model (GBM), a general linear model (GLM), a random forest (RF), the multivariate adaptive regression splines (MARS) method, classification tree analysis (CTA), an artificial neural network (ANN), the surface range envelope (SRE) method, flexible discriminant analysis (FDA), and the maximum entropy (Maxent) method. Before constructing the model, it was necessary to process the species distribution data. Biomod2 provides several methods to generate nonexistence (pseudoabsence) points from background research data [39]. The "random" command was used to randomly generate 1200 pseudoabsence data points for model simulation. Then, the "biomod_tuning" command was used to optimize the model parameters and select 70% of the sample data for training. The remaining 30% of the sample data were used to verify the performance of the model [40]. The resulting single models were evaluated using 3 metrics: the true skill statistic (TSS), AUC, and kappa coefficient [41].

Single models with accuracies that met the selected standard were integrated into an ensemble model using a weighted average approach [42]. First, the results of the single models used in the construction of the ensemble model were normalized so that the predictions of single SDMs were in the range of [0, 1]. This process was repeated 10 times to avoid random errors associated with the use of a single model. Then, the weights for model combination were determined based on the AUC and TSS values of each model, and the single models used to construct the ensemble model were determined with fixed cutoffs of TSS > 0.7 and AUC > 0.8. The higher the average AUC and TSS values were after multiple runs, the greater the weight assigned to the corresponding single model was when it was incorporated into the ensemble model.

2.4. Changes in the Spatial Pattern of the Suitable Distribution Ranges of Species

The 0/1 probability value cutoff of "suitable" or "unsuitable" was obtained by running the model. The spatial units with values below the cutoff were considered unsuitable habitats, and the spatial units with values above the cutoff were divided into 3 equal parts, corresponding to minimally, moderately, and highly suitable habitats [43]. Two time nodes, namely, the middle of this century and the end of this century, were selected to analyze the future suitable area changes of *V. peregusna*, and the average values from 2040–2060 (2050s) and 2080–2100 (2090s) were calculated [44]. Based on the "binary_meth" operation in Biomod2, we obtained the results of the suitable/unsuitable (0/1) simulation and used the "biomod_rangesize" function to calculate the changes in the spatial pattern of the suitable areas of *R. opimus* under future climate change scenarios [45]. Finally, the results, in matrix format, were loaded into ArcGIS v10.4.1 for visual representation.

3. Results

3.1. Model Accuracy

Among the 10 models evaluated, only the GAM failed to run successfully due to the difficulty in obtaining parameter values; the other 9 models were all run successfully, and a total of 90 sets of results were obtained. The different models were compared (Figure S1), and the GBM displayed the highest accuracy and the best stability for the three evaluation metrics, suggesting that it was the best choice for assessing suitable habitats for *V. peregusna* using a single model. Although higher scores were obtained for the FDA and RF models, they each yielded one data anomaly in the calculation process. Moreover, low scores were obtained for the other models, which failed to reach acceptable performance levels. From the 90 sets of results, we selected a total of 39 eligible models to construct the ensemble model. The final ensemble model yielded a kappa coefficient of 0.91, a TSS value of 0.94, and an AUC value of 0.96, indicating excellent results.

3.2. Current Distribution Range

Suitable habitats for *V. peregusna* were found over almost all of Eurasia, except in a few tropical areas in Southeast Asia. Based on the current climate scenario, the suitable habitats for *V. peregusna* covered 3935.92×10^4 km^2, of which moderately suitable habitats accounted for 2415.17×10^4 km^2, followed by low-suitability habitats (867.99×10^4 km^2). Highly suitable habitats accounted for the smallest area (652.76×10^4 km^2). The highly suitable habitats were mainly found in the following regions: (1) the plain area at the border of China and Mongolia in the east, (2) the plateau area from the Orkhon River to Khangai Mountain in the west–central part of Mongolia, (3) the area from 40 to 50° N from Bulgaria and Ukraine in the west to the Junger Basin in China and the Siberian Plain in Russia in the east, and (4) sporadic highly suitable habitats in Spain, Italy, Hungary, Poland, Lithuania, Latvia, and the far east of Russia (Figure 2).

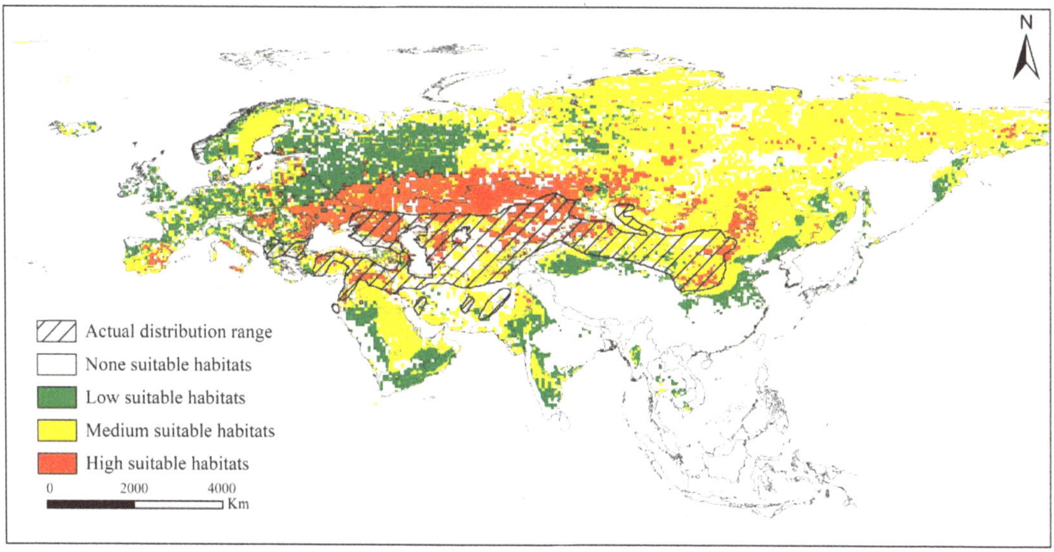

Figure 2. The current suitable distribution range of *V. peregusna*.

3.3. *Future Changes in Suitable Habitat Area*

Based on the three future climate scenarios, the geographical distributions of moderately and highly suitable habitats for *V. peregusna* were predicted to decrease to varying degrees (Figure 3). Although the percentages of the increases and decreases differed in various scenarios, the decrease in the area of suitable habitats for *V. peregusna* was much larger than the corresponding increase in each scenario. The lost suitable habitats were mainly in Europe, and most areas from northern to southern Europe were predicted to no longer be suitable for *V. peregusna* survival, especially in several countries around the Black Sea, where many *V. peregusna* have been recorded. Increases and decreases in the area of suitable habitats were variable in other regions, with decreases occurring in the eastern part of Saudi Arabia, in the border area between northeastern China and Russia, and on the Western Siberian Plain. Regions with increases in the area of suitable habitats were concentrated around the Ural Mountains in Russia, in the central and southern parts of the Arabian Peninsula, in North China, and in central Xinjiang, China.

The predicted area changes in each scenario indicated that both SSP126 and SSP585 in the middle of this century will result in extensive losses of moderately and highly suitable habitats for *V. peregusna* (Table 2), with the lost area exceeding 900×10^4 km^2, accounting for 30% of the area of existing suitable habitats. In comparison, the case of SSP245 was relatively optimistic, but the lost area still reached 646×10^4 km^2, which was 21.06% less than that in the current period. At the end of this century, SSP245 is projected to be the scenario with the most severe decrease in the area of suitable habitats for *V. peregusna*, and the species range is predicted to decrease to only two-thirds of the existing distribution area, with a loss of 35.83% of suitable habitat. SSP126 is the most optimistic scenario for habitat suitability based on the predicted results. Notably, the percentages of area gain and loss for *V. peregusna* habitats are projected to be 5.89% and 11.91%, respectively, with an overall change of only 11.91%. The change in the area of suitable habitat under SSP585 is predicted to be stable at the end of this century, and the increases and decreases in suitable habitat area are expected to be consistent with those at 50° S.

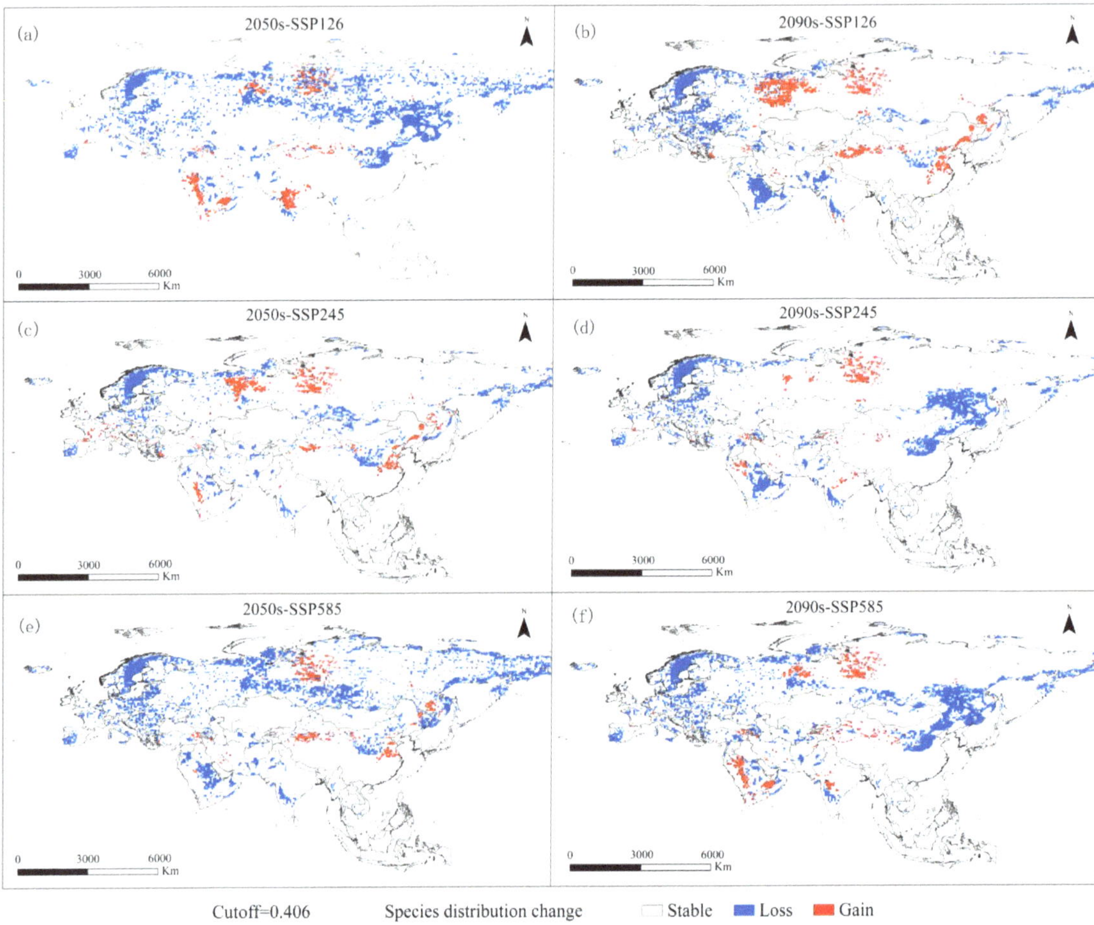

Figure 3. The spatial pattern changes in potential moderately and highly suitable areas for *V. peregusna* in different periods.

Table 2. Changes in moderately and highly suitable areas for *V. peregusna* in different climate scenarios in the future.

Periods	Climate Scenario	Suitable Habitat Area ($\times 10^4$ km^2)	Loss ($\times 10^4$ km^2)	Stable ($\times 10^4$ km^2)	Gain ($\times 10^4$ km^2)	Species Range Change (%)	Percentage Loss (%)	Percentage Gain (%)
Current		3067.93						
2050	SSP126	2226.43	978.64	2089.29	137.14	−27.43	31.90	4.47
	SSP245	2553.79	646.18	2421.75	132.04	−16.76	21.06	4.30
	SSP585	2226.66	926.29	2141.64	85.02	−27.42	30.19	2.77
2090	SSP126	2702.43	546.16	2521.77	180.66	−11.91	17.80	5.89
	SSP245	2038.78	1099.21	1968.72	70.06	−33.55	35.83	2.28
	SSP585	2213.48	978.59	2089.34	124.14	−27.85	31.90	4.05

4. Discussion

In general, species ecological niches evolve at a much slower rate than climate change [46], and species respond to rapid climate change by dispersing to new suitable habitats, adapting, or becoming extinct [47]. Despite the widespread distribution of *V. peregusna* in Eurasia, it is alarming that our assessment indicated that a 12 to 34% decrease in the suitable habitats for *V. peregusna* may occur in the coming decades, implying that the survival of *V. peregusna* may be severely affected by climate change; this trend is largely associated with the extremely low number of *V. peregusna* per unit area of distribution and the poor migratory capacity of the species [19]. However, the future predictions obtained with the model in this study are relatively uncertain. Our results only indicate the probability of potential occurrence and do not represent real changes in the species distribution or habitat area [48]. In addition to the environment, the main factors affecting the distribution of the species include various biological and nonbiological factors, such as competition, disease, and human disturbances [49].

In contrast to the results of ecological niche modeling studies for other species [50,51], the modeling results for *V. peregusna* in this study did not show a clear pattern of response to different climatic scenarios, i.e., no shift of suitable habitats to higher latitudes or a significant decrease in the area of suitable habitats with increasing temperature. Surprisingly, the moderate carbon emission concentration at the end of this century is projected to result in the largest decrease in the suitable area for *V. peregusna*. We speculate that the reason for this result is directly related to the unique living habits of *V. peregusna* and the selection of environmental factors retained in the model. *V. peregusna* is the only mustelid animal that hibernates [25], and its food sources are predominantly rodents and lizards. Precipitation and soil factors accounted for more than 80% of the environmental influence in the modeling process, and temperature parameters, which were most influenced by changes in climate scenarios, accounted for only 0.66% of the influence in the ensemble model. Although it did not have a direct impact, climate change still had a substantial effect on the suitable habitats of *V. peregusna*, indicating that current wildlife conservation efforts face serious challenges and that it is necessary to pay increased attention to climate change to avoid species extinction.

To enhance biodiversity conservation, we recommend the following two measures to ensure that *V. peregusna* will not become an endangered species in the future. First, for regions with low climate impacts, the protection of *V. peregusna* should be strengthened, human interference and the use of anticoagulant rodenticides should be reduced, in situ protection should be enhanced, nature reserves should be established as soon as possible, and the hunting and trading of wild animals should be closely monitored. Second, for regions with relatively fragile climates, it is necessary to strengthen captive management, tentatively adopt ex situ conservation measures, promote domestication and breeding, and implement overall population resource monitoring.

The IUCN Red List is the most widely used wildlife conservation standard and reference for prioritizing conservation and ecological research [52], and the accuracy of Red List assessments has global implications [53]. IUCN assessments primarily consider the current drivers of species declines, such as population fluctuations and human pressure on populations and their suitable habitats, without adequately identifying potential future risks, such as threats posed by climate change [54,55]. Combined with studies of other listed species [56–58], we suggest that the IUCN consider the threats posed by climate change in future assessment efforts and incorporate distribution changes resulting from climate change into the assessment metrics for Red List species.

With the ensemble model, we effectively mitigated overfitting and improved the accuracy of predictions, but it is undeniable that the use of models to study species distributions has certain limitations. First, the results of the model are species distribution predictions, not the actual distribution of a species [59]. Second, the reproduction and migration of species are complex and dynamic processes. Natural disasters, human activities, and intraspecific competition are also important factors that influence species distributions. No existing prediction model can capture interspecies competition well. In this study, we only considered environmental factors, and this limitation may have affected the prediction accuracy to a certain extent. Third, too sparse a sample size can lead to errors in the prediction results of ecological niche models; therefore, further global co-operation is required to improve the ability to assess and protect wildlife through data sharing.

5. Conclusions

This study applied the ensemble model to evaluate the spatial distribution of the potential habitat of the endangered species *V. peregusna* in Eurasia and the effects of future climate change on its habitat suitability. The results showed that GBM is the best single model for predicting suitable habitats for *V. peregusna*, and the ensemble model showed higher accuracy and stability than single models. Under current conditions, the area of suitable habitats for *V. peregusna* has reached 3935.92×10^4 km^2, making it a species with a wide distribution range. In the future, climate change will severely affect the distribution and substantially reduce the area of suitable habitats for *V. peregusna*, thus posing an extremely serious challenge to the conservation of *V. peregusna*. These findings are expected to support the development of practical solutions to prevent the extinction of *V. peregusna* populations.

Supplementary Materials: The following supporting information can be downloaded at: https://www.mdpi.com/article/10.3390/ani13142341/s1, Supplementary Table S1: Distribution points of *Vormela peregusna*; Supplementary Figure S1: Comparison of AUC, kappa coefficient, and TSS evaluations of 9 models.

Author Contributions: Conceptualization, X.W.; methodology, X.C. and X.W.; software, X.C. and Y.H.; validation, J.L. and Q.C.; formal analysis, Y.S. and D.C.; investigation, F.J.; writing—original draft preparation, X.C. and Y.H.; writing—review and editing, J.L. and X.W.; visualization, F.J., J.L. and Q.C.; supervision, X.W.; project administration, X.W.; funding acquisition, X.W. All authors have read and agreed to the published version of the manuscript.

Funding: This work was supported by the National Forestry and Grassland Administration Forestry plant new varieties and patent protection application [KJZXXP202212].

Institutional Review Board Statement: Not applicable.

Informed Consent Statement: Not applicable.

Data Availability Statement: The original contributions presented in this study are included in the article/Supplementary Material, and further inquiries can be directed to the corresponding author.

Conflicts of Interest: The authors declare no conflict of interest.

References

1. Thapa, A.; Wu, R.; Hu, Y.; Nie, Y.; Singh, P.B.; Khatiwada, J.R.; Yan, L.; Gu, X.; Wei, F. Predicting the potential distribution of the endangered red panda across its entire range using MaxEnt modeling. *Ecol. Evol.* **2018**, *8*, 10542–10554. [CrossRef]
2. Scherrer, D.; Christe, P.; Guisan, A. Modelling bat distributions and diversity in a mountain landscape using focal predictors in ensemble of small models. *Divers. Distrib.* **2019**, *25*, 770–782. [CrossRef]
3. International Union for Conservation of Nature. The IUCN Red List of Threatened Species. Version 2022-2. Available online: https://www.iucnredlist.org (accessed on 15 December 2022).
4. Hoekstra, J.M.; Boucher, T.M.; Ricketts, T.H.; Roberts, C. Confronting a biome crisis: Global disparities of habitat loss and protection. *Ecol. Lett.* **2005**, *8*, 23–29. [CrossRef]
5. He, F.; Hubbell, S.P. Species-area relationships always overestimate extinction rates from habitat loss. *Nature* **2011**, *473*, 368–371. [CrossRef] [PubMed]
6. Coristine, L.E.; Kerr, J.T. Habitat loss, climate change, and emerging conservation challenges in Canada. *Can. J. Zool.* **2011**, *89*, 435–451. [CrossRef]
7. Galante, P.J.; Triguero, S.C.; Paz, A.; Aiello-Lammens, M.; Gerstner, B.E.; Johnson, B.A.; Kass, J.M.; Merow, C.; Noguera-Urbano, E.A.; Pinilla-Buitrago, G.E.; et al. changeRangeR: An R package for reproducible biodiversity change metrics from species distribution estimates. *Conserv. Sci. Pract.* **2022**, *5*, e12863. [CrossRef]
8. Brodie, S.; Smith, J.A.; Muhling, B.A.; Barnett, L.A.K.; Carroll, G.; Fiedler, P.; Bograd, S.J.; Hazen, E.L.; Jacox, M.G.; Andrews, K.S.; et al. Recommendations for quantifying and reducing uncertainty in climate projections of species distributions. *Glob. Chang. Biol.* **2022**, *28*, 6586–6601. [CrossRef] [PubMed]
9. Hodel, R.G.J.; Soltis, D.E.; Soltis, P.S. Hindcast-validated species distribution models reveal future vulnerabilities of mangroves and salt marsh species. *Ecol. Evol.* **2022**, *12*, e9252. [CrossRef]
10. Norberg, A.; Abrego, N.; Blanchet, F.G.; Adler, F.R.; Anderson, B.J.; Anttila, J.; Araújo, M.B.; Dallas, T.; Dunson, D.; Elith, J.; et al. A comprehensive evaluation of predictive performance of 33 species distribution models at species and community levels. *Ecol. Monogr.* **2019**, *89*, e01370. [CrossRef]

11. Tez, C.; Gündüz, I.; Kefelioğlu, H. Contributions to distribution, reproduction biology and ecology of *Vormela peregusna* (Güldenstaedt, 1770) (Mammalia: Carnivora) in Turkey. *Pak. J. Biol. Sci.* **2001**, *4*, 74–76. [CrossRef]
12. Gama, M.; Crespo, D.; Dolbeth, M.; Anastácio, P. Predicting global habitat suitability for *Corbicula fluminea* using species distribution models: The importance of different environmental datasets. *Ecol. Model.* **2016**, *319*, 163–169. [CrossRef]
13. Di Cola, V.; Broennimann, O.; Petitpierre, B.; Breiner, F.T.; D'Amen, M.; Randin, C.; Engler, R.; Pottier, J.; Pio, D.; Dubuis, A.; et al. ecospat: An R package to support spatial analyses and modeling of species niches and distributions. *Ecography* **2017**, *40*, 774–787. [CrossRef]
14. Hao, T.; Elith, J.; Lahoz-Monfort, J.J.; Guillera-Arroita, G. Testing whether ensemble modelling is advantageous for maximising predictive performance of species distribution models. *Ecography* **2020**, *43*, 549–558. [CrossRef]
15. Schickele, A.; Leroy, B.; Beaugrand, G.; Goberville, E.; Hattab, T.; Francour, P.; Raybaud, V. Modelling European small pelagic fish distribution: Methodological insights. *Ecol. Model.* **2020**, *416*, 108902. [CrossRef]
16. Puzachenko, A.Y.; Abramov, A.V.; Rozhnov, V.V. Cranial variation and taxonomic content of the marbled polecat *Vormela peregusna* (Mustelidae, Carnivora). *Mamm. Biol.* **2017**, *83*, 10–20. [CrossRef]
17. Ben-David, M.; Pellis, S.M.; Pellis, V.C. Feeding habits and predatory behaviour in the marbled polecat (*Vormela peregusna syriaca*): I. Killing methods in relation to prey size and prey behaviour. *Behaviour* **1991**, *118*, 127–143. [CrossRef]
18. Liu, G.; Zhao, S.; Tan, W.; Hornok, S.; Yuan, W.; Mi, L.; Wang, S.; Liu, Z.; Zhang, Y.; Hazihan, W.; et al. Rickettsiae in red fox (*Vulpes vulpes*), marbled polecat (*Vormela peregusna*) and their ticks in northwestern China. *Parasit. Vectors* **2021**, *14*, 204. [CrossRef]
19. Abramov, A.V.; Kranz, A.; Maran, T. Vormela peregusna. In *The IUCN Red List of Threatened Species 2016*; International Union for Conservation of Nature and Natural Resources: Gland, Switzerland, 2016. [CrossRef]
20. Zidarova, S. Present distribution of the marble polecat *Vormela peregusna* (Güldenstädt, 1770) (Carnivora: Mustelidae) in central Western Bulgaria, with an observation on its defensive behavior. *Acta Zool. Bulg.* **2022**, *74*, 403–408.
21. Wright, P.G.R.; Croose, E.; Hunter, S.B.; MacPherson, J.; Çoraman, E.; Yarotskiy, V.; Moisieieva, V.; Karapandža, B.; Hoxha, B.; Madalina, P.; et al. Can social media be used to inform the distribution of the marbled polecat, *Vormela peregusna*? *Mammal Res.* **2023**, *68*, 295–304. [CrossRef]
22. Ivanov, V.; Spassov, N. Some new data on the distribution, habitats and ecology of the threatened European mustelids Mustela eversmanii and *Vormela peregusna* in Bulgaria. *Hist. Nat. Bulg.* **2015**, *21*, 267–271.
23. AL-Sheikhly, O.S.; Ararat, K.; Ahmed, S.H. New records for the first time in six decades of the Vulnerable Marbled Polecat *Vormela peregusna* in Iraq. *Small Carniv. Conserv.* **2022**, *60*, e60004.
24. Dulamtseren, S.; Shar, S.; Murdoch, J.; Reading, R.P.; Gantulga, J.; Usukhjrargal, D.; Suuri, B. Contributions to the distribution of the marbled polecat (*Vormela peregusna* Guldenstaedt, 1770) in Mongolia. *Small Carniv. Conserv.* **2009**, *40*, 29–32.
25. Wen, X.; Cheng, X.; Dong, Y.; Wang, Q.; Lin, X. Analysis of the activity rhythms of the great gerbil (*Rhombomys opimus*) and its predators and their correlations based on infrared camera technology. *Glob. Ecol. Conserv.* **2020**, *24*, e01337. [CrossRef]
26. Werner, N.Y. Small carnivores, big database—Inferring possible small carnivore distribution and population trends in Israel from over 30 years of recorded sightings. *Small Carniv. Conserv.* **2012**, *47*, 17–25.
27. He, B.; Tuya, W.; Qinchaoketu, S.; Nanzad, L.; Yong, M.; Kesi, T.; Sun, C. Climate Change Characteristics of Typical Grassland in the Mongolian Plateau from 1978 to 2020. *Sustainability* **2022**, *14*, 16529. [CrossRef]
28. Vale, C.G.; Brito, J.C. Desert-adapted species are vulnerable to climate change: Insights from the warmest region on Earth. *Glob. Ecol. Conserv.* **2015**, *4*, 369–379. [CrossRef]
29. Iknayan, K.J.; Beissinger, S.R. Collapse of a desert bird community over the past century driven by climate change. *Proc. Natl. Acad. Sci. USA* **2018**, *115*, 8597–8602. [CrossRef]
30. Chen, T.; Bao, A.; Jiapaer, G.; Guo, H.; Zheng, G.; Jiang, L.; Chang, C.; Tuerhanjiang, L. Disentangling the relative impacts of climate change and human activities on arid and semiarid grasslands in central asia during 1982–2015. *Sci. Total Environ.* **2019**, *53*, 1311–1325. [CrossRef]
31. Gherman, C.M.; Sándor, A.D.; Kalmár, Z.; Marinov, M.; Mihalca, A.D. First report of *Borrelia burgdorferi* sensu lato in two threatened carnivores: The *Marbled polecat, Vormela peregusna* and the European mink, *Mustela lutreola* (Mammalia: Mustelidae). *BMC Vet. Res.* **2012**, *8*, 137. [CrossRef]
32. Liu, X.; Yang, M.; Liu, G.; Zhao, S.; Yuan, W.; Xiao, R.; Hazihan, W.; Hornok, S.; Wang, Y. Molecular evidence of *Rickettsia raoultii*, "*Candidatus* Rickettsia barbariae" and a novel Babesia genotype in marbled polecats (*Vormela peregusna*) at the China-Kazakhstan border. *Parasit. Vectors* **2018**, *11*, 450. [CrossRef]
33. Hadad, E.; Charter, M.; Kosicki, J.Z.; Yosef, R. Prey-base does not influence breeding success in eagle owls (*Bubo bubo*) in Judea, Israel. *Animals* **2022**, *12*, 1280. [CrossRef]
34. Gray, A.; Brito, J.C.; Edwards, C.W.; Figueiró, H.V.; Koepfli, K.P. First complete mitochondrial genome of the Saharan striped polecat (*Ictonyx libycus*). *Mitochondrial DNA Pt. B Resour.* **2022**, *7*, 1957–1960. [CrossRef] [PubMed]
35. Wen, X.; Zhao, G.; Cheng, X.; Chang, G.; Dong, X.; Lin, X. Prediction of the potential distribution pattern of the great gerbil (*Rhombomys opimus*) under climate change based on ensemble modelling. *Pest Manag. Sci.* **2022**, *78*, 3128–3134. [CrossRef] [PubMed]
36. Li, W.; Zhang, Y.; Shi, X.; Zhou, W.; Huang, A.; Mu, M.; Qiu, B.; Ji, J. Development of land surface model BCC_AVIM2.0 and its preliminary performance in LS3MIP/CMIP6. *J. Meteorol. Res.* **2019**, *33*, 851–869. [CrossRef]
37. Graham, M.H. Confronting multicollinearity in ecological multiple regression. *Ecology* **2003**, *84*, 2809–2815. [CrossRef]

38. Rose, E.L. Multivariate analysis of categorical data: Theory. *Struct. Equ. Model.* **1995**, *2*, 274–276. [CrossRef]
39. Thuiller, W.; Lafourcade, B.; Engler, R.; Araújo, M.B. BIOMOD—A platform for ensemble forecasting of species distributions. *Ecography* **2009**, *32*, 369–373. [CrossRef]
40. Singh, K.; McClean, C.J.; Büker, P.; Hartley, S.E.; Hill, J.K. Mapping regional risks from climate change for rainfed rice cultivation in India. *Agric. Syst.* **2017**, *156*, 76–84. [CrossRef] [PubMed]
41. Allouche, O.; Tsoar, A.; Kadmon, R. Assessing the accuracy of species distribution models: Prevalence, kappa and the true skill statistic (TSS). *J. Appl. Ecol.* **2006**, *43*, 1223–1232. [CrossRef]
42. Fourcade, Y.; Besnard, A.G.; Secondi, J. Paintings predict the distribution of species, or the challenge of selecting environmental predictors and evaluation statistics. *Glob. Ecol. Biogeogr.* **2018**, *27*, 245–256. [CrossRef]
43. Liu, C.; White, M.; Newell, G. Selecting thresholds for the prediction of species occurrence with presence-only data. *J. Biogeogr.* **2013**, *40*, 778–789. [CrossRef]
44. Patasaraiya, M.K.; Devi, R.M.; Sinha, B.; Bisaria, J. Predicting impacts of climate change on teak and sal forests in central India using maximum entropy modeling: An approach for future conservation and silvicultural strategies. *For. Sci.* **2023**, fxad014. [CrossRef]
45. Guisan, A.; Thuiller, W.; Zimmermann, N.E. *Habitat Suitability and Distribution Models: With Applications in R*; Ecology, Biodiversity and Conservation); Cambridge University Press: Cambridge, UK, 2017.
46. Quintero, I.; Wiens, J.J. Rates of projected climate change dramatically exceed past rates of climatic niche evolution among vertebrate species. *Ecol. Lett.* **2013**, *16*, 1095–1103. [CrossRef] [PubMed]
47. Christmas, M.J.; Breed, M.F.; Lowe, A.J. Constraints to and conservation implications for climate change adaptation in plants. *Conserv. Genet.* **2016**, *17*, 305–320. [CrossRef]
48. Blair, M.E.; Le, M.D.; Xu, M. Species distribution modeling to inform transboundary species conservation and management under climate change: Promise and pitfalls. *Front. Biogeogr.* **2022**, *14*, e54662. [CrossRef]
49. Ikeda, T.; Higashide, D.; Shichijo, T. Impact of human disturbance in Japan on the distribution and diel activity pattern of terrestrial mammals. *J. Nat. Conserv.* **2022**, *70*, 126293. [CrossRef]
50. Kufa, C.A.; Bekele, A.; Atickem, A. Impacts of climate change on predicted habitat suitability and distribution of Djaffa Mountains Guereza (*Colobus guereza gallarum*, Neumann 1902) using MaxEnt algorithm in Eastern Ethiopian Highland. *Glob. Ecol. Conserv.* **2022**, *35*, e02094. [CrossRef]
51. Baral, K.; Adhikari, B.; Bhandari, S.; Kunwar, R.M.; Sharma, H.P.; Aryal, A.; Ji, W. Impact of climate change on distribution of common leopard (*Panthera pardus*) and its implication on conservation and conflict in Nepal. *Heliyon* **2023**, *9*, e12807. [CrossRef]
52. Pimm, S.L.; Jenkins, C.N.; Abell, R.; Brooks, T.M.; Gittleman, J.L.; Joppa, L.N.; Raven, P.H.; Roberts, C.M.; Sexton, J.O. The biodiversity of species and their rates of extinction, distribution, and protection. *Science* **2014**, *344*, 1246752. [CrossRef]
53. Guan, T.; Owens, J.R.; Kong, Y.; Jian, Y.; Gu, X.; Song, Y. The need for IUCN species distribution update—The case from takin (*Budorcas taxicolor*) in Southwest China. *Ecol. Evol.* **2022**, *12*, e9222. [CrossRef]
54. Foden, W.B.; Butchart, S.H.; Stuart, S.N.; Vié, J.C.; Akçakaya, H.R.; Angulo, A.; DeVantier, L.M.; Gutsche, A.; Turak, E.; Cao, L.; et al. Identifying the world's most climate change vulnerable species: A systematic trait-based assessment of all birds, amphibians and corals. *PLoS ONE* **2013**, *8*, e65427. [CrossRef] [PubMed]
55. Wheatley, C.J.; Beale, C.M.; Bradbury, R.B.; Pearce-Higgins, J.W.; Critchlow, R.; Thomas, C.D. Climate change vulnerability for species-assessing the assessments. *Glob. Chang. Biol.* **2017**, *23*, 3704–3715. [CrossRef] [PubMed]
56. Akçakaya, H.R.; Butchart, S.H.M.; Watson, J.E.M.; Pearson, R.G. Preventing species extinctions resulting from climate change. *Nat. Clim. Chang.* **2014**, *4*, 1048–1049. [CrossRef]
57. Wessely, J.; Hülber, K.; Gattringer, A.; Kuttner, M.; Moser, D.; Rabitsch, W.; Schindler, S.; Dullinger, S.; Essl, F. Habitat-based conservation strategies cannot compensate for climate-change-induced range loss. *Nat. Clim. Chang.* **2017**, *7*, 823–827. [CrossRef]
58. Bonnot, T.W.; Cox, W.A.; Thompson, F.R.; Millspaugh, J.J. Threat of climate change on a songbird population through its impacts on breeding. *Nat. Clim. Chang.* **2018**, *8*, 718–722. [CrossRef]
59. Liu, C.; White, M.; Newell, G. Measuring and comparing the accuracy of species distribution models with presence—Absence data. *Ecography* **2011**, *34*, 232–243. [CrossRef]

animals

MDPI

Article

Comparative Analysis of Gut Microbiota between Wild and Captive Golden Snub-Nosed Monkeys

Yunting Wang †, Xuanyi Yang †, Mingyi Zhang and Huijuan Pan *

School of Ecology and Nature Conservation, Beijing Forestry University, Beijing 100083, China
* Correspondence: phjjanine@bjfu.edu.cn; Tel.: +86-10-6233-6801
† These authors contributed equally to this work.

Simple Summary: Golden snub-nosed monkey is a critically endangered primate. Determining the gut microbial diversity, construction, and function is vital for protecting the golden snub-nosed monkey. The gut microbiota plays an essential role in regulating the physiological health of wild animals. The dominant phyla in the gut microbiota of captive and wild golden snub-nosed monkeys are *Bacteroidetes*, *Firmicutes*, and *Kiritimatiellaeota*. There are significant differences in the alpha and beta diversities of the gut microbiota between wild and captive golden snub-nosed monkeys, with the captive population having higher alpha diversity than the wild population. Functional predictions related to the Kyoto Encyclopedia of Genes and Genomes (KEGG) database showed that the most significant functional pathway at the second level between captive and wild monkeys was carbohydrate metabolism.

Abstract: Environmental shifts and dietary habits could directly affect the gut microbiota of animals. In this study, we investigated the gut microbiota of golden snub-nosed monkeys under two different conditions: captive and wild. Our study adopted a non-invasive sampling method, using full-length 16S rRNA Pacbio SMAT sequencing technology to compare the gut microbiota of wild and captive golden snub-nosed monkeys. The results showed that the captive populations had higher alpha diversity than the wild populations, and there were also significant differences in beta diversity. The linear discriminant analysis effect size (LEfSe) analysis showed 39 distinctly different taxonomic units. At the phylum level, the most dominant bacteria under captive and wild conditions were Bacteroidetes and Firmicutes. This study revealed that the different fiber intake between wild and captive populations might be the main reason for the difference in the gut microbiota. We found that captive golden snub-nosed monkeys had less beneficial bacteria and more potentially pathogenic bacteria than wild ones. Functional predictions showed that the most significant functional pathway at the second level between the captive and wild monkeys was carbohydrate metabolism. Therefore, our results indicate that diet changes caused by captivity could be the main reason impacting the gut microbiota of captive golden snub-nosed monkeys. We further highlight the potential impact of diet changes on the health of captive golden snub-nosed monkeys and offer some suggestions for the feeding of captive golden snub-nosed monkeys.

Keywords: golden snub-nosed monkey; captive; wild; gut microbiota

Citation: Wang, Y.; Yang, X.; Zhang, M.; Pan, H. Comparative Analysis of Gut Microbiota between Wild and Captive Golden Snub-Nosed Monkeys. *Animals* **2023**, *13*, 1625. https://doi.org/10.3390/ani13101625

Academic Editor: Chunwang Li

Received: 6 April 2023
Revised: 7 May 2023
Accepted: 10 May 2023
Published: 12 May 2023

1. Introduction

Golden snub-nosed monkey (*Rhinopithecus roxellana*) is one of the most endangered species in China [1]. It is only distributed in a remote mountainous area at elevations between 1500 and 3500 m in Sichuan, Gansu, Shaanxi, and Hubei provinces, with approximately 22,000–23,000 individuals in the wild [2]. They usually occupy large home ranges, live in family units, and have complex social relationships [3,4]. As an iconic endangered species and flagship protected species in China, the government has adopted both in situ and ex situ conservation strategies of the golden snub-nosed monkey. Ex situ conservation has been widely applied in protecting endangered animals that have difficulty surviving and reproducing in the wild [5,6]. However, living in a human-controlled environment under ex situ conservation could cause problems in animal health, such as gut microbial disorder, which currently causes significant concerns in wildlife conservation [7–9].

The gut microbiome is composed of bacteria, archaea, viruses, and eukaryotic microbes, and they have tremendous potential to impact our physiology, both in healthy and disease conditions [10,11]. The gut microbiota and the host interact through a long-term coevolutionary process to form a complex and relatively stable microbial environment [12]. The gut microbiota plays a vital role in the maintenance of the normal life activities of the host. It has been demonstrated that the gut microbiota is essential in the host's ecological adaptation, such as immunity digestion, diet behavior, and metabolism [13–21].

Significant differences exist between wild and captive lifestyles. The latter includes contact with humans, antibiotic exposure, altered diet composition, and increased stress, which may lead to microbiome disruption in captive animals [22]. Multiple studies have proved that captivity leads to diet alternations, and the living environment can significantly affect the animal's gut microbiota. Some studies on mammals demonstrated significant differences in the relative abundance of Firmicutes and Bacteroidetes between captive and wild individuals [9,23]. However, the influence of the external environment on the gut microbiota is not limited to the changes in the abundance of specific species or genera. For example, a previous study on wild and captive sika deer (*Cervus nippon*) has proved that captivity could also affect the diversity of the gut microbiota [24]. In studies of non-human primates (NHPs), captivity altered the original structure of the gut microbiome shaped by natural dietary sources [9,25–27] and increased the abundance and diversity of antibiotic genes [28]. Previous studies on humans and animals have suggested that many gastrointestinal (GI) diseases and metabolic diseases are strongly connected with gut microbiome disruption [29–32]. In brief, captivity affects the host's health by disrupting the gut microbiome. Such disruption causes alterations in the gut microbiota and can lead to pathogen colonization [33–36], which makes captive animals more susceptible to disease. In addition, the gut microbiota can also significantly affect the metabolism of the host [37]. The metabolite production of the gut microbiota from dietary sources will ultimately affect host health [38]. As a typical folivorous primate, the main food for golden snub-nosed monkeys in the wild environment is a large number of leaves [2], while in captivity, they have less intake of cellulose and protein and more intake of carbohydrates and fat [39,40]. Close contact with humans has increased the probability of infection with pathogenic bacteria. All of these factors will lead to changes in the intestinal tract microorganisms and affect health [17].

Therefore, studying the gut microbiota of golden snub-nosed monkeys in the wild and captivity is essential for protecting this species. In our study, we adopted a non-invasive sampling method that is harmless to animals, using full-length 16S rRNA Pacbio SMAT sequencing technology to compare the gut microbiota of 19 healthy wild and captive golden snub-nosed monkeys. This can help us better understand the adverse effects of captivity on the health status of golden snub-nosed monkeys and provide some feasible suggestions for managing captive golden snub-nosed monkeys.

2. Materials and Methods

2.1. Samples Collection

A total of 19 fecal samples were collected from 9 wild and 10 captive healthy golden snub-nosed monkeys. The captive samples were collected from Shanghai Wild Animal Park; the wild samples were collected from Mianyang, Sichuan Province. All fecal samples were collected and preserved in 15 mL centrifugal tubes immediately after defecation, snap-frozen in liquid nitrogen, and stored at −80 °C. DNA samples were stored frozen (−20 °C) until use.

2.2. DNA Extraction

Following the protocols provided by the manufacturer, the QIAamp DNA Stool Mini Kit (Qiagen, Valencia, CA, USA) was utilized to extract DNA. The concentration of DNA was assessed utilizing the Qubit dsDNA HS Assay Kit and Qubit 3.0 Fluorometer (Invitrogen, Thermo Fisher Scientific, Waltham, MA, USA). To determine the quality and quantity of the DNA, both a Nanodrop (ND-1000) spectrophotometer (Nanodrop Technologies, Wilmington, DE, USA) and agarose gel electrophoresis methods were utilized.

2.3. 16S rRNA Genes Amplicon Sequencing and Bioinformatics Analysis

The 16S rRNA gene was amplified using barcoded 27 forward and 1492 reverse primers (27F:5′-AGAGTTTGATCCTGGCTCAG-3′; 1492R:5′-CTACGGCTACCTTGTTACGA-3′) and sequenced using PacBio Sequel.

PCR amplification was carried out by performing 25 cycles using a KOD One PCR Master Mix (from TOYOBOLife Science, Shanghai, China). Initial denaturation was at 95 °C for 2 min, followed by denaturation at 98 °C for 10 s, annealing at 55 °C for 30 s, extension at 72 °C for 1 min and 30 s, and a final extension at 72 °C for 10 min. Following purification with Agencourt AMPure XP Beads (supplied by Beckman Coulter, Indianapolis, IN, USA), the total number of PCR amplicons was measured using the Qubit dsDNA HS Assay Kit and Qubit 3.0 Fluorometer (manufactured by Invitrogen, Thermo Fisher Scientific, Waltham, MA, USA). Following individual quantification, the amplicons were combined in equal quantities, and then SMRTbell libraries were generated from the amplified DNA using the SMRTbell Express Template Prep Kit 2.0 (from Pacific Biosciences, Menlo Park, CA, USA), as per the manufacturer's guidelines. After purification, the SMRTbell libraries from the pooled and barcoded samples were sequenced on a PacBio Sequel II 8M cell utilizing the Sequel II Sequencing Kit 2.0 (from Pacific Bioscience, Menlo Park, CA, USA).

Sequencing data were processed using SMRTlink software (version 8.0) to obtain circular consensus sequencing (CCS) reads through the filtering and demultiplexing of raw reads. To increase the reliability of CCS reads, which were generated following the demultiplexing of raw reads using SMRTlink software (version 8.0), we performed quality filtering in Cutadapt (version 2.7) to identify and discard CCS reads that did not match the forward and reverse primers and those falling outside the specified length range of 1200–1650 bp.

Microbiome bioinformatic analysis was performed with QIIME2 v.2020.11 [41]. Sequences were then quality-filtered, denoised, and merged, and chimeras were removed; then, amplicon sequence variants (ASVs) were output using DADA 2 v.1.8 [42,43], and the ASVs with abundance <0.001% were filtered. To generate the taxonomic table, we aligned ASV sequences against the SILVA reference database (version 132) pretrained at 99% sequence identity using the feature-classifier plugin available in QIIME2 [44].

2.4. Statistical Analysis

We evaluated the alpha diversity of the microbial communities using Chao1, Shannon, Simpson, Faith's, and Pielou's evenness indices, calculated with QIIME2. Additionally, we performed Kruskal–Wallis tests to examine differences in alpha diversity between the two groups [45–49]. We assessed the similarity of community structure among different groups by conducting principal coordinate analysis (PCoA) for which we utilized a Bray–Curtis dissimilarity matrix to estimate beta diversity. Then, we used the pair group method with arithmetic means (UPGMAs) and non-metric multidimensional scaling (NMDS) to analyze the beta diversity [50–52]. The ANOSIM (analysis of similarities) was used to evaluate the similarity between different individuals [53]. We graphically represented the relationship between captive and wild samples using a hierarchical clustering method. The samples were clustered using the average method, based on the distance matrix.

Kruskal–Wallis tests were used to detect the abundant differential features between captive and wild groups. LEfSe analysis (https://huttenhower.sph.harvard.edu/galaxy/ (accessed on 15 July 2022)) was performed to reveal the different taxa identified between captive and wild groups. A size-effect threshold of 3.0 on the logarithmic LDA score was used to discriminate functional biomarkers [54]. Student's *t*-test was used to determine the microbial communities and functional compositions that were significantly different between the two groups (*p*-value < 0.05). To control for multiple comparisons, we performed false discovery rate correction for the obtained *p*-values [55].

PICRUSt2 was used to predict the functional profiles of microbial communities [56]. The functional profiles of the identified taxa were predicted using the KEGG database. The Kruskal–Wallis test was used to calculate the differences in the gut microbiota between the captive and wild monkeys (*p*-value < 0.05).

3. Results

3.1. Microbial Community Profiles

A total of 174,012 clean reads were obtained from the 16S rRNA of the 19 snub-nosed monkeys (10 captive; 9 wild) (Table S1). We identified 1012 unique ASVs from the 19 fecal samples based on taxonomic annotation, distributed across 14 phyla, 21 classes, 30 orders, 54 families, and 142 genera (Table S2). The results indicated that the detection of bacteria in the samples was comprehensive, with a Good's coverage of nearly 99% (means ± SD = 98.5 ± 1.80%). The rarefaction curves in Figure S1 indicate that, as sequencing depth increased, the number of observed species also increased; however, the curve eventually plateaued, indicating that the sequencing depth met the requirement for subsequent analysis.

The analysis result showed that the top three dominant phyla in both captive and wild groups were *Bacteroidetes* (captive 48.89% vs. wild 42.47%; 45.68% on average); *Firmicutes* (captive 21.74% vs. wild 15.23%; 18.49% on average); and *Kiritimatiellaeota* (captive 13.05% vs. wild 22.21%; 17.63% on average) (Figure 1A,B). At the phylum level, *Bacteroidetes*, *Firmicutes*, *Proteobacteria*, *Tenericutes*, *Cyanobacteria*, and *Lentisphaerae* had higher relative abundance in the captive group, while *Kiritimatiellaeota*, *Verrucomicrobia*, *Spirochaetes*, and *Planctomycetes* had higher relative abundance in the wild group (Figure 2A). At the genus level, the relative abundances of *Clostridiales_vadinBB60_group* and *Prevotella_1* in the captive group were significantly higher than those in the wild group ($p < 0.05$). The relative abundances of *Rikenellaceae_RC9_gut_group*, *WCHB1-41*, *Akkermansia*, *p-2534-18B5_gut_group*, *Muribaculaceae*, *Ruminococcaceae_UCG-010*, *Treponema_2*, and *Lachnospiraceae_NK4A136_group* in the wild group were significantly higher than those in the captive group ($p < 0.05$) (Figure 2B).

3.2. Diversity Analysis of Microbiota in Captive and Wild Golden Snub-Nosed Monkey

We compared the alpha diversity of the microbiota between the captive group and the wild group based on the abundance at the genus level across all cohorts. The results showed that the captive environment greatly changed the gut microbiome's alpha diversity of snub-nosed monkeys. We observed a significant difference in the Shannon index (accounts for species richness and evenness) ($p = 0.022$) and the observed species index (accounts for the number of species contained in a community) ($p = 0.034$) between captive and wild monkeys (Figure 3A), indicating high richness, evenness and species number of the microbiota in captive monkeys.

PCoA and NMDS based on Bray–Curtis showed distinct differences in the diversity of the gut microbiota between captive and wild monkeys (Figure 3B,C). The result indicated that the microbial communities from captive monkeys clustered together and were separated from wild monkeys along the principal coordinate axis, which suggests significant differences in the microbial community composition between captive and wild monkeys.

3.3. Microbial Taxa Differences in the Gut Microbiota between Wild and Captive Golden Snub-Nosed Monkeys

The Venn diagram demonstrates the differences in gut microbiota composition between the two groups. A total of 1012 ASVs were detected in the captive and wild monkeys.

The captive and wild groups had 35 ASV overlaps, and the captive group had 611 unique ASVs, whereas the wild group had 366 unique ASVs (Figure 4).

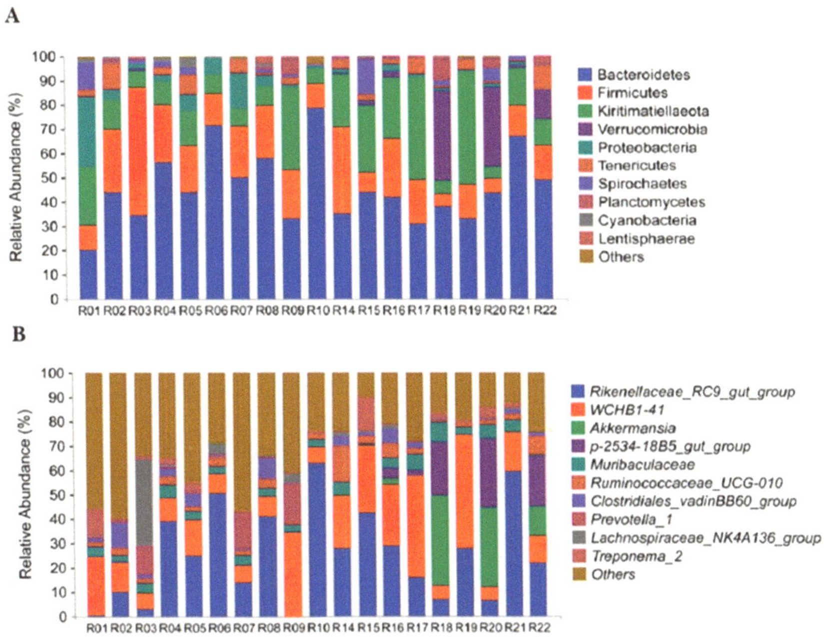

Figure 1. The basic structure of the bacterial community composition of each sample at the phylum (**A**) and genus (**B**) level. Stacked bar graphs illustrate the abundances and the x-axis represents the sample names.

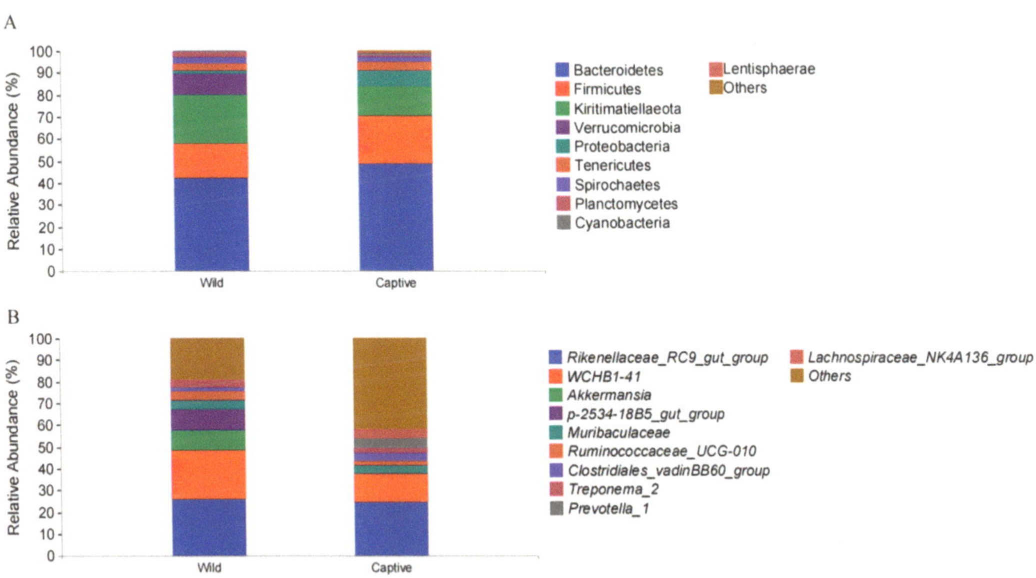

Figure 2. The bacterial community distributions and relative abundances comparison of the two groups at the phylum (**A**) and genus (**B**) level.

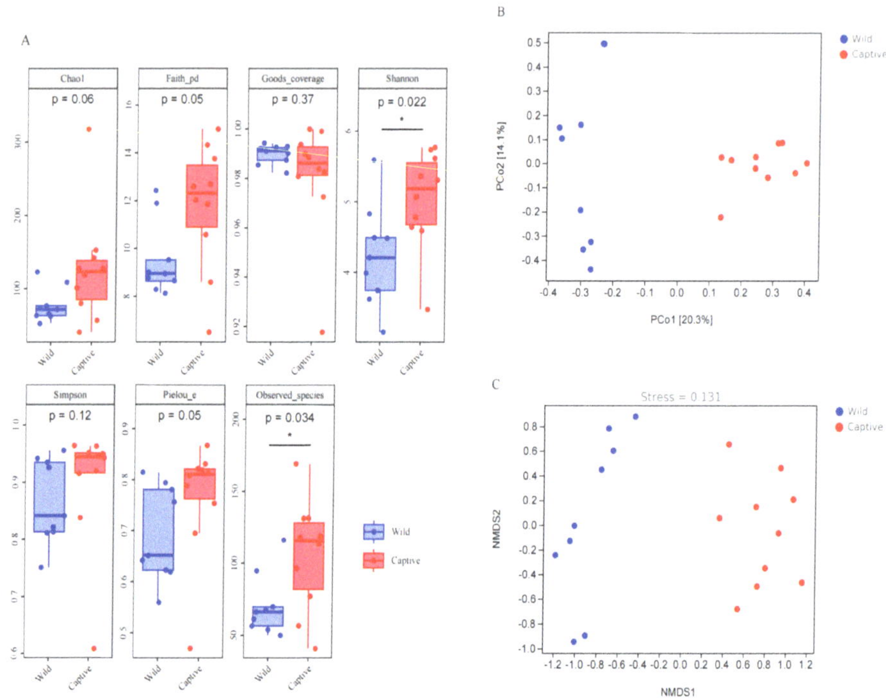

Figure 3. Boxplots showing α-diversity indices of microbiota in the captive and wild groups. The indices of Chao 1, Simpson, Shannon, Faith's PD, Pielou's evenness, and Good's coverage. The Chao 1 index is commonly used in ecology to estimate the total number of species; larger Chao 1 values represent the total number of species. The Shannon index and Simpson's diversity index are standard diversity measures, reflecting the samples' richness and evenness. * $p < 0.05$ (**A**). The principal coordinate analysis (PCoA) plot of Bray–Curtis distances shows the ecological distance between gut microbial communities in captive and wild monkeys. Blue dots represent wild monkeys, and red dots represent captive monkeys. A closer distance between two points infers a higher similarity (**B**). Using the Bray–Curtis distance, the NMDS plot displays the calculated distance between two groups based on dissimilarity in ASV composition. Blue dots represent wild monkeys, and red dots represent captive monkeys (**C**).

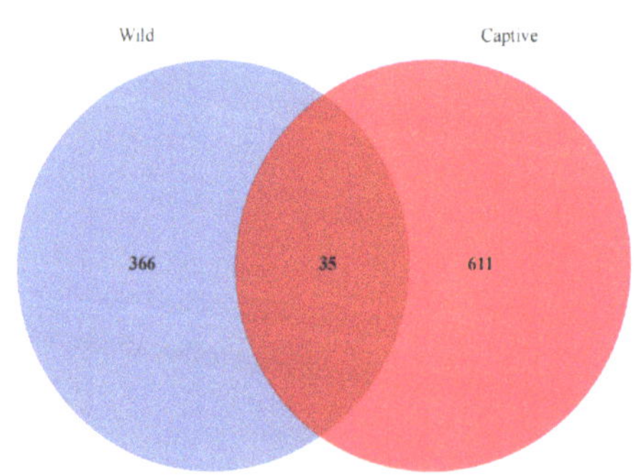

Figure 4. Venn diagram illustrating the number of ASVs shared by the captive and wild groups. The shared taxa by all individuals were to represent the core microbiota in two groups. The 2 groups shared 35 common ASVs, and within each group, there were 611 ASVs in the captive and 366 in the wild group.

To investigate the potential differences in microbial community composition between captive and wild populations, we utilized LEfSe tests to detect variations in the relative abundance of bacterial taxa. Figure 5 displays the microbial communities with significant differences in the relative abundances between the captive and wild groups. At the phylum level, the relative abundances of *Proteobacteria*, *Cyanobacteria*, and *Epsilonbacteraeota* in the captive golden snub-nosed monkeys were significantly higher than those in wild golden snub-nosed monkeys. In contrast, the relative abundance of *Verrucomicrobia* in wild golden snub-nosed monkeys was significantly higher than that in captive golden snub-nosed monkeys. At the genus levels, the relative abundances of *2534_18B5_gut_group*, *Ruminococcaceae_UCG_002*, *Oxalobacter*, and *Akkermansia* in wild golden snub-nosed monkeys were significantly higher than in captive golden snub-nosed monkeys. At the same time, *Prevotella_1*, *Prevotellaceae_UCG_003*, *Candidatus_Stoquefichus*, *Tyzzerella*, *Phascolarctobacterium*, *Gastranaerophilales*, *Coprococcus_2*, *Parasutterella*, *Prevotella_2*, and *Prevotellaceae_UCG_003* were more abundant in captive golden snub-nosed monkeys.

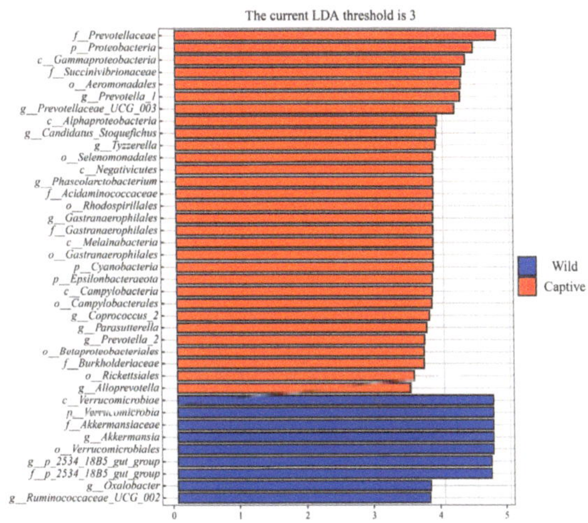

(A)

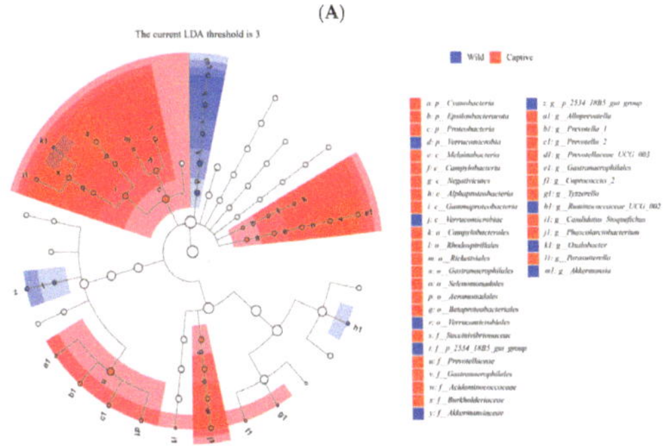

(B)

Figure 5. Using the LEfSe analysis to identify the bacterial taxa with significant differences between the two groups. The cladogram shows the evolutionary clades of different species. The circles radiating from the inside to the outside represent the taxonomic level from phylum to genus. Each small circle at a different taxonomic level represents a taxonomy at that level, and the diameter of the circles is proportional to the relative abundance. Red nodes represent microbial groups that play an important role in the captive group, blue nodes represent microbial groups that play an important role in the wild group, and yellow nodes mean no significant difference (**A**). Bacterial taxa with a significant difference and an LDA score greater than the estimated value (3.0); the histogram length represents the LDA score (**B**).

3.4. Functional Differences in Predicted Metagenomic between the Gut Microbiota of Wild and Captive Golden Snub-Nosed Monkeys

We performed a variance analysis of KEGG metabolic pathways in both captive and wild golden snub-nosed monkeys. A total of 129 KEGG Level 3 pathways were annotated, and the Kruskal–Wallis test was utilized to check the significance of K-numbers, where 28 of them were found to be significantly different ($p < 0.05$) between the wild and captive groups (Table S3). The majority of KEGG categories were found to be associated with metabolic processes, genetic information processing, cellular processes, environmental information processing, organismal systems, and human diseases (Figure 6A). The comparative analysis of the second-level pathways showed seven significantly different pathways ($p < 0.05$) (Figure 6B), including carbohydrate metabolism (ko00620), endocrine system (ko04910), membrane transport (ko03070), digestive system (ko04974), cellular community prokaryotes (ko05111), the metabolism of cofactors and vitamins (ko00760), and folding, sorting, and degradation (ko04122) (Table S3). The captive group had a higher abundance in the digestive system, cellular community prokaryotes, and metabolism of cofactors and vitamins. The wild group had a higher abundance of carbohydrate metabolism, endocrine system, and membrane transport (Figure 6B).

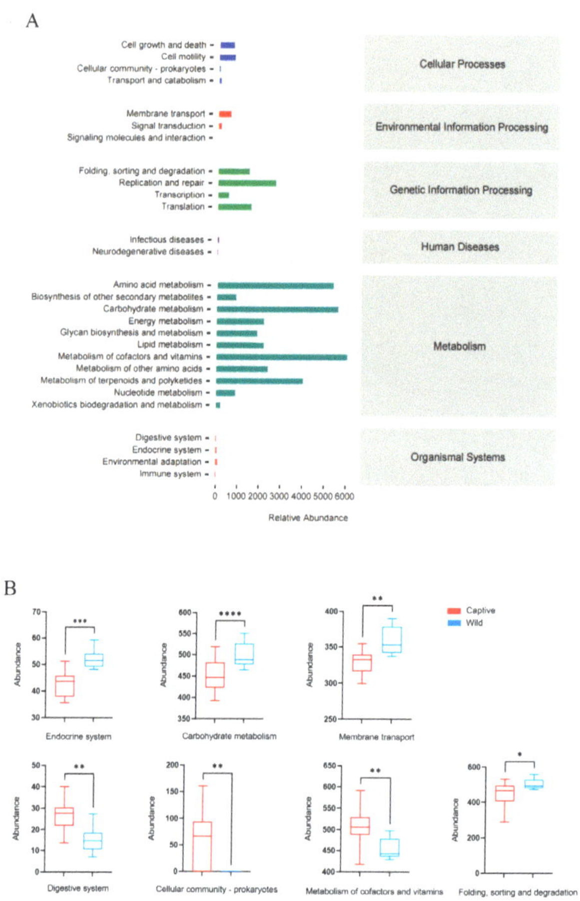

Figure 6. The prediction of the abundance of KEGG pathway classifications for all samples (**A**). Differential functional abundance in RPKM of KEGG Level 2 pathways in the gut microbial between captive and wild monkeys (**B**), * $p < 0.05$, ** $p < 0.01$, *** $p < 0.001$, **** $p < 0.0001$.

4. Discussion

In this study, we analyzed the difference in the gut microbiota between wild and captive golden snub-nosed monkeys using full-length 16S rRNA PacBio SMAT sequencing technology. We found that captive life may alter the community structure of the gut microbiota in golden snub-nosed monkeys. In addition, captive golden snub-nosed monkeys have more potential pathogens, which could cause GI problems, indicating that captive life might affect the gastrointestinal health of golden snub-nosed monkeys. Furthermore, our results suggest that diet might be the main cause affecting the gut microbiota of golden snub-nosed monkeys under different circumstances.

In our study, the dominant phyla in both captive and wild golden snub-nosed monkeys were *Bacteroidetes* and *Firmicutes* (Figure 1A), the same as other studies on primate gut microbiota [57–61]. *Bacteroidetes* can help degrade simple sugars, proteins, and carbohydrates, while *Firmicutes* are the main cellulolytic bacteria that can degrade fiber and cellulose [62–65]. Free-ranging golden snub-nosed monkeys consume a fiber-rich diet in the wild, and their primary food is more diverse, including lichens, leaves, seeds, fruits, buds, and bark [66]. Meanwhile, the diet of captive golden snub-nosed monkeys contains lower crude fiber than natural diets, and captive monkeys only have a mean of 15% crude fiber intake, while that of wild monkeys is up to 52% [39,67]. We found that *Bacteroidetes* were more abundant than *Firmicutes* in captive monkeys (Figure S2), which could be due to captive golden snub-nosed monkeys consuming more carbohydrates than wild monkeys.

We found the taxa with significant differences between the two groups using LEfSe analysis (Figure 5A,B). There are several notable bacterial taxa; some are potential pathogens, some are beneficial for hosts' health, and some are associated with diet. The genus *Prevotella* was significantly more abundant in captive golden snub-nosed monkeys. The high abundance of *Prevotella* often corresponds to an increased ability to digest simple carbohydrates [68–70]. This suggests that in captive individuals, the ability to digest cellulose is reduced as a result of food changes which in turn increases the ability to digest simple carbohydrates. In addition, *Prevotella* has a beneficial impact on glucose metabolism [71,72], and researchers have confirmed the role of *Prevotella* in regulating host health. Therefore, we could speculate that the increased abundance of *Prevotella* in captive monkeys is related to adaptation to captivity. *Proteobacteria* were also observed in the captive group. It is considered to be a marker of microbial dysbiosis and potential diagnostic criteria for disease [73] and is closely correlated with inflammatory bowel disease (IBD) [33,74] and colorectal cancer [75]. A higher *Proteobacteria* ratio is usually connected with poor health conditions [76,77]. *Proteobacteria* also play a role in intestinal inflammation. Studies on immunodeficient mice have pointed to the disturbance of the gut microbiome in diseased immunodeficient mice with a higher proportion of *Proteobacteria* species [78,79]. Another study of Crohn's disease also showed an increased relative abundance of Proteobacteria [80]. In addition, studies have shown that a high intake of sugars may increase the relative abundance of *Proteobacteria* in the gut [81,82]. Therefore, we can speculate that captive golden snub-nosed monkeys' high sugar intake (such as fructose and starch) could be the reason for the increased abundance of *Proteobacteria* in the captive group's gut microbiota. We found that the genus *Akkermansia* was abundant in the wild group. *Akkermansia* is considered a marker of intestinal health [83] and is essential in enhancing glucose tolerance, reducing insulin resistance, and regulating pathways in establishing basal metabolic homeostasis [84]. Studies have confirmed that diet significantly affects *Akkermansia*, and a decreased abundance of *Akkermansia* could be a sign of malnutrition [85].

In the results of our diversity analysis, captive monkeys' gut microbiota was richer and more diverse than that of wild monkeys. As typical folivorous primates, golden snub-nosed monkeys can obtain a more homogeneous diet in the wild but obtain richer food types in a captive environment; thus, we suggest that this factor may result in a significant increase in their gut microbial species and abundance in their gut [86]. In addition, animals in captivity have more frequent contact with humans and live in a more complex environment. The veterinary treatment of their diseases and the use of drugs may also lead to significant changes in their gut microbiota [87].

Though the findings showed that there was no significant difference in the makeup of functional pathways at the first level of KEGG analysis between the two groups (Figure 6A), gene function predictions showed that the second-level pathway was mainly related to metabolism, suggesting that the gut microbiota are closely related to their natural environments, especially for the host diet [88]. Thus, our study results indicate that gut microbiota plays an essential role in host physiology, and more studies are needed to investigate the mechanism of functional pathways further.

5. Conclusions

In summary, our study suggests that the gut microbiota of golden snub-nosed monkeys could be affected by a captive environment, especially due to changes in diet. By comparing the differences in the gut microbiota of both groups, we found a divergence in the diversity of the captive and wild monkeys' gut microbiota, which could be due to the captive monkeys consuming different food than the wild monkeys. On the one hand, we found that the wild monkeys had unique beneficial bacteria (*Akkermansia*), while the captive monkeys had more potentially pathogenic bacteria, suggesting that captive monkeys have a higher potential to get infected with diseases and suffer from poor health. On the other hand, *Prevotella*, which has a positive effect on glucose metabolism, was found to have a higher proportion in the gut of captive monkeys, indicating adaptation to captivity. The functional prediction analysis further confirmed the functional differences between the microbiota of the captive and wild monkeys. Our study could have implications for the implementation of instructions on how to feed animals in captivity. Overall, we suggest that captivity could disrupt the gut microbiota, but on the other hand, this disruption might help the host adjust to captive life.

Supplementary Materials: The following supporting information can be downloaded at: https://www.mdpi.com/article/10.3390/ani13101625/s1, Supplementary Figure S1: Rarefaction curves of the observed species index (A) and rank abundance curves of the ASVs (B); Supplementary Figure S2: Bar chart illustrating the relative abundance of Firmicutes and Bacteroidetes in the gut microbiota of captive and wild groups; Supplementary Table S1: Metadata of the samples and data information. Supplementary Table S2: ASVs of each group's taxonomy. The relative abundance of each taxonomy. Supplementary Table S3: Relative abundance of K-numbers and KO abundance of captive and wild monkeys. KO differs between captive and wild monkeys calculated by the Kruskal–Wallis test. Supplementary Table S4: Numbers of taxonomic units of the samples and alpha diversity index.

Author Contributions: Conceptualization, H.P.; methodology, Y.W. and X.Y.; software, Y.W. and X.Y.; validation, Y.W. and X.Y.; formal analysis, Y.W. and X.Y.; investigation, Y.W.; writing—original draft preparation, Y.W. and X.Y.; writing—review and editing, M.Z. and H.P.; visualization, M.Z., Y.W. and X.Y.; supervision, H.P.; project administration, H.P.; funding acquisition, H.P. All authors have read and agreed to the published version of the manuscript.

Funding: This study was supported by the National Natural Science Foundation of China (32070404) and the State Forestry Administration of China.

Institutional Review Board Statement: The procedures and protocols of all the experiments in this study have been approved by the Institution of Animal Care and the Ethics Committee of Beijing Forestry University, Approval No. EAWC_BJFU_2022012.

Data Availability Statement: Sequencing data can be found in the Sequence Read Archive (http://www.ncbi.nlm.nih.gov/bioproject/971738) under BioProject PRJNA971738.

Acknowledgments: The authors thank Xiaochen Wang and Ming Li for their help in review and editing. We also thank Ming Li for providing constructive suggestions for this paper.

Conflicts of Interest: The authors have no conflict of interest to declare.

References

1. Sung, W.; Yan, X. *China Species Red List*; Science Press: Beijing, China, 2004.
2. Zhou, X.; Wang, B.; Pan, Q.; Zhang, J.; Kumar, S.; Sun, X.; Liu, Z.; Pan, H.; Lin, Y.; Liu, G.; et al. Whole-genome sequencing of the snub-nosed monkey provides insights into folivory and evolutionary history. *Nat. Genet.* **2014**, *46*, 1303–1310. [CrossRef]
3. Qi, X.-G.; Garber, P.A.; Ji, W.; Huang, Z.-P.; Huang, K.; Zhang, P.; Guo, S.-T.; Wang, X.-W.; He, G.; Zhang, P.; et al. Satellite telemetry and social modeling offer new insights into the origin of primate multilevel societies. *Nat. Commun.* **2014**, *5*, 5296. [CrossRef]
4. Huang, Y.-P.; Xiang, J.-T.; Wang, C.-H.; Ren, D.; David, J.; Xu, T. Lichen as a biomonitor for vehicular emission of metals: A risk assessment of lichen consumption by the Sichuan snub-nosed monkey (*Rhinopithecus roxellana*). *Ecotoxicol. Environ. Saf.* **2019**, *180*, 679–685. [CrossRef]
5. Crees, J.J.; Collins, A.C.; Stephenson, P.; Meredith, H.M.; Young, R.P.; Howe, C.; Price, M.R.S.; Turvey, S.T. A comparative approach to assess drivers of success in mammalian conservation recovery programs. *Conserv. Biol.* **2016**, *30*, 694–705. [CrossRef]
6. Shan, L.; Hu, Y.; Zhu, L.; Yan, L.; Wang, C.; Li, D.; Jin, X.; Zhang, C.; Wei, F. Large-scale genetic survey provides insights into the captive management and reintroduction of giant pandas. *Mol. Biol. Evol.* **2014**, *31*, 2663–2671. [CrossRef] [PubMed]
7. Trevelline, B.K.; Fontaine, S.S.; Hartup, B.K.; Kohl, K.D. Conservation biology needs a microbial renaissance: A call for the consideration of host-associated microbiota in wildlife management practices. *Proc. R. Soc. B Boil. Sci.* **2019**, *286*, 20182448. [CrossRef]
8. Wei, F.; Wu, Q.; Hu, Y.; Huang, G.; Nie, Y.; Yan, L. Conservation metagenomics: A new branch of conservation biology. *Sci. China Life Sci.* **2019**, *62*, 168–178. [CrossRef] [PubMed]
9. McKenzie, V.J.; Song, S.J.; Delsuc, F.; Prest, T.L.; Oliverio, A.M.; Korpita, T.M.; Alexiev, A.; Amato, K.R.; Metcalf, J.L.; Kowalewski, M.; et al. The effects of captivity on the mammalian gut microbiome. *Integr. Comp. Biol.* **2017**, *57*, 690–704. [CrossRef]
10. Ahrodia, T.; Das, S.; Bakshi, S.; Das, B. Structure, functions, and diversity of the healthy human microbiome. *Hum. Microbiome Health Dis. Part A* **2022**, *191*, 53.
11. Clayton, J.B.; Gomez, A.; Amato, K.; Knights, D.; Travis, D.A.; Blekhman, R.; Knight, R.; Leigh, S.; Stumpf, R.; Wolf, T.; et al. The gut microbiome of nonhuman primates: Lessons in ecology and evolution. *Am. J. Primatol.* **2018**, *80*, e22867. [CrossRef]
12. Ley, R.E.; Hamady, M.; Lozupone, C.; Turnbaugh, P.J.; Ramey, R.R.; Bircher, J.S.; Schlegel, M.L.; Tucker, T.A.; Schrenzel, M.D.; Knight, R.; et al. Evolution of mammals and their gut microbes. *Science* **2008**, *320*, 1647–1651. [CrossRef]
13. Clayton, J.B.; Al-Ghalith, G.A.; Long, H.T.; Tuan, B.V.; Cabana, F.; Huang, H.; Vangay, P.; Ward, T.; Minh, V.V.; Tam, N.A.; et al. Associations between nutrition, gut microbiome, and health in a novel nonhuman primate model. *Sci. Rep.* **2018**, *8*, 11159. [CrossRef] [PubMed]
14. Chung, H.; Pamp, S.J.; Hill, J.A.; Surana, N.K.; Edelman, S.M.; Troy, E.B.; Reading, N.C.; Villablanca, E.J.; Wang, S.; Mora, J.R.; et al. Gut immune maturation depends on colonization with a host-specific microbiota. *Cell* **2012**, *149*, 1578–1593. [CrossRef] [PubMed]
15. Round, J.L.; Mazmanian, S.K. The gut microbiota shapes intestinal immune responses during health and disease. *Nat. Rev. Immunol.* **2009**, *9*, 313–323. [CrossRef] [PubMed]
16. Cho, I.; Blaser, M.J. The human microbiome: At the interface of health and disease. *Nat. Rev. Genet.* **2012**, *13*, 260–270. [CrossRef]
17. Amato, K.R.; Metcalf, J.L.; Song, S.J.; Hale, V.L.; Clayton, J.; Ackermann, G.; Humphrey, G.; Niu, K.; Cui, D.; Zhao, H.; et al. Using the gut microbiota as a novel tool for examining colobine primate GI health. *Glob. Ecol. Conserv.* **2016**, *7*, 225–237. [CrossRef]
18. Chevalier, C.; Stojanović, O.; Colin, D.J.; Suarez-Zamorano, N.; Tarallo, V.; Veyrat-Durebex, C.; Rigo, D.; Fabbiano, S.; Stevanović, A.; Hagemann, S.; et al. Gut microbiota orchestrates energy homeostasis during cold. *Cell* **2015**, *163*, 1360–1374. [CrossRef] [PubMed]
19. Tardiolo, G.; Romeo, O.; Zumbo, A.; Di Marsico, M.; Sutera, A.M.; Cigliano, R.A.; Paytuví, A.; D'Alessandro, E. Characterization of the Nero Siciliano Pig Fecal Microbiota after a Liquid Whey-Supplemented Diet. *Animals* **2023**, *13*, 642. [CrossRef] [PubMed]
20. Wang, X.-C.; Zhang, J.-L.; Pan, H.-J.; Chen, Y.-X.; Mao, S.-X.; Qi, J.-W.; Shen, Y.; Zhang, M.-Y.; Xiang, Z.-F.; Li, M. Unique characteristics of gut microbiota in black snub-nosed monkeys (*Rhinopithecus strykeri*) reveal an enzymatic mechanism of adaptation to dietary vegetation. *Zool. Res.* **2023**, *44*, 357. [CrossRef]
21. Zhang, M.; Wang, X.; Wang, Z.; Mao, S.; Zhang, J.; Li, M.; Pan, H. Metatranscriptomic Analyses Reveal Important Roles of the Gut Microbiome in Primate Dietary Adaptation. *Genes* **2023**, *14*, 228. [CrossRef]
22. Dallas, J.W.; Warne, R.W. Captivity and animal microbiomes: Potential roles of microbiota for influencing animal conservation. *Microb. Ecol.* **2022**, *85*, 820–838. [CrossRef] [PubMed]
23. Eigeland, K.A.; Lanyon, J.M.; Trott, D.J.; Ouwerkerk, D.; Blanshard, W.; Milinovich, G.J.; Gulino, L.-M.; Martinez, E.; Merson, S.; Klieve, A.V. Bacterial community structure in the hindgut of wild and captive dugongs (*Dugong dugon*). *Aquat. Mamm.* **2012**, *38*, 402–411. [CrossRef]
24. Guan, Y.; Yang, H.; Han, S.; Feng, L.; Wang, T.; Ge, J. Comparison of the gut microbiota composition between wild and captive sika deer (*Cervus nippon hortulorum*) from feces by high-throughput sequencing. *AMB Express* **2017**, *7*, 212. [CrossRef] [PubMed]
25. Clayton, J.B.; Vangay, P.; Huang, H.; Ward, T.; Hillmann, B.M.; Al-Ghalith, G.A.; Travis, D.A.; Long, H.T.; Tuan, B.V.; Minh, V.V.; et al. Captivity humanizes the primate microbiome. *Proc. Natl. Acad. Sci. USA* **2016**, *113*, 10376–10381. [CrossRef]
26. Malukiewicz, J.; Cartwright, R.A.; Dergam, J.A.; Igayara, C.S.; Kessler, S.E.; Moreira, S.B.; Nash, L.T.; Nicola, P.A.; Pereira, L.C.; Pissinatti, A.; et al. The gut microbiome of exudivorous marmosets in the wild and captivity. *Sci. Rep.* **2022**, *12*, 5049. [CrossRef] [PubMed]

27. Hale, V.L.; Tan, C.L.; Niu, K.; Yang, Y.; Zhang, Q.; Knight, R.; Amato, K.R. Gut microbiota in wild and captive Guizhou snub-nosed monkeys, *Rhinopithecus brelichi*. *Am. J. Primatol.* **2019**, *81*, e22989. [CrossRef] [PubMed]
28. Jia, T.; Chang, W.-S.; Marcelino, V.R.; Zhao, S.; Liu, X.; You, Y.; Holmes, E.C.; Shi, M.; Zhang, C. Characterization of the gut microbiome and resistomes of wild and zoo-captive macaques. *Front. Vet. Sci.* **2022**, *8*, 1664. [CrossRef] [PubMed]
29. Kamada, N.; Seo, S.-U.; Chen, G.Y.; Núñez, G. Role of the gut microbiota in immunity and inflammatory disease. *Front. Vet. Sci.* **2013**, *13*, 321–335. [CrossRef]
30. Brown, C.T.; Davis-Richardson, A.G.; Giongo, A.; Gano, K.A.; Crabb, D.B.; Mukherjee, N.; Casella, G.; Drew, J.C.; Ilonen, J.; Knip, M.; et al. Gut microbiome metagenomics analysis suggests a functional model for the development of autoimmunity for type 1 diabetes. *PLoS ONE* **2011**, *6*, e25792. [CrossRef]
31. Boerner, B.P.; Sarvetnick, N.E. Type 1 diabetes: Role of intestinal microbiome in humans and mice. *Ann. N. Y. Acad. Sci.* **2011**, *1243*, 103–118. [CrossRef]
32. Guo, W.; Mishra, S.; Wang, C.; Zhang, H.; Ning, R.; Kong, F.; Zeng, B.; Zhao, J.; Li, Y. Comparative study of gut microbiota in wild and captive giant pandas (*Ailuropoda melanoleuca*). *Genes* **2019**, *10*, 827. [CrossRef] [PubMed]
33. David, L.A.; Maurice, C.F.; Carmody, R.N.; Gootenberg, D.B.; Button, J.E.; Wolfe, B.E.; Ling, A.V.; Devlin, A.S.; Varma, Y.; Fischbach, M.A.; et al. Diet rapidly and reproducibly alters the human gut microbiome. *Nature* **2014**, *505*, 559–563. [CrossRef] [PubMed]
34. O'Mahony, S.M.; Marchesi, J.R.; Scully, P.; Codling, C.; Ceolho, A.-M.; Quigley, E.M.; Cryan, J.F.; Dinan, T.G. Early life stress alters behavior, immunity, and microbiota in rats: Implications for irritable bowel syndrome and psychiatric illnesses. *Biol. Psychiatry* **2009**, *65*, 263–267. [CrossRef]
35. Stecher, B.; Robbiani, R.; Walker, A.W.; Westendorf, A.M.; Barthel, M.; Kremer, M.; Chaffron, S.; Macpherson, A.J.; Buer, J.; Parkhill, J.; et al. *Salmonella enterica* serovar typhimurium exploits inflammation to compete with the intestinal microbiota. *PLoS Biol.* **2007**, *5*, e244. [CrossRef] [PubMed]
36. Stecher, B.; Hardt, W.-D. The role of microbiota in infectious disease. *Trends Microbiol.* **2008**, *16*, 107–114. [CrossRef]
37. Qin, J.; Li, R.; Raes, J.; Arumugam, M.; Burgdorf, K.S.; Manichanh, C.; Nielsen, T.; Pons, N.; Levenez, F.; Yamada, T.; et al. A human gut microbial gene catalogue established by metagenomic sequencing. *Nature* **2010**, *464*, 59–65. [CrossRef]
38. Ursell, L.K.; Haiser, H.J.; Van Treuren, W.; Garg, N.; Reddivari, L.; Vanamala, J.; Dorrestein, P.C.; Turnbaugh, P.J.; Knight, R. The intestinal metabolome: An intersection between microbiota and host. *Gastroenterology* **2014**, *146*, 1470–1476. [CrossRef]
39. Chen, S.T.; Luo, X.; Hou, R.; Raubenheimer, D.; Ji, W.; Jin, X.; Jiang, Z.; Yu, X.; Wang, J.; Li, M.; et al. Nutrient balancing by captive golden snub-nosed monkeys (*Rhinopithecus roxellana*). *Int. J. Primatol.* **2018**, *39*, 1124–1138. [CrossRef]
40. Guo, S.T.; Hou, R.; Garber, P.A.; Raubenheimer, D.; Righini, N.; Ji, W.H.; Jay, O.; He, S.J.; Wu, F.; Li, F.F.; et al. Nutrient-specific compensation for seasonal cold stress in a free-ranging temperate colobine monkey. *Funct. Ecol.* **2018**, *32*, 2170–2180. [CrossRef]
41. Bolyen, E.; Rideout, J.R.; Dillon, M.R.; Bokulich, N.A.; Abnet, C.C.; Al-Ghalith, G.A.; Alexander, H.; Alm, E.J.; Arumugam, M.; Asnicar, F.; et al. Reproducible, interactive, scalable and extensible microbiome data science using QIIME 2. *Nat. Biotechnol.* **2019**, *37*, 852–857. [CrossRef]
42. Callahan, B.J.; McMurdie, P.J.; Rosen, M.J.; Han, A.W.; Johnson, A.J.A.; Holmes, S.P. DADA2: High-resolution sample inference from Illumina amplicon data. *Nat. Methods* **2016**, *13*, 581–583. [CrossRef] [PubMed]
43. Callahan, B.J.; Wong, J.; Heiner, C.; Oh, S.; Theriot, C.M.; Gulati, A.S.; McGill, S.K.; Dougherty, M.K. High-throughput amplicon sequencing of the full-length 16S rRNA gene with single-nucleotide resolution. *Nucleic Acids Res.* **2019**, *47*, e103. [CrossRef] [PubMed]
44. Bokulich, N.A.; Kaehler, B.D.; Rideout, J.R.; Dillon, M.; Bolyen, E.; Knight, R.; Huttley, G.A.; Gregory Caporaso, J. Optimizing taxonomic classification of marker-gene amplicon sequences with QIIME 2's q2-feature-classifier plugin. *Microbiome* **2018**, *6*, 90. [CrossRef]
45. Shannon, C.E. A mathematical theory of communication. *Bell Syst. Tech. J.* **1948**, *27*, 379–423. [CrossRef]
46. Simpson, E.H. Measurement of diversity. *Nature* **1949**, *163*, 688. [CrossRef]
47. Faith, D.P. Conservation evaluation and phylogenetic diversity. *Biol. Conserv.* **1992**, *61*, 1–10. [CrossRef]
48. Pielou, E.C. The measurement of diversity in different types of biological collections. *J. Theor. Biol.* **1966**, *13*, 131–144. [CrossRef]
49. Good, I.J. The population frequencies of species and the estimation of population parameters. *Biometrika* **1953**, *40*, 237–264. [CrossRef]
50. Lozupone, C.A.; Hamady, M.; Kelley, S.T.; Knight, R. Quantitative and qualitative β diversity measures lead to different insights into factors that structure microbial communities. *Appl. Environ. Microbiol.* **2007**, *73*, 1576–1585. [CrossRef]
51. Lozupone, C.; Knight, R. UniFrac: A new phylogenetic method for comparing microbial communities. *Appl. Environ. Microbiol.* **2005**, *71*, 8228–8235. [CrossRef] [PubMed]
52. Bray, J.R.; Curtis, J.T. An ordination of the upland forest communities of southern Wisconsin. *Ecol. Monogr.* **1957**, *27*, 326–349. [CrossRef]
53. Anderson, M.J.; Walsh, D.C. PERMANOVA, ANOSIM, and the Mantel test in the face of heterogeneous dispersions: What null hypothesis are you testing? *Ecol. Monogr.* **2013**, *83*, 557–574. [CrossRef]
54. Segata, N.; Izard, J.; Waldron, L.; Gevers, D.; Miropolsky, L.; Garrett, W.S.; Huttenhower, C. Metagenomic biomarker discovery and explanation. *Genome Biol.* **2011**, *12*, R60. [CrossRef]

55. Benjamini, Y.; Hochberg, Y. Controlling the false discovery rate: A practical and powerful approach to multiple testing. *J. R. Stat. Soc. Ser. B* **1995**, *57*, 289–300. [CrossRef]
56. Langille, M.G.; Zaneveld, J.; Caporaso, J.G.; McDonald, D.; Knights, D.; Reyes, J.A.; Clemente, J.C.; Burkepile, D.E.; Vega Thurber, R.L.; Knight, R.; et al. Predictive functional profiling of microbial communities using 16S rRNA marker gene sequences. *Nat. Biotechnol.* **2013**, *31*, 814–821. [CrossRef]
57. McKenney, E.A.; Rodrigo, A.; Yoder, A.D. Patterns of gut bacterial colonization in three primate species. *PLoS ONE* **2015**, *10*, e0124618. [CrossRef] [PubMed]
58. Hale, V.L.; Tan, C.L.; Knight, R.; Amato, K.R. Effect of preservation method on spider monkey (*Ateles geoffroyi*) fecal microbiota over 8 weeks. *J. Microbiol. Methods* **2015**, *113*, 16–26. [CrossRef]
59. Ma, J.; Prince, A.L.; Bader, D.; Hu, M.; Ganu, R.; Baquero, K.; Blundell, P.; Alan Harris, R.; Frias, A.E.; Grove, K.L.; et al. High-fat maternal diet during pregnancy persistently alters the offspring microbiome in a primate model. *Nat. Commun.* **2014**, *5*, 3889. [CrossRef]
60. Xu, B.; Xu, W.; Li, J.; Dai, L.; Xiong, C.; Tang, X.; Yang, Y.; Mu, Y.; Zhou, J.; Ding, J.; et al. Metagenomic analysis of the Rhinopithecus bieti fecal microbiome reveals a broad diversity of bacterial and glycoside hydrolase profiles related to lignocellulose degradation. *BMC Genomics* **2015**, *16*, 174. [CrossRef]
61. Barelli, C.; Albanese, D.; Donati, C.; Pindo, M.; Dallago, C.; Rovero, F.; Cavalieri, D.; Tuohy, K.M.; Hauffe, H.C.; De Filippo, C. Habitat fragmentation is associated to gut microbiota diversity of an endangered primate: Implications for conservation. *Sci. Rep.* **2015**, *5*, 14862. [CrossRef]
62. Dooley, H.M.; Judge, D.S. Kloss gibbon (*Hylobates klossii*) behavior facilitates the avoidance of human predation in the Peleonan forest, Siberut Island, Indonesia. *Am. J. Primatol.* **2015**, *77*, 296–308. [CrossRef]
63. Hon, N.; Behie, A.M.; Rothman, J.M.; Ryan, K.G. Nutritional composition of the diet of the northern yellow-cheeked crested gibbon (*Nomascus annamensis*) in northeastern Cambodia. *Primates* **2018**, *59*, 339–346. [CrossRef]
64. Bäckhed, F.; Ding, H.; Wang, T.; Hooper, L.V.; Koh, G.Y.; Nagy, A.; Semenkovich, C.F.; Gordon, J.I. The gut microbiota as an environmental factor that regulates fat storage. *Proc. Natl. Acad. Sci. USA* **2004**, *101*, 15718–15723. [CrossRef]
65. Chen, H.; Wang, C.; Huasai, S.; Chen, A. Effects of dietary forage to concentrate ratio on nutrient digestibility, ruminal fermentation and rumen bacterial composition in Angus cows. *Sci. Rep.* **2021**, *11*, 17023. [CrossRef]
66. Guo, S.; Li, B.; Watanabe, K. Diet and activity budget of *Rhinopithecus roxellana* in the Qinling Mountains, China. *Primates* **2007**, *48*, 268–276. [CrossRef]
67. Hou, R.; He, S.; Wu, F.; Chapman, C.A.; Pan, R.; Garber, P.A.; Guo, S.; Li, B. Seasonal variation in diet and nutrition of the northern-most population of *Rhinopithecus roxellana*. *Am. J. Primatol.* **2018**, *80*, e22755. [CrossRef] [PubMed]
68. De Filippo, C.; Cavalieri, D.; Di Paola, M.; Ramazzotti, M.; Poullet, J.B.; Massart, S.; Collini, S.; Pieraccini, G.; Lionetti, P. Impact of diet in shaping gut microbiota revealed by a comparative study in children from Europe and rural Africa. *Proc. Natl. Acad. Sci. USA* **2010**, *107*, 14691–14696. [CrossRef] [PubMed]
69. Flint, H.J.; Scott, K.P.; Duncan, S.H.; Louis, P.; Forano, E. Microbial degradation of complex carbohydrates in the gut. *Gut Microbes* **2012**, *3*, 289–306. [CrossRef] [PubMed]
70. Wu, G.D.; Chen, J.; Hoffmann, C.; Bittinger, K.; Chen, Y.-Y.; Keilbaugh, S.A.; Bewtra, M.; Knights, D.; Walters, W.A.; Knight, R.; et al. Linking long-term dietary patterns with gut microbial enterotypes. *Science* **2011**, *334*, 105–108. [CrossRef]
71. Kovatcheva-Datchary, P.; Nilsson, A.; Akrami, R.; Lee, Y.S.; De Vadder, F.; Arora, T.; Hallen, A.; Martens, E.; Björck, I.; Bäckhed, F. Dietary fiber-induced improvement in glucose metabolism is associated with increased abundance of Prevotella. *Cell Metab.* **2015**, *22*, 971–982. [CrossRef]
72. Koeth, R.A.; Wang, Z.; Levison, B.S.; Buffa, J.A.; Org, E.; Sheehy, B.T.; Britt, E.B.; Fu, X.; Wu, Y.; Li, L.; et al. Intestinal microbiota metabolism of L-carnitine, a nutrient in red meat, promotes atherosclerosis. *Nat. Med.* **2013**, *19*, 576–585. [CrossRef] [PubMed]
73. Shin, N.-R.; Whon, T.W.; Bae, J.-W. Proteobacteria: Microbial signature of dysbiosis in gut microbiota. *Trends Biotechnol.* **2015**, *33*, 496–503. [CrossRef] [PubMed]
74. Seksik, P.; Rigottier-Gois, L.; Gramet, G.; Sutren, M.; Pochart, P.; Marteau, P.; Jian, R.; Dore, J. Alterations of the dominant faecal bacterial groups in patients with Crohn's disease of the colon. *Gut* **2003**, *52*, 237–242. [CrossRef]
75. Wang, T.; Cai, G.; Qiu, Y.; Fei, N.; Zhang, M.; Pang, X.; Jia, W.; Cai, S.; Zhao, L. Structural segregation of gut microbiota between colorectal cancer patients and healthy volunteers. *ISME J.* **2012**, *6*, 320–329. [CrossRef] [PubMed]
76. Mizrahi, I.; Wallace, R.J.; Moraïs, S. The rumen microbiome: Balancing food security and environmental impacts. *Nat. Rev. Genet.* **2021**, *19*, 553–566. [CrossRef]
77. Monira, S.; Nakamura, S.; Gotoh, K.; Izutsu, K.; Watanabe, H.; Alam, N.H.; Endtz, H.P.; Cravioto, A.; Ali, S.I.; Nakaya, T.J.; et al. Gut microbiota of healthy and malnourished children in Bangladesh. *Front. Microbiol.* **2011**, *2*, 228. [CrossRef]
78. Garrett, W.S.; Lord, G.M.; Punit, S.; Lugo-Villarino, G.; Mazmanian, S.K.; Ito, S.; Glickman, J.N.; Glimcher, L.H. Communicable ulcerative colitis induced by T-bet deficiency in the innate immune system. *Cell* **2007**, *131*, 33–45. [CrossRef]
79. Devkota, S.; Wang, Y.; Musch, M.W.; Leone, V.; Fehlner-Peach, H.; Nadimpalli, A.; Antonopoulos, D.A.; Jabri, B.; Chang, E.B. Dietary-fat-induced taurocholic acid promotes pathobiont expansion and colitis in $Il10^{-/-}$ mice. *Nature* **2012**, *487*, 104–108. [CrossRef]
80. Gevers, D.; Kugathasan, S.; Denson, L.A.; Vázquez-Baeza, Y.; Van Treuren, W.; Ren, B.; Schwager, E.; Knights, D.; Song, S.J.; Yassour, M.; et al. The treatment-naive microbiome in new-onset Crohn's disease. *Cell Host Microbe* **2014**, *15*, 382–392. [CrossRef]

81. Satokari, R. High intake of sugar and the balance between pro- and anti-inflammatory gut bacteria. *Nutrients* **2020**, *12*, 1348. [CrossRef]
82. Roca-Saavedra, P.; Mendez-Vilabrille, V.; Miranda, J.M.; Nebot, C.; Cardelle-Cobas, A.; Franco, C.M.; Cepeda, A. Food additives, contaminants and other minor components: Effects on human gut microbiota—A review. *J. Physiol. Biochem.* **2018**, *74*, 69–83. [CrossRef]
83. Frugé, A.D.; Van der Pol, W.; Rogers, L.Q.; Morrow, C.D.; Tsuruta, Y.; Demark-Wahnefried, W. Fecal Akkermansia muciniphila is associated with body composition and microbiota diversity in overweight and obese women with breast cancer participating in a presurgical weight loss trial. *J. Acad. Nutr. Diet.* **2020**, *120*, 650–659. [CrossRef] [PubMed]
84. Derrien, M.; Belzer, C.; de Vos, W.M. Akkermansia muciniphila and its role in regulating host functions. *Microb. Pathog.* **2017**, *106*, 171–181. [CrossRef] [PubMed]
85. Everard, A.; Belzer, C.; Geurts, L.; Ouwerkerk, J.P.; Druart, C.; Bindels, L.B.; Guiot, Y.; Derrien, M.; Muccioli, G.G.; Delzenne, N.M.; et al. Cross-talk between *Akkermansia muciniphila* and intestinal epithelium controls diet-induced obesity. *Proc. Natl. Acad. Sci. USA* **2013**, *110*, 9066–9071. [CrossRef] [PubMed]
86. Campbell, T.P.; Sun, X.; Patel, V.H.; Sanz, C.; Morgan, D.; Dantas, G. The microbiome and resistome of chimpanzees, gorillas, and humans across host lifestyle and geography. *ISME J.* **2020**, *14*, 1584–1599. [CrossRef] [PubMed]
87. Grassotti, T.T.; de Angelis Zvoboda, D.; da Fontoura Xavier Costa, L.; De Araujo, A.J.G.; Pereira, R.I.; Soares, R.O.; Wagner, P.G.C.; Frazzon, J.; Frazzon, A.P.G. Antimicrobial Resistance Profiles in *Enterococcus* spp. Isolates from Fecal Samples of Wild and Captive Black Capuchin Monkeys (*Sapajus nigritus*) in South Brazil. *Front. Microbiol.* **2018**, *9*, 2366. [CrossRef] [PubMed]
88. Muegge, B.D.; Kuczynski, J.; Knights, D.; Clemente, J.C.; González, A.; Fontana, L.; Henrissat, B.; Knight, R.; Gordon, J.I. Diet drives convergence in gut microbiome functions across mammalian phylogeny and within humans. *Science* **2011**, *332*, 970–974. [CrossRef]

animals

MDPI

Communication

Acoustic Presence of Cetaceans in the Miaodao Archipelago, China

Zhaolong Cheng [1,2], Yongtao Li [1,2], Matthew Keith Pine [3], Xiaoling Wan [4], Tao Zuo [1,2], Mingxiang Niu [1,2] and Jun Wang [1,2,*]

[1] Yellow Sea Fisheries Research Institute, Chinese Academy of Fishery Sciences, Qingdao 266071, China
[2] Laboratory for Marine Ecology and Environmental Science, Pilot National Laboratory for Marine Science and Technology (Qingdao), Qingdao 266237, China
[3] Department of Biology, University of Victoria, Victoria, BC V8P 5C2, Canada
[4] School of Animal Science and Nutritional Engineering, Wuhan Polytechnic University, Wuhan 430023, China
* Correspondence: wangjun@ysfri.ac.cn

Simple Summary: Little effort has been made to conserve cetaceans in the Miaodao Archipelago, which is hindered by a lack of baseline data on their species and distribution patterns. Using a passive acoustic monitoring technique, we found a decrease in cetacean diversity; the East Asian finless porpoise is the sole cetacean species that can be reliability detected in this area, and their distribution exhibits seasonally patterns. Further research and conservation measures are needed to protect cetaceans in this area.

Abstract: Once an important cetacean habitat, the Miaodao Archipelago has been altered by human-induced disturbances over several decades. While cetacean diversity is known to have decreased, no recent data on species diversity around Miaodao are known to exist. Capitalizing on the high vocal activity of cetaceans, three passive acoustic surveys, including towed and stationary types, were undertaken to detect the presence of species-specific vocalizations in May 2021, October 2021, and July 2022, as most cetacean sightings occurred during May and August in recent years. The results revealed that the East Asian finless porpoise is the sole cetacean species that can be reliably observed around the archipelago, as no other species were detected. The acoustic data also revealed potentially clumped distributions of finless porpoises with some seasonal variation. While not acoustically detected during any of the surveys, humpback whales, minke whales, and killer whales have been visually sighted in the region. The lack of acoustic detection of these species suggests that they are likely to be temporary visitors to the region, or at least exhibit strong seasonality in their presence within the region. These new data provide the latest snapshot of cetacean presence around the Miaodao Archipelago that can help inform future research and conservation.

Keywords: passive acoustic monitoring; species diversity; distribution; conservation

Citation: Cheng, Z.; Li, Y.; Pine, M.K.; Wan, X.; Zuo, T.; Niu, M.; Wang, J. Acoustic Presence of Cetaceans in the Miaodao Archipelago, China. *Animals* **2023**, *13*, 1306. https://doi.org/10.3390/ani13081306

Academic Editor: Chunwang Li

Received: 15 March 2023
Revised: 5 April 2023
Accepted: 10 April 2023
Published: 11 April 2023

1. Introduction

The Miaodao Archipelago is made up of 32 small islands and is located at the intersection of the Yellow Sea and the Bohai Sea, between the Shandong and Liaodong Peninsulas. This region is characterized by a temperate, semi-humid continental and ocean climate, and supports economically important commercial fisheries and agriculture [1]. Historical whaling and stranding records show that the Miaodao Archipelago is utilized by both mysticetes and odontocetes, including four baleen species (humpback whale (*Megaptera novaeangliae*), minke whale (*Balaenoptera acutorostrata*), fin whale (*Balaenoptera physalus*), and gray whale (*Eschrichtius robustus*)) and five odontocete species (killer whale (*Orcinus orca*), false killer whale (*Pseudorca crassidens*), short-beaked common dolphin (*Delphinus delphis*), common bottlenose dolphin (*Tursiops truncatus*), and the marine subspecies of narrow-ridged finless porpoise (*Neophocaena asiaeorientalis*)-the

East Asian finless porpoise (*N. a. sunameri*)) [2]. However, cetacean species in this region have been suffering sharp declines in population size due to detrimental influences from human activities, such as commercial whaling, fishery bycatch, chemical and noise pollution, and overfishing [3]. Based on the recent sighting records, only two baleen and two odontocete species have been found in these waters (such as humpback whale, http://www.jiaodong.net/travel/system/2018/05/07/013670986.shtml (accessed on 9 April 2023); minke whale, https://ishare.ifeng.com/c/s/7nzD8FqWGIS (accessed on 9 April 2023); killer whale, https://www.sohu.com/a/329178594_120044938 (accessed on 9 April 2023), and https://v.youku.com/v_show/id_XMzk0MTk4NjUwNA==.html (accessed on 9 April 2023); East Asian finless porpoise, https://www.toutiao.com/article/6663770826057187844/?wid=1660867925355 (accessed on 9 April 2023), https://sjb.qlwb.com.cn/qlwb/content/20220526/ArticelA01003FM.htm (accessed on 9 April 2023), and http://w.dzwww.com/p/5172554.html (accessed on 9 April 2023)). All of the four species are listed in the National Key Protected Animals in the newly modified list of key protected wild animals in China in 2021, of which humpback whale, minke whale, and killer whale are Grade 1 protected animals and the East Asian finless porpoise is a Grade 2 protected animal, meaning it is critically necessary for them to be protected. There is thus a need to further understand basic information on the population of these species within and around the Miaodao Archipelago, such as their distribution patterns and dynamics, and behaviors, so as to push for appropriate conservation practices.

Despite this, little scientific attention has been given to study the cetacean presence within the Miaodao Archipelago. A key reason may be that effective monitoring of these species is challenging. The classical approach to monitoring cetaceans, being visual-based methodologies, continues to provide valuable information on cetaceans, including their abundance and distribution patterns. However, despite their benefits, visual surveys are expensive (particularly vessel-based surveys), time consuming, labor intensive, and limited to good weather and visibility conditions [4]. In addition, some species, in particular the finless porpoise, can sometimes be missed by observers due to their fast swimming and lack of dorsal fin [5]. The passive acoustic method counters these limitations and has thus become a fast-evolving tool in marine mammal monitoring. Passive acoustic monitoring (PAM) is cost-effective and has been widely used in cetacean studies [5–7]. Cetaceans rely heavily on their vocalizations in order to adapt to their underwater environment, making them particularly suitable for PAM approaches [6].

Humpback whales are commonly known for their singing, made up of repeated series of complex vocalizations [8]. They also produce vast repertoires of non-song social calls. Males sing during the breeding season, whereas both males and females produce social calls when in groups of three or more adults [9]. The fundamental frequency of sounds produced by humpback whales ranges between tens of Hz and 4 kHz, with higher harmonics extending beyond 24 kHz [10]. Minke whales in the North Pacific produce unique vocalizations (commonly referred to as boings) during their breeding season from fall to spring [11]. The fundamental frequencies of boings ranged between 1 and 1.8 kHz, with harmonics extending to approximately 9 kHz [12]. Killer whales produce broad-band echolocation clicks up to 85 kHz, whistles between 1.5 kHz and 18 kHz, and pulse calls that appear tonal but are a rapid series of pulses with peak energies between 1 and 6 kHz [13–15]. East Asian finless porpoises produce narrowband high-frequency echolocation clicks well over 100 kHz that can be easily identified from other bio- and abiotic sounds [16].

Considering that cetacean sightings occur mostly during May and August, we investigated the acoustic presence of the species and distribution patterns of cetaceans in the Miaodao Archipelago using a PAM system in May 2021, October 2021, and July 2022 in this study. As all of the potentially existing species in this region are considered endangered and data deficient, the information provided here could be helpful for future conservation efforts.

2. Materials and Methods

2.1. Data Collection

Data were collected using a SoundTrap 300HF (Ocean Instruments, Auckland, New Zealand) recorder that was set to record continuously at a 576 kHz sampling rate in the Miaodao Archipelago (Figure 1).

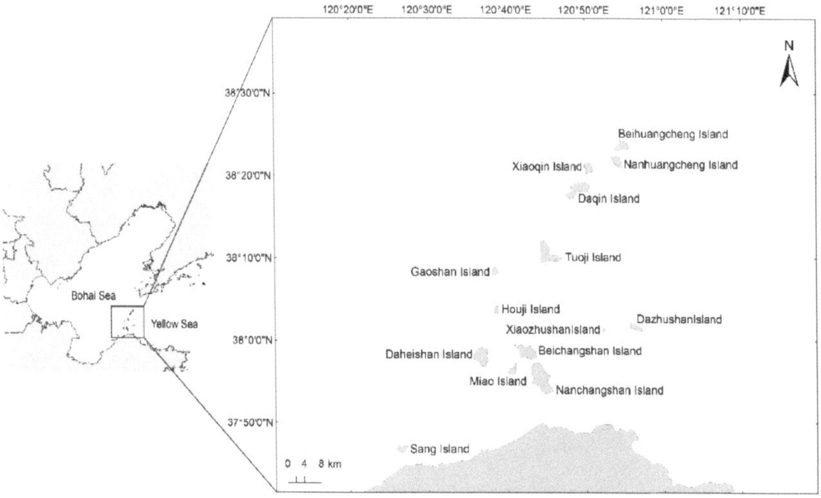

Figure 1. Map of the passive acoustic monitoring area where the humpback whale, minke whale, killer whale, and East Asian finless porpoise may be encountered.

To cover the study area evenly, towed PAM survey transects were spaced at 10 km. A 26 m long fishing vessel (at approximately 12 km/h) was used to tow the recorder 100 m behind the boat using a nylon rope. An iron bar (approximately 300 g) was affixed to the front of the recorder, and five floats were also attached to the nylon rope to suspend the recorder in the seawater without touching the sea bed. A time-stamped GPS receiver (Garmin GPSMAP 639sc; Garmin International Ltd., Olathe, KS, USA) was used to track the survey routes.

Stationary PAM surveys were conducted during the nighttime of the towed PAM surveys when the fishing vessel was anchored and the engine was turned off. The recorder was attached with plastic bands to a rope, and was deployed vertically into the water at 2.5 m depth using a 7 kg weight.

2.2. Data Analysis

Acoustic recordings were downloaded and converted to WAV files using SoundTrap host software (Ocean Instruments, Auckland, New Zealand). The WAV files were then uploaded to Raven Pro 1.6 software (the Cornell Lab of Ornithology, Ithaca, NY, USA) to manually identify potential cetacean vocalizations.

To pick out the sound of humpback whales and minke whales, acoustic recordings were inspected in the range of 0–5 kHz with the following parameters: window type: Hanning; window size: 10,000; 3 dB filter bandwidth: 82.8 Hz; time grid overlap: 50%; hop size: 8.68 ms; frequency grid DFT size: 16,384; grid spacing: 35.2 Hz. To locate the sounds from the killer whale and East Asian finless porpoises, acoustic recordings were inspected in full bandwidth with the following parameters: window type: Hanning; window size: 2048; 3 dB filter bandwidth: 405 Hz; time grid overlap: 50%; hop size: 1.78 ms; frequency grid DFT size: 2048; grid spacing: 281 Hz.

A detection was defined as a series of vocalizations within 5 min following the last vocalization. In towed PAM, the acoustic detection density was calculated by dividing the survey distance into the numbers of detection.

3. Results

Surveys were conducted in May 2021, October 2021, and July 2022. In towed PAM, the recorder was towed 356.49 km in 30.08 h, 423.52 km in 32.58 h, and 504.76 km in 44.75 h, respectively. In stationary PAM, the recorder was deployed for 124.62 h, 87.85 h, and 103.28 h, respectively. No humpback whale, minke whale, or killer whale vocalizations were detected. Only narrow band high frequency clicks from the East Asian finless porpoises were detected (Figure 2).

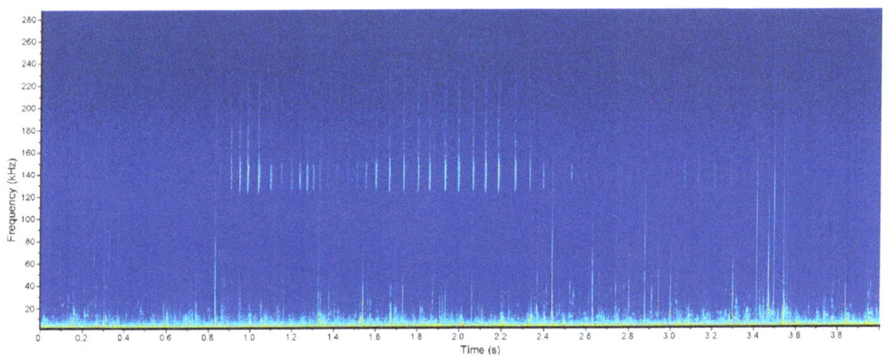

Figure 2. Spectrogram of the narrow band high frequency clicks recorded from the East Asian finless porpoises in Miaodao Archipelago waters (hanning; window size: 2048; 3 dB filter bandwidth: 405 Hz; time grid overlap: 50%; hop size: 1.78 ms; frequency grid DFT size: 2048; grid spacing: 281 Hz).

In towed PAM, 169 East Asian finless porpoise detections (56 in May 2021, 64 in October 2021, and 49 in July 2022) were found, which were throughout the Miaodao Archipelago, but displayed a seasonal distribution pattern. Dividing the survey area into four parts by N38°05′ and E120°45′, the details of the porpoise detection, survey distance, and porpoise detection density in each part are listed in Table 1. Survey routes and detection locations are shown in Figures 3–5.

Table 1. Details of the results of the towed passive acoustic monitoring in the Miaodao Archipelago waters in May 2021, October 2021, and July 2022.

		Survey		
		May 2021	**October 2021**	**July 2022**
Part I	Detection	33	27	33
	Distance (km)	200.65	154.76	176.44
	Density (/km)	0.16	0.17	0.19
Part II	Detection	8	21	4
	Distance (km)	59.63	146.26	183.75
	Density (/km)	0.13	0.14	0.02
Part III	Detection	8	7	8
	Distance (km)	54.66	32.96	57.58
	Density (/km)	0.15	0.21	0.14
Part IV	Detection	7	9	4
	Distance (km)	41.55	89.54	86.99
	Density (/km)	0.17	0.10	0.05

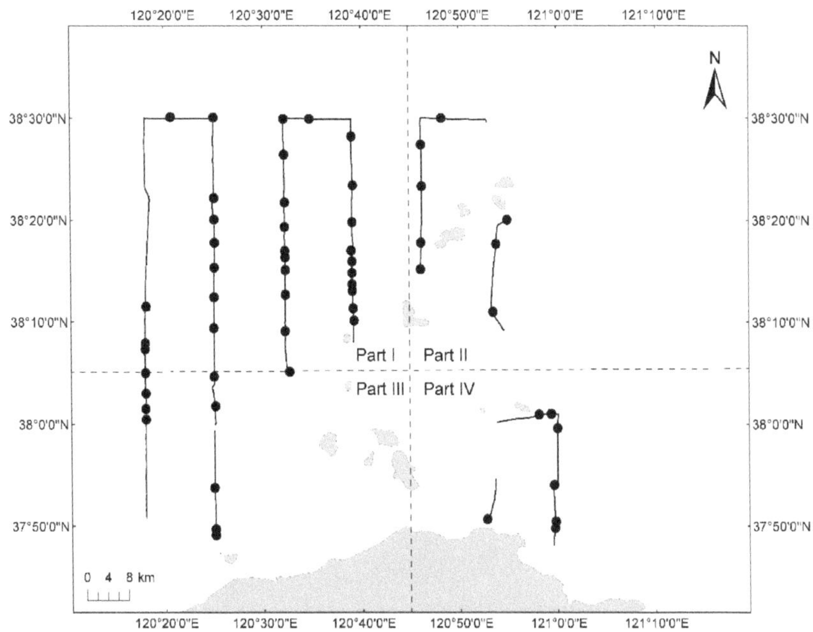

Figure 3. Survey routes (black line) and acoustic detection location (black dot) for the East Asian finless porpoises in Miaodao Archipelago waters in May 2021.

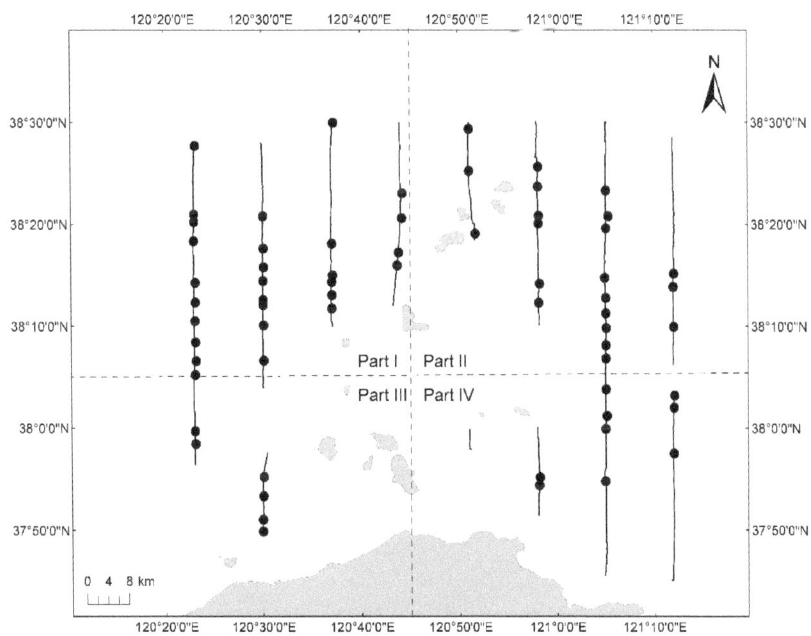

Figure 4. Survey routes (black line) and acoustic detection location (black dot) for the East Asian finless porpoises in Miaodao Archipelago waters in October 2021.

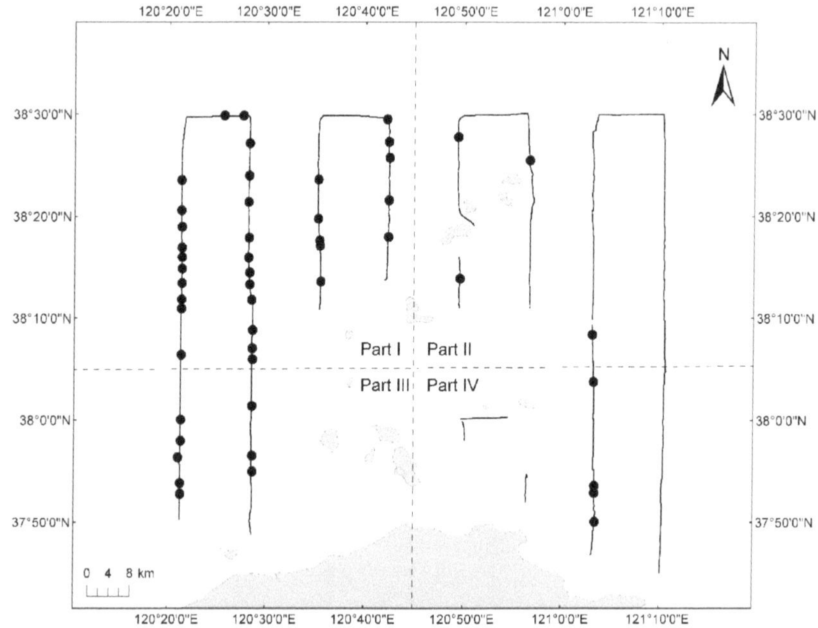

Figure 5. Survey routes (black line) and acoustic detection location (black dot) for the East Asian finless porpoises in Miaodao Archipelago waters in July 2022.

In stationary PAM, 99 East Asian finless porpoise detections (54 in May 2021, 24 in October 2021, and 21 in July 2022) were found. The duration of the porpoise acoustic detection recorded from each site is showed in Table 2 and Figure 6. In summary, the proportions of times in which porpoises were detected were 32.14% in May 2021, 11.64% in October 2021, and 5.02% in July 2022.

Table 2. Details of the results of the stationary passive acoustic monitoring in Miaodao Archipelago waters in May 2021, October 2021, and July 2022.

Site	Monitoring Time	Detection Duration	Proportion
20210518	810 min	164 min	20.25%
20210519	820 min	805 min	98.18%
20210520	645 min	505 min	78.29%
20210522	2500 min	682 min	27.29%
20210524	662 min	221 min	33.33%
20210527	2040 min	30 min	1.45%
20211012	894 min	178 min	19.89%
20211013	768 min	339 min	44.16%
20211014	1001 min	0 min	0%
20211020	755 min	69 min	9.09%
20211021	1117 min	15 min	1.34%
20211022	736 min	15 min	2.01%
20220717	775 min	0 min	0%
20220718	645 min	86 min	13.28%
20220720	1352 min	40 min	2.94%
20220721	2180 min	0 min	0%
20220723	630 min	34 min	5.47%
20220724	615 min	79 min	12.90%

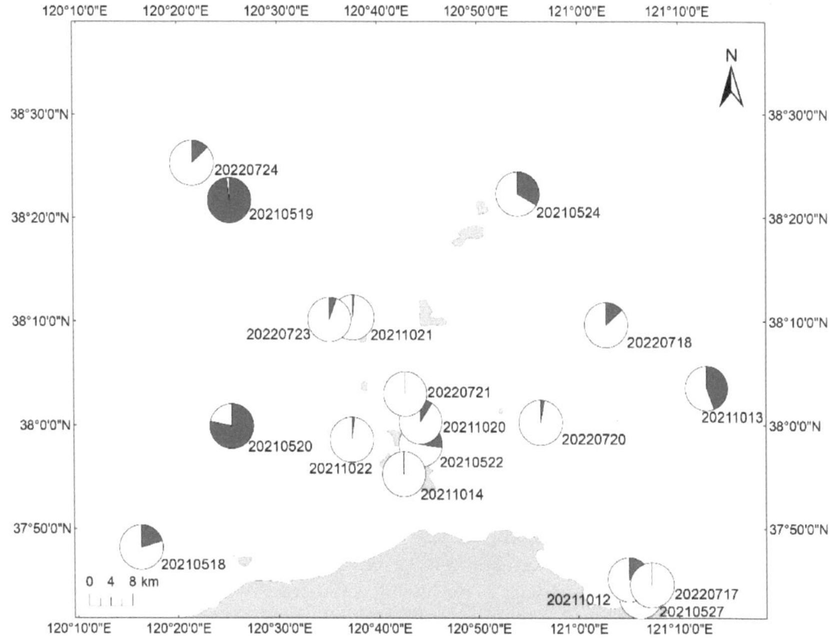

Figure 6. Map showing the proportions of the duration of porpoise acoustic detection in each site (dark in pie charts) in Miaodao Archipelago waters in May 2021, October 2021, and July 2022.

4. Discussion

Based on historical records and recent citizen information, it was expected that at least vocalizations from humpback whale, minke whale, killer whale, and East Asian finless porpoise would be found in our recordings. However, in the present study, only narrow band high frequency clicks, i.e., sounds produced by East Asian finless porpoise, were detected.

Our results suggest that the diversity of cetaceans around the Miaodao Archipelago waters have declined: only four species were sighted in recent years, and only East Asian finless porpoises were detected in our recordings. Cheng et al. [3] suggest that whaling (stopped since 1980 when China joined the International Whaling Commission), food deficiency caused by overfishing, bycatch, and environmental pollution were their main threats. The minke whales in Miaodao Archipelago were previously believed to originate from Japan, while the humpback whales were thought to come from Sea of Okhotsk, and killer whales were believed to be from the Chukchi Peninsula [17]. Their sightings in Miaodao Archipelago mostly occurred between May and August, when several fish species migrate through this area [18]. Notwithstanding, the precise origins and the motivations behind some patterns still require further study.

The present study further suggests that the Miaodao Archipelago is a vital habitat for the East Asian finless porpoise, concurring with previous data from Cheng et al. [3] who obtained baseline information of the East Asian finless porpoises in south Bohai Sea using local ecological knowledge. In addition, the present study updates the baseline distribution information for the porpoise in Miaodao Archipelago: in May and October, their distribution was relative even, as the detection rates in all four parts were above or equal 0.10 per km; however, in July, a heterogeneous distribution pattern was found, and their detections were concentrated in west part of the Miaodao Archipelago, i.e., part I and part III in the present study. Human disturbance and prey spatio-temporal distribution are the key factors influencing the fine-scale habitat use in cetacean [19,20]. In this study, the porpoises spent more time in areas distant from islands and lands. It is possible that porpoises preferred these habitats owing to the low level of disturbance from human activities. The distribution and movement patterns of East Asian finless porpoises in the adjacent waters were found to be closely correlated with prey [5,21]. Seasonal porpoise distribution variation in the present study may indicate the area changes for fish aggregation.

There are some limitations in the present study. Firstly, only one recorder was used, which made it impossible to determine the orientation of the sound source, i.e., we could not directly count the animals the recorder recorded. Then, shellfish aquaculture and fixed fishing net flourish in the Miaodao Archipelago. This was particularly the case in part III, part IV, and the area close to the islands, which hindered towing a recorder behind the survey boat. Thirdly, there is still the possibility that humpback, minke, and killer whales were missed due to our limited sampling efforts. For example, mammal-eating killer whales often travel or forage without discernibly echolocating [22]. In the future, PAM towing hydrophone array, long-term stationary PAM, and visual surveys, are recommended to be carried out to clarify the movement pattern of the humpback, minke, and killer whales, as well as the population status, size, age class composition, habitat range, and annual or seasonal present pattern of the East Asian finless porpoises in this water.

The conservation of cetaceans in the Miaodao Archipelago water is insufficient due to a previous lack of baseline knowledge. The findings of the present study underscore the importance of conducting further research, taking conservation measures to protect cetaceans in this area. Marine anthropogenic noise and fish resource management are key issues in protecting cetaceans [5]. Therefore, it is important to regulate human activities such as commercial shipping, oil extracting, and wind turbine construction in the Miaodao Archipelago water. The use of fishing methods such as bottom trawling and drift gillnets should be prohibited in this area in order to protect the marine ecosystem. There is a need for increased education initiatives in communities and schools to raise public awareness about marine mammal conservation. Establishing rescue organizations to save marine mammals injured in bycatch events is also important. In the future, it will be necessary to establish a marine protected area covering these regions.

Author Contributions: Conceptualization, Z.C. and J.W.; methodology, Z.C. and J.W.; validation, T.Z., M.N. and J.W.; formal analysis, Z.C. and M.K.P.; investigation, Z.C. and Y.L.; data curation, Z.C.; writing—original draft preparation, Z.C.; writing—review and editing, M.K.P., X.W. and J.W.; supervision, T.Z., M.N. and J.W.; funding acquisition, J.W. and Z.C. All authors have read and agreed to the published version of the manuscript.

Funding: This research was funded by the Shandong Provincial Natural Science Foundation (grant number ZR2022QC099); the Bureau of Fisheries, The Ministry of Agriculture and Rural Affairs of the People's Republic of China (grant number 125162002000210008); and the China National Offshore Oil Corporation foundation (grant number CF-MEEC/TR/2022-10).

Institutional Review Board Statement: Not applicable.

Informed Consent Statement: Not applicable.

Data Availability Statement: The data presented in this study are available on request from the corresponding author.

Acknowledgments: We would like to thank the captain of Baidi Li for his valuable help during data collection. We also thank Jiaqiang Chen for his help in data collection.

Conflicts of Interest: The authors declare no conflict of interest.

References

1. Wang, X.L.; Zhang, J. A nonlinear model for assessing multiple probabilistic risks: A case study in South five-island of Miaodao National Nature Reserve in China. *J. Environ. Manag.* **2007**, *85*, 1101–1108. [CrossRef] [PubMed]
2. Jefferson, T.A.; Webber, M.A.; Pitman, R.L.; Gorter, U. *Marine Mammals of the World*; Elsevier: London, UK, 2015.

3. Cheng, Z.; Pine, M.K.; Li, Y.; Zuo, T.; Niu, M.; Wan, X.; Zhao, X.; Wang, K.; Wang, J. Using local ecological knowledge to determine ecological status and threats of the East Asian finless porpoise, *Neophocaena asiaeorientalis sunameri*, in south Bohai Sea, China. *Ocean. Coast Manag.* **2021**, *203*, 105516. [CrossRef]
4. Kimura, S.; Akamatsu, T.; Wang, K.; Wang, D.; Li, S.; Dong, S.; Arai, N. Comparison of stationary acoustic monitoring and visual observation of finless porpoises. *J. Acoust. Soc. Am.* **2009**, *125*, 547–553. [CrossRef] [PubMed]
5. Cheng, Z.; Li, Y.; Pine, M.K.; Zuo, T.; Niu, M.; Wang, J. Association between porpoise presence and fish choruses: Implications for feeding strategies and ecosystem-based conservation of the East Asian finless porpoise. *Integr. Zool.* **2023**, *18*, 169–182. [CrossRef] [PubMed]
6. Zimmer, W.M.X. *Passive Acoustic Monitoring of Cetacean*; Cambridge University Press: Cambridge, UK, 2011; pp. 239–258.
7. Barlow, J.; Cheeseman, T.; Trickey, J.S. Acoustic detections of beaked whales, narrow-band high-frequency pulses and other odontocete cetaceans in the Southern Ocean using an autonomous towed hydrophone recorder. *Deep. Sea Res. Part II Top. Stud. Oceanogr.* **2021**, *193*, 104973. [CrossRef]
8. Ross-Marsh, E.C.; Elwen, S.H.; Prinsloo, A.S.; James, B.X.; Gridley, T. Singing in South Africa: Monitoring the occurrence of humpback whale (*Megaptera novaeangliae*) song near the Western Cape. *Bioacoustics* **2021**, *30*, 163–179. [CrossRef]
9. Silber, G.K. The relationship of social vocalizations to surface behavior and aggression in the Hawaiian humpback whale (*Megaptera novaeangliae*). *Can. J. Zool.* **1986**, *64*, 2075–2080. [CrossRef]
10. Tsujii, K.; Akamatsu, T.; Okamoto, R.; Mori, K.; Mitani, Y.; Umeda, N. Change in singing behavior of humpback whales caused by shipping noise. *PLoS ONE* **2018**, *13*, e0204112. [CrossRef] [PubMed]
11. Nikolich, K.; Towers, J. Vocalizations of common minke whales (*Balaenoptera acutorostrata*) in an eastern North Pacific feeding ground. *Bioacoustics* **2018**, *29*, 97–108. [CrossRef]
12. Oswald, J.N.; Au, W.W.; Duennebier, F. Minke whale (*Balaenoptera acutorostrata*) boings detected at the Station ALOHA Cabled Observatory. *J. Acoust. Soc. Am.* **2011**, *129*, 3353–3360. [CrossRef] [PubMed]
13. Ford, J.K. Acoustic behaviour of resident killer whales (*Orcinus orca*) off Vancouver Island, British Columbia. *Can. J. Zool.* **1989**, *67*, 727–745. [CrossRef]
14. Wellard, R.; Erbe, C.; Fouda, L.; Blewitt, M. Vocalisations of Killer Whales (*Orcinus orca*) in the Bremer Canyon, Western Australia. *PLoS ONE* **2015**, *10*, e136535.
15. Reyes Reyes, M.V.; Baumann-Pichering, S.; Simonis, A.; Melcón, M.L.; Trickey, J.; Hildebrand, J.; Iñíguez, M. High-frequency modulated signals recorded off the Antarctic Peninsula area: Are killer whales emitting them? *Acoust. Aust.* **2017**, *45*, 253–260. [CrossRef]
16. Li, S.; Wang, D.; Wang, K.; Akamatsu, T.; Ma, Z.; Han, J. Echolocation click sounds from wild inshore finless porpoise (*Neophocaena phocaenoides sunameri*) with comparisons to the sonar of riverine *N. p. asiaeorientalis*. *J. Acoust. Soc. Am.* **2007**, *121*, 3938–3946. [CrossRef] [PubMed]
17. Wang, P. *Chinese Cetaceans*; Chemical Industry Press: Beijing, China, 2012; pp. 27–376.
18. Chen, D. *Fishery Ecology of the Bohai Sea and the Yellow Sea*; Ocean Press: Beijing, China, 1991; pp. 143–463.
19. Dransfield, A.; Hines, E.; McGowan, J.; Holzman, B.; Nu, N.; Elliott, M.; Howar, J.; Jahncke, J. Where the whales are: Using habitat modeling to support changes in shipping regulations within National Marine Sanctuaries in Central California. *Endang. Species Res.* **2014**, *26*, 39–57. [CrossRef]
20. Brough, T.; Rayment, W.; Slooten, E.; Dawson, S. Spatiotemporal distribution of foraging in a marine predator: Behavioural drivers of hotspot formation. *Mar. Ecol. Prog. Ser.* **2020**, *635*, 187–202. [CrossRef]
21. Cheng, Z.; Yu, G.; Li, Y.; Zuo, T.; Niu, M.; Wang, J. Distribution pattern of the East Asian finless porpoise in the Huanghe River estuary and its adjacent waters in spring. *Oceanol. Limnol. Sin.* **2022**, *53*, 505–512.
22. Barrett-Lennard, L.G.; Ford, J.K.B.; Heise, K.A. The mixed blessing of echolocation: Differences in sonar use by fish-eating and mammal-eating killer whales. *Anim. Behav.* **1996**, *51*, 553–565. [CrossRef]

animals

MDPI

Review

The Science Underlying Giant Panda Conservation Translocations

Yue Wang [1,2,†], Wei Wei [1,2,†], Feiyun Yuan [3], Dandan Cao [1,2] and Zejun Zhang [1,2,4,*]

1 Key Laboratory of Southwest China Wildlife Resources Conservation (Ministry of Education), China West Normal University, Nanchong 637001, China; wangyue@stu.cwnu.edu.cn (Y.W.); weiw@cwnu.edu.cn (W.W.); cxy9@stu.cwnu.edu.cn (D.C.)
2 Liziping Giant Panda's Ecology and Conservation Observation and Research Station of Sichuan Province, China West Normal University, Nanchong 637001, China
3 Sichuan Lushi Expressway Co., Ltd., Chengdu 610041, China; chenyx@stu.cwnu.edu.cn
4 Chengdu Normal University, Chengdu 611130, China
* Correspondence: zhangzj@ioz.ac.cn
† These authors contributed equally to this work.

Simple Summary: Pandas are a flagship species for global animal conservation and wild individuals are scarce. They are now segregated into 33 local populations, and 25 of them are too small to be self-sustaining. For these small populations, in addition to preservation of pandas in situ, translocation is one of the options for the species' recovery. This paper reviews the scientific progress in conservation translocation of pandas, with the aim of providing theoretical guidance to improve the success rate of released pandas, and uses pandas as a model species to provide reference for the global conservation translocation of rare and endangered species.

Abstract: The giant panda (*Ailuropoda melanoleuca*) is the flagship species of animal conservation worldwide, and the number of captive pandas reached 673 in 2021. According to the Fourth National Survey Report on the Giant Panda, there are 1864 wild pandas, segregated into 33 local populations, and 25 of these populations are too small to be self-sustaining. In addition to the conservation and restoration of panda habitats, conservation translocations, an approach that has been shown to be effective in slowing or reversing biodiversity loss, are highly desirable for panda conservation. The captive-bred panda population has grown rapidly, laying the foundation for releasing captive-bred pandas into the wild. This paper reviews the scientific advances in conservation translocations of pandas. Studies have shown that before translocation conservation programs are implemented, we should determine what factors are causing the depletion of the original population at the release site. The selection of suitable release sites and individuals will help to improve the survival rate of released individuals in the wild. Pre-release training and post-release monitoring are essential to ensure successful releases. We also see the great potential for increasing applications of Adaptive Management to improve the success of giant panda conservation translocation programs. This review provides theoretical guidance for improvement of the success rate in conservation translocations for captive pandas, and uses the panda as a model species to provide a global reference for the conservation translocations of rare and endangered species.

Keywords: artificial intervention; conservation translocation; giant panda; local population; self-sustaining

Citation: Wang, Y.; Wei, W.; Yuan, F.; Cao, D.; Zhang, Z. The Science Underlying Giant Panda Conservation Translocations. *Animals* **2023**, *13*, 3332. https://doi.org/10.3390/ani13213332

Academic Editor: Brian L. Cypher

Received: 26 September 2023
Revised: 18 October 2023
Accepted: 24 October 2023
Published: 26 October 2023

1. Introduction

The Earth is experiencing its sixth mass extinction, which, unlike the previous five, is largely human induced [1]. Biodiversity conservation has implications for the sustainable development of human societies and has become one of the most widespread concerns of the international community [2]. China is one of the world's richest countries in terms of biodiversity and also one of the most threatened [3].

The giant panda (*Ailuropoda melanoleuca*) is undoubtedly the most discussed and represented species in China's biodiversity conservation efforts. It is known as China's national treasure because of its lovely appearance and the scarcity of individuals in the wild. China has made countless efforts to protect the species. From the establishment of China's first giant panda reserves in 1963, 67 nature reserves (NRs) were established by 2015, covering approximately 54% of the giant panda's habitat and protecting approximately 67% of the wild giant panda population [4]. Meanwhile, the Chinese government announced the official establishment of the Giant Panda National Park (GPNP) at the COP 15 meeting in 2021, which spans three provinces, Sichuan, Shaanxi, and Gansu, and covers a total area of 27,134 km^2, with the aim of protecting the habitat of the giant panda and its sympatric species [5], marking a new phase in panda conservation. Although giant pandas have been downlisted from Endangered to Vulnerable in IUCN red lists, the Chinese government's protection of pandas has increased rather than decreased [6], certainly setting an example for most of the world's protected species.

According to The Fourth National Survey Report on the Giant Panda, there are 1864 pandas in the wild across China, separated into 33 local populations (Figure 1). Based on minimum area requirement (MAR) and minimal viable population (MVP), some of these populations are too small to be self-sustaining and, therefore, are at high risk of extinction [4]. Sustainable populations are those that meet the two conditions of panda population reaching its MVP and a habitat area greater than 114.7 km^2 [7], of which there are 8 in total; unsustainable populations mean populations that do not meet both of these conditions, of which there are 25 in total. Clearly, the Chinese government is aware that the conservation of giant pandas is still a serious challenge. In parallel to building a more comprehensive conservation system for wild giant panda populations and their habitats, many active human interventions have been undertaken, including conservation translocation [8], damaged habitat restoration [9], and ecological corridor construction [10]. Of particular interest is conservation translocation, which refers to the intentional movement of organisms from one place to another.

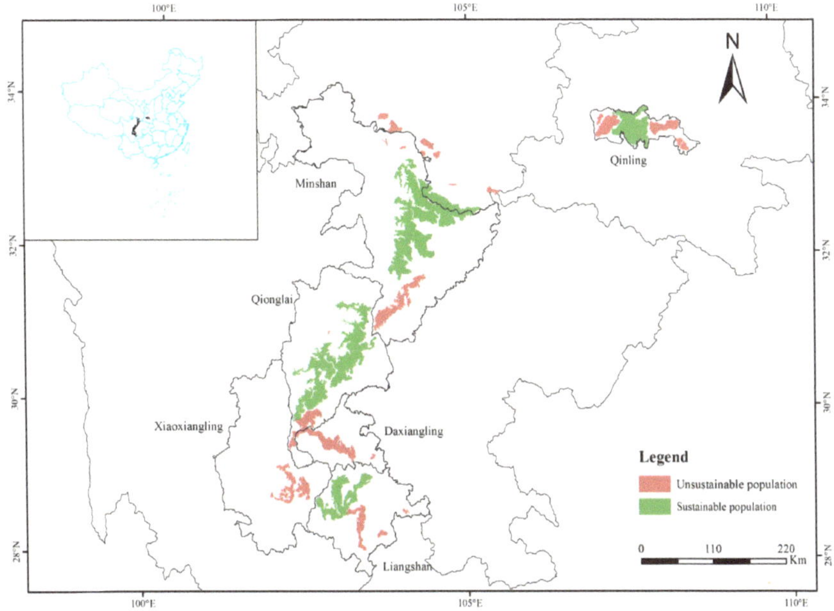

Figure 1. 33 isolated populations of giant pandas in the six mountains.

There are many types of conservation translocations, including introduction, reintroduction, and reinforcement. Conservation translocation projects of charismatic species include the reintroduction of the California condor (*Gymnogyps californianus*) [11] and Lord Howe Island woodhen (*Hypotaenidia sylvestris*) [12]. However, available data for wildlife reintroductions suggest that the majority of programs suffer from weak planning, often insufficient monitoring, habitat-related issues, and failure to establish viable populations [13]. These drawbacks have led to the development of 'reintroduction biology' to bridge the research–management gap. Successful translocations should be guided by explicit theoretical frameworks based on clearly defined objectives and rigorous scientific studies [14].

Benefitting from the resolution of the so-called 'trilemma' of captive breeding, that is, difficult to rut, difficult to conceive, and difficult to raise young and survive, the captive-bred giant panda population has grown rapidly over the last 20 years, reaching 673 individuals by 2021 (Figure 2), laying the foundation for conservation translocation of captive-bred giant pandas into the wild. In fact, an international symposium on the feasibility of captive-bred giant panda reinforcement was held in Wolong in 1997; however, experts concluded that the conditions for releasing giant pandas into the wild were not yet met. In 2006, Zhang et al. (2006) concluded that the conditions for the release of captive giant pandas into the wild in China were primarily met, and the time was ripe for release [15]. However, the actual release of giant pandas in China began in 2005 with the rescue of "Shenglin 1" in the wild.

Figure 2. The population of captive giant pandas worldwide from 2010 to 2021.

To date, a total of 15 releases have been recorded (Table 1), of which 3 were wild-born and 12 were captive-born; 3 individuals died (20%), and 12 survived (80%). The fact that all of the wild-born individuals survived and 3 of the captive-born individuals died seems to indicate that the former had a much higher chance of survival than the latter, although the sample size is too small to confirm this. The only wild-born individual with a reliable record is Luxin, who was released in 2009 at Liziping NR in Shimian. In the future, we should record reintroductions in the same way to increase the sample and knowledge base. Overall, conservation translocation of giant pandas in China is proceeding slowly. Before launching this complex and difficult procedure, managers and decision-makers need to clearly understand the rationale and technical challenges of conservation translocations, especially in the face of so many small populations with high extinction risk.

Table 1. The release of giant pandas in China to date.

Name	Gender	Source	Release Year	Release Site	Survival Status
Shenglin 1	Female	Wild-born	2005	Longxi-hongkou NR	Still alive
Xiangxiang	Male	Captive-born	2006	Wolong NR	Died (fights and injuries)
Luxin	Female	Wild-born	2009	Liziping NR	Still alive (produced offspring)
Taotao	Male	Captive-born	2012	Liziping NR	Still alive
Zhangxiang	Female	Captive-born	2013	Liziping NR	Still alive
Xuexue	Female	Captive-born	2014	Liziping NR	Died (conditional pathogenic infections)
Huajiao	Female	Captive-born	2015	Liziping NR	Still alive
Hesheng	Male	Captive-born	2016	Liziping NR	Died (attacked by unknown animal)
Huayan	Female	Captive-born	2016	Liziping NR	Still alive
Zhangmeng	Female	Captive-born	2016	Liziping NR	Still alive
Yingxue	Female	Captive-born	2017	Liziping NR	Still alive
Baxi	Male	Captive-born	2017	Liziping NR	Still alive
Qinxin	Female	Captive-born	2018	Longxi-hongkou NR	Still alive
Xiaohetao	Female	Captive-born	2018	Longxi-hongkou NR	Still alive
Tangtang	Female	Wild-born	2021	Foping NR	Still alive

Note: In brackets are the cause of death of the dead individuals and the current status of the surviving individuals. Survival status of giant pandas until October 2023.

Fortunately, owing to the unique scientific and conservation value of the species, scientific research on giant pandas has garnered numerous studies, such as studies on habitat quality [16] and population dynamics [17] of giant pandas, covering both field and captive populations and different levels, from individual to landscape (Table 2). With the support Iof many scientific papers, the conservation biology of giant pandas can be said to have become a model case-study in the field of wildlife research. This article will introduce the selection of release sites and individuals for release, as well as pre-release and post-release work to explore the science underlying giant panda conservation translocations.

Table 2. Research topics on giant pandas and published papers examined from 2000 to 2022.

Research Topic	Number of Papers from Different Sources		
	Science Direct	Web of Science	CNKI
Wild giant panda	1471	282	653
Captive giant panda	551	234	637
Giant panda	3330	1398	9462
Giant panda population	2036	347	1144
Giant panda landscape	625	98	46

CNKI: China National Knowledge Infrastructure.

2. Selecting Suitable Release Sites

When identifying suitable sites for translocation, it is apparent that 'species should never be released blindly without extensive assessment of habitat quality' [18]. It is extremely important to select a suitable habitat for the released species in the process of conservation translocation [19]. Today, wild giant pandas are fragmented across six mountain systems in southwest China: the Qinling, Minshan, Qionglai, Daxiangling, Xiaoxiangling, and Liangshan mountains, and are separated into 33 small populations. To release captive giant pandas into the wild, we first need to understand the habitat requirements of giant pandas. According to existing studies, giant pandas usually choose old growth or secondary forests with medium to high densities of bamboo, medium altitudes and gentle slopes, and usually choose to avoid human disturbance and livestock [20–23]. In addition, forest age, topography, and the presence of bamboo were key predictor variables determining habitat selection by giant pandas at different scales [24,25].

Monitoring the released captive-bred giant panda Zhangxiang, Lei et al. (2015) found differences in dietary composition and microhabitat usage between Zhangxiang and local wild pandas. The percentage of bitten-not-broken bamboo was also larger from Zhangxiang than for wild pandas [26]. Thus, the ecological conditions of the release enclosure should be as similar as possible to the habitat of the population to be joined by the released individuals, otherwise the released individuals will face a process of behavioural adjustment and adaptation while integrating into the population. This is, in fact, a soft release process to maximise the chances of survival of the released individuals in the wild.

In addition to the above habitat requirements, these two other factors need to be a focused: MAR and MVP. MAR for a population is the area required for the long-term survival of that population. Qing et al. (2016) concluded that the MAR of giant panda populations is approximately 114.7 km^2 [7]. Therefore, we need to upgrade the area of each habitat to at least this level to ensure that release sites are sufficient to sustain the growth of reintroduced populations in the long term [27]. A piece of counterevidence is provided by Xiangxiang; although it was released into an area with a favorable habitat, there was already a stable population of wild pandas. Xiangxiang may have died from competition with other wild individuals for territory and resources. Therefore, releasing pandas into an area which already has a dense existing population of pandas will be riskier because of competition and territoriality. Areas where wild populations have been extirpated or depleted should be given priority for receiving translocated individuals, but the reason for depletion must first be controlled before release.

Small populations are susceptible to large fluctuations caused by various levels of stochastic processes (e.g., genetics, environment, disasters, population statistics randomness) and are thus at risk of small population extinction [28]. Among the genetic problems for populations are inbreeding depression and loss of genetic diversity from genetic drift. Loss of genetic diversity can in turn be expected to cause a decreased ability in the population to adapt to environmental change and to survive outbreaks of disease, which is a huge blow to maintaining the dynamic balance of populations. Therefore, the number of pandas in each small population should reach a minimum value to cope with possible negative impacts, that is, the need to reach MVP. Further research is still needed on the calculation of MVP for wild giant panda populations. According to the Fourth National Survey Report on the Giant Panda, only 8 of the 33 populations meet the sustainability criteria; the remaining 25 populations are located in habitats where the risk of extinction can be reduced by releasing captive individuals or creating ecological corridors (Figure 1). The status of 25 local populations at risk of extinction and the recommended measures to prevent their extinction are as follows (Table 3). These measures are implemented on the basis that, as far as possible, the wild giant panda populations meet the MVP while their habitats meet the MAR.

Table 3. Local populations at risk of extinction and their mitigation measures.

Mountain Ranges	Populations Name	Number of Extant Giant Pandas	Habitat Area (km^2)	Suggested Mitigation Measures
Qinling	Qinling A	7	211.41	1
	Qinling B	20	594.22	1
	Qinling D	36	641.01	1
	Qinling E	3	90.47	1 + 2
	Qinling F	4	128.45	1 + 2

Table 3. *Cont.*

Mountain Ranges	Populations Name	Number of Extant Giant Pandas	Habitat Area (km^2)	Suggested Mitigation Measures
Minshan	Minshan A	4	135.93	1 + 2
	Minshan B	9	238.81	1
	Minshan C	3	204.89	1
	Minshan D	1	56.00	2
	Minshan E	2	93.39	2
	Minshan F	2	36.07	2
	Minshan H	2	34.06	2
	Minshan I	1	33.69	2
	Minshan L	35	1363.99	1
Qionglai	Qionglai D	29	800.44	1
	Qionglai E	1	17.27	2
Daxiangling	Daxiangling A	4	205.75	1
	Daxiangling B	32	979.16	1
	Daxiangling C	2	43.78	2
Xiaoxiangling	Xiaoxiangling A	21	442.31	1
	Xiaoxiangling B	9	751.33	1
Liangshan	Liangshan B	22	528.41	1
	Liangshan C	3	173.60	1
	Liangshan D	4	107.59	1 + 2
	Liangshan E	3	48.12	2

Note: 1 represents the release of captive-bred individuals; 2 represents the establishment of ecological corridors. All data and populations name were obtained from the Fourth National Survey Report on the Giant Panda.

3. Selecting Suitable Release Candidacy

3.1. Individual Source

Captive-bred animals fare relatively poorly in reintroduction programs [29]. They potentially have poor health and abnormal behaviour due to captivity [30] or a lack of important capacities, such as avoiding predators, foraging, and mating. A complete repertoire of survival-critical behaviours of free-ranging wild giant pandas should be assembled. Another avenue of interest is whether it is possible to translocate more wild individuals. The purpose of translocation is not to off-load captive animals or to justify captive breeding expenditures. Giant pandas are translocated to promote conservation. It may even be more conservation-friendly to move wild-caught animals from populations with a surplus of juveniles to populations below MVP, rather than reintroducing captive-bred animals. However, assessing the reasonableness of this idea requires more experimental studies.

3.2. Body Condition

The first thing to consider is the health of the release candidate. Pre-translocation parasite screening of wild populations and risk assessments are indispensable. For captive populations, selection for host tolerance can enhance the success of reintroduction or translocation [31]. The giant panda is known to be susceptible to natural infection with canine distemper virus (CDV). Bronson et al. (2007) recommend that giant pandas be vaccinated annually using the canarypox-vectored recombinant distemper vaccine [32]. Geng et al. (2020) found that giant panda cubs had the strongest immune response after the second vaccination [33]. Recently, Dai et al. (2021) also identified the complete genome sequence of a novel circovirus, giant panda-associated circovirus, in giant panda blood, but its pathogenesis in giant pandas needs to be further explored [34]. It is therefore essential that captive-bred individuals are thoroughly examined prior to release to select individuals more suitable for release, and be properly vaccinated to reduce the chance of infection by the virus.

3.3. Parasites and Viruses

Parasitic infection is another noteworthy issue for candidates. *Baylisascaris schroederi* is the most common parasite in wild and captive giant pandas, and the visceral larval migrants caused by *B. schroederi* infection are identified as the most significant threat to the survival of the giant panda, with the probability of death from this disease in the wild increasing significantly between 1971 and 2005 [35]. Zhu et al. (2020) used seven polymorphic functional Major Histocompatibility Complex (MHC) genes and found that heterozygotes and certain MHC variants were less likely to be infected with *B. schroederi*. When selecting individuals for release, this method can be used to select pandas that are less susceptible to disease, supplemented by the previous selection criteria, thus improving the chances of survival of released individuals in the wild [36]. Xinyuan, a two-year-old female who died from respiratory and renal failure before her release to Liziping NR, reminded us of the need for increased surveillance and research on the infectious diseases of giant pandas and the development of suitable vaccines [37]. In addition, there is always a serious risk that translocated animals will carry novel diseases into the wild and infect healthy populations [38]. This is one of the main risks that must be considered before translocation is carried out.

3.4. Sex and Age

Next, the sex and age of the release candidacy need to be considered. One genetic study suggest that female-biased dispersal occurs in giant pandas, most likely due to competition for birth dens among females, inbreeding avoidance, and enhanced inclusive fitness among related males [39]. This finding is supported by direct observation of a GPS-collared subadult female [40]. Moderate post-release dispersal is necessary for released individuals to survive in a changing environment and to promote gene flow among metapopulations [41]. This observation seems to suggest that the effect of releasing females is better than the effect of releasing males. Indeed, female pandas will directly take part in production and nurturing offspring. More females released can be expected to make a greater contribution to population recovery [42].

However, we should also consider the community structure of the target population for release, and that if there is a significant female bias in the community, then releasing males will be necessary. In theory, a single male panda could mate with several females and easily spread his genes widely among the target population. Therefore, mixed-sex combinations with female bias have better results than female-only or male-only combinations [8].

In addition, we should release subadult individuals at 3 to 4 years of age to keep the age structure of the target population for release growing and capable of normal reproductive activity. This is because they have, firstly, the basic ability to survive; secondly, sufficient time to establish home ranges, integrate into the community, and engage in learning about reproductive behaviour; and thirdly, the greatest contribution to population reinforcement [17].

3.5. Behaviour

Next, we also need to consider the behavioural factors of the release candidants. For behaviour, we focus on chemical communication behaviour and antipredatory behaviour. Chemical communication is important in pandas and there are multiple ways of marking, this includes anogenital gland secretion (AGS), urine, etc. The marks left behind contain information on individual status, sex, age, oestrus status, etc. [43,44]. Zhou et al. (2019) demonstrated that there is a difference in the chemical composition of AGS between captive and wild giant pandas and that this difference is most likely one of the important reasons for the low natural reproductive ability of captive giant pandas [45]. In this regard, we suggest that wild giant panda feces and marks from the release site can be collected for the release individuals to acclimatize to in advance.

Natural predators of giant pandas are mainly medium and large carnivores living in the same area, such as leopards (*Panthera pardus*), dholes (*Cuon alpinus*) and wolves (*Canis lupus*). Although Li et al. (2020) showed that the number of large carnivores in giant panda habitats is decreasing [46], mortality due to predation is a major cause of failure, and captive-bred giant pandas are isolated from predators throughout their lifetime and may no longer express antipredator behaviour [47]. Even though there is likely an innate component to predator recognition in captive giant pandas based on experiments using predator urine, antipredator training in prerelease preparation procedures is indispensable [48]. van Heezik et al. (1999) demonstrated that using a model predator was not an effective conditioning stimulus to houbara bustards (*Chlamydotis [undulata] macqueenii*), but pre-release training with a live predator significantly improved post-release survival [49]. Therefore, future research could explore whether live predators should be included in the pre-release training of giant pandas and whether they are more effective than model predators.

3.6. Genetic Consideration

Last but not least, for potential reintroduction, the maintenance of genetic diversity of captive-bred species through pedigree management has been increasingly addressed [50], but improving the exchange of genetic materials among institutions will be necessary for captive giant pandas because of lower levels of allelic diversity and heterozygosity in captive-bred populations compared to isolated wild populations [51]. For wild, isolated, and smaller populations of giant pandas with low genetic diversity which are facing a high level of inbreeding, genetic rescue can prevent the negative consequences of disrupted gene flow and isolation by increasing population size at population establishment or by gene flow as the population expands and connects with neighbouring populations [52,53]. For example, exchange between populations of Chengdu and Wanglang should be encouraged because of similar wild founder sources [54]. Whole genome sequencing of 34 wild giant pandas led to the classification of pandas into three distinct populations, Qinling (QIN), Minshan (MIN), and Qionglai–Daxiangling–Xiaoxiangling–Liangshan (QXL) [55], suggesting that different feasible schemes of recovery programs against three management units should be brought forwards. Individuals from different populations should not be cross-released to avoid the potential risk of outbreeding depression [54].

4. Pre-Release Training of Candidate

There are usually two ways of releasing captive animals, namely hard releasing and soft releasing [56]. At present, conservation translocations of giant pandas usually adopt soft release, which consists of these four main steps: prerelease training, acclimatization in enclosures, release, and post-release monitoring. Among these steps for release, pre-release training is widely considered to be the key to successful soft release [57]. Zhang et al. (2017) found that the inclusion of manual intervention in pr-release training caused differences in the behavioural development of released individuals compared to those individuals without manual intervention, things like habitat selection and home range [58]. To increase the success rate of soft release, we should minimize manual intervention in pre-release training.

In a study of the released individual Zhangxiang, Lei et al. (2015) found that there were significant differences in food habits between Zhangxiang and wild giant pandas during reintroduction training; it was less capable of handling bamboo than wild individuals, suggesting that Zhangxiang had to readapt to a new environment after release into the wild even after pre-release training, which inevitably reduced its chance of survival [26]. In view of this consideration, in the future, when building wild acclimatization pens, management should try to ensure that the site is similar to the wild environment that the released pandas will enter. In the end, it is important for captive pandas preparing for release to form as little dependence on people as possible [58]. For example, the recently popular 'panda costume', in which giant panda keepers disguise themselves as pandas to get closer to captive individuals, seems to be useful in preventing the imprinting of humans on panda cubs.

5. Post-Release Monitoring of Released Individuals

Post-release monitoring of released individuals is critical to ensure adequate population establishment, growth, and viability [14], and these metrics are commonly used to assess the success of translocation [59]. Monitoring of individual giant pandas released into the wild has commonly been carried out, especially in the Liziping NR [42]. To date, many advanced technologies have been used for post-release monitoring, such as GPS collars, camera trapping technology, genetic monitoring, and automatic giant panda identification systems (a new Feature-Fusion Network with Patch Detector) [8,60,61]. Post-release monitoring actually includes these two aspects: monitoring of released individuals and monitoring of population dynamics, but most of the current monitoring is directed at the former, and research on the latter is lacking. The purpose of releasing captive giant pandas in the wild is to form self-sustaining populations to avoid the extinction of small populations [14,38], so monitoring population dynamics is essential, and research in this area should be strengthened in the future. Bubac et al. (2019) also found that most translocation studies were conducted for only 1–4 years of post-release monitoring, with the highest proportion of failures occurring in the first 4 years [13]. Based on this finding, and in conjunction with existing research on the release of giant pandas, we should monitor captive giant pandas for a longer period of time after their release into the wild.

6. Future Directions

Although we are currently able to select suitable release sites for individuals about to be released, one more thing we should do in the future is to determine the factors causing the depletion of the original population at the release site. Habitat degradation and fragmentation are well known to threaten the sustainable survival of giant pandas [62], but we should determine which specific factors are causing pandas to become unsuitable for distribution in this area, such as changes in food (bamboo), breeding dens, climate, and forest cover [10,63–65]. Only after this determination can we take measures to make the release site suitable for releasing individuals to form self-sustaining populations. In addition, the multiscale Maxent approach can help conservationists model the habitat suitability of giant pandas, which can guide the selection of release sites [66]. Surveys between releases are therefore necessary, and in some areas, some active human interventions may have to be undertaken, such as habitat expansion, construction of giant panda breeding dens, improving bamboo forest growth, etc.

Reintroduction by releasing captive-bred individuals to recover wild giant panda populations within their historical distribution range is the most common approach [15]. Applications of other conservation translocations of giant pandas need to be explored by reliable model predictions, even though it is extremely challenging. At present, before the conservation translocations of giant pandas, we should first focus on restoring suitable habitats and building ecological corridors so that existing giant panda populations have sufficient resources to sustain themselves, which is the primary way to protect wild giant panda populations.

Morris et al. (2021) summarized what is known about conservation translocation projects and concluded that factors such as the number of individuals released, the geographical location of the release, and the characteristics of the released species can affect the success of conservation translocations of terrestrial vertebrates [67]. Facing the complex ecological requirements (habitat quality, food resources, predator control, breeding dens, and climate change) of giant pandas, the best methods for selecting suitable release sites are often uncertain. Adaptive management (AM) can be helpful in the face of such uncertainty by balancing the benefits of improved information against the goals of management [68]. We see great potential for increasing applications of AM to improve the success of giant panda conservation translocation programs. AM can recognise and combat uncertainty and help us make better decisions [69].

7. Conclusions

Overall, the conservation of giant pandas is changing from passive in situ conservation to active conservation with human intervention, and is moving from crude to more precise approaches. In the foreseeable future, scientific research will become increasingly prominent in the development of giant panda conservation. As one of the key instruments for the maintenance of wild giant panda populations, conservation translocation should be actively promoted and continuously improved. As a core issue, improving the survival rate of released individuals in the wild is crucial. This study provides a theoretical basis for improving the survival of captive giant panda individuals in the wild by reviewing aspects such as the selection of individuals for release and release sites. However, before the conservation translocations of pandas, we should first focus on restoring suitable habitats and building ecological corridors so that existing populations have sufficient resources to sustain themselves, which is the primary way to protect wild populations. As a model species, the giant panda also provides a reference for the global conservation of rare and endangered species in conservation translocation, and we are actively providing Chinese wisdom towards global biodiversity conservation.

Author Contributions: Conceptualization, Z.Z.; Methodology, Y.W. and F.Y.; Software, Y.W., W.W., D.C., F.Y. and Z.Z.; Investigation, Y.W., W.W. and Z.Z.; Data curation, Y.W., W.W., D.C., F.Y. and Z.Z.; Writing—original draft, Y.W. and D.C.; Writing—review & editing, Y.W., W.W. and Z.Z.; Supervision, Z.Z.; Project administration, Z.Z. Funding acquisition, Z.Z. and F.Y. All authors have read and agreed to the published version of the manuscript.

Funding: This study was supported by the National Natural Science Foundation of China (32070524, 32270552), Giant Panda International Cooperation Foundation (2023-7) and the Second Qinghai-Tibet Plateau Comprehensive Scientific Survey (2019QZKK05010502).

Institutional Review Board Statement: Not applicable.

Informed Consent Statement: Not applicable.

Data Availability Statement: Not applicable.

Conflicts of Interest: The authors declare no conflict of interest.

References

1. Wagler, R. 6th mass extinction. In *Encyclopedia of the Anthropocene*; Dellasala, D.A., Goldstein, M.I., Eds.; Elsevier: Oxford, UK, 2018; pp. 9–12.
2. Cafaro, P. Three ways to think about the sixth mass extinction. *Biol. Conserv.* **2015**, *192*, 387–393. [CrossRef]
3. Zhang, S.; Zhou, Y.; Yu, R.; Xu, X.; Xu, M.; Li, G.; Wang, W.; Yang, Y. China's biodiversity conservation in the process of implementing the sustainable development goals (sdgs). *J. Clean. Prod.* **2022**, *338*, 130595. [CrossRef]
4. State Forestry Administration of China. *The Fourth National Giant Panda Survey*; State Forestry Administration of China Press: Beijing, China, 2015.
5. Wei, F.W.; Costanza, R.; Dai, Q.; Stoeckl, N.; Gu, X.D.; Farber, S.; Nie, Y.G.; Kubiszewski, I.; Hu, Y.B.; Swaisgood, R.; et al. The value of ecosystem services from giant panda reserves. *Curr. Biol.* **2018**, *28*, 2174. [CrossRef]
6. Swaisgood, R.R.; Wang, D.; Wei, F. Panda downlisted but not out of the woods. *Conserv. Lett.* **2018**, *11*, e12355. [CrossRef]
7. Qing, J.; Yang, Z.; He, K.; Zhang, Z.; Gu, X.; Yang, X.; Zhang, W.; Yang, B.; Qi, D.; Dai, Q. The minimum area requirements (mar) for giant panda: An empirical study. *Sci. Rep.* **2016**, *6*, 37715. [CrossRef]
8. Yang, Z.; Gu, X.; Nie, Y.; Huang, F.; Huang, Y.; Dai, Q.; Hu, Y.; Yang, Y.; Zhou, X.; Zhang, H. Reintroduction of the giant panda into the wild: A good start suggests a bright future. *Biol. Conserv.* **2018**, *217*, 181–186. [CrossRef]
9. Hong, M.S.; Wei, W.; Tang, J.F.; Zhou, H.; Han, H.; Zhang, Z.J. Positive responses from giant pandas to the natural forest conservation programme based on slope utilisation. *Glob. Ecol. Conserv.* **2021**, *27*, e01616. [CrossRef]
10. Bai, W.; Huang, Q.; Zhang, J.; Stabach, J.; Huang, J.; Yang, H.; Songer, M.; Connor, T.; Liu, J.; Zhou, S.; et al. Microhabitat selection by giant pandas. *Biol. Conserv.* **2020**, *247*, 108615. [CrossRef]
11. Walters, J.R.; Derrickson, S.R.; Michael Fry, D.; Haig, S.M.; Marzluff, J.M.; Wunderle, J.M., Jr. Status of the california condor (*Gymnogyps californianus*) and efforts to achieve its recovery. *Auk* **2010**, *127*, 969–1001. [CrossRef]
12. Frith, C. *The Woodhen: A Flightless Island Bird Defying Extinction*; CSIRO Publishing: Canberra, Australia, 2013.
13. Bubac, C.M.; Johnson, A.C.; Fox, J.A.; Cullingham, C.I. Conservation translocations and post-release monitoring: Identifying trends in failures, biases, and challenges from around the world. *Biol. Conserv.* **2019**, *238*, 108239. [CrossRef]
14. Armstrong, D.P.; Seddon, P.J. Directions in reintroduction biology. *Trends Ecol. Evol.* **2008**, *23*, 20–25. [CrossRef] [PubMed]

15. Zhang, Z.; Zhang, S.; Wei, F.; Wang, H.; Li, M.; Hu, J. Translocation and discussion on reintroduction of captive giant panda. *Acta Theriol. Sin.* **2006**, *26*, 292.
16. Kang, D. A review of the impacts of four identified major human disturbances on the habitat and habitat use of wild giant pandas from 2015 to 2020. *Sci. Total Environ.* **2021**, *763*, 142975. [CrossRef]
17. Kong, L.Q.; Xu, W.H.; Xiao, Y.; Pimm, S.L.; Shi, H.; Ouyang, Z.Y. Spatial models of giant pandas under current and future conditions reveal extinction risks. *Nat. Ecol. Evol.* **2021**, *5*, 1309. [CrossRef] [PubMed]
18. Lomolino, M.V.; Channell, R. Range collapse, re-introductions, and biogeographic guidelines for conservation. *Conserv. Biol.* **1998**, *12*, 481–484. [CrossRef]
19. Baling, M.; Stuart-Fox, D.; Brunton, D.H.; Dale, J. Habitat suitability for conservation translocation: The importance of considering camouflage in cryptic species. *Biol. Conserv.* **2016**, *203*, 298–305. [CrossRef]
20. Hu, J.C.; Schaller, G.B.; Pan, W.S.; Zhu, J. *The Giant Panda of Wolong*; Sichuan Science and Technology Press: Chengdu, China, 1985.
21. Hull, V.; Zhang, J.D.; Huang, J.Y.; Zhou, S.Q.; Vina, A.; Shortridge, A.; Li, R.G.; Liu, D.A.; Xu, W.H.; Ouyang, Z.Y.; et al. Habitat use and selection by giant pandas. *PLoS ONE* **2016**, *11*, e0162266. [CrossRef]
22. Liu, Z.; Dayananda, B.; Jeffree, R.A.; Tian, C.; Zhang, Y.; Yu, B.; Zheng, Y.; Jing, Y.; Si, P.; Li, J. Giant panda distribution and habitat preference: The influence of sympatric large mammals. *Glob. Ecol. Conserv.* **2020**, *24*, e1221. [CrossRef]
23. Wei, F.; Swaisgood, R.; Hu, Y.; Nie, Y.; Yan, L.; Zhang, Z.; Qi, D.; Zhu, L. Progress in the ecology and conservation of giant pandas. *Conserv. Biol.* **2015**, *29*, 1497–1507. [CrossRef] [PubMed]
24. Zhang, Z.; Swaisgood, R.R.; Zhang, S.; Nordstrom, L.A.; Wang, H.; Gu, X.; Hu, J.; Wei, F. Old-growth forest is what giant pandas really need. *Biol. Lett.* **2011**, *7*, 403–406. [CrossRef]
25. Zhiyun, O.; Jianguo, L.; Han, X.; Yingchun, T.; Hemin, Z. An assessment of giant panda habitat in wolong nature reserve. *Acta Ecol. Sin.* **2001**, *21*, 1869–1874.
26. Lei, M.; Yuan, S.; Yang, Z.; Hong, M.; Yang, X.; Gu, X.; Huang, F.; Zhang, Z. Comparison of microhabitats and foraging strategies between the captive-born zhangxiang and wild giant pandas: Implications for future reintroduction. *Environ. Sci. Pollut. Res.* **2015**, *22*, 15089–15096. [CrossRef] [PubMed]
27. Rodrigues, M.; Beck, B.; Travis, D.; Unwin, S.; Walkup, K. *Best Practice Guidelines for the Re-Introduction of Great Apes*; IUCN: Gland, Switzerland, 2007.
28. Guo, J.; Chen, Y.; Hu, J. Population viability analysis of giant pandas in the yele nature reserve. *J. Nat. Conserv.* **2002**, *10*, 35–40. [CrossRef]
29. Griffith, B.; Scott, J.M.; Carpenter, J.W.; Reed, C. Translocation as a species conservation tool: Status and strategy. *Science* **1989**, *245*, 477–480. [CrossRef] [PubMed]
30. Mathews, F.; Orros, M.; Mclaren, G.; Gelling, M.; Foster, R. Keeping fit on the ark: Assessing the suitability of captive-bred animals for release. *Biol. Conserv.* **2005**, *121*, 569–577. [CrossRef]
31. Venesky, M.D.; Mendelson Iii, J.R.; Sears, B.F.; Stiling, P.; Rohr, J.R. Selecting for tolerance against pathogens and herbivores to enhance success of reintroduction and translocation. *Conserv. Biol.* **2012**, *26*, 586–592. [CrossRef]
32. Bronson, E.; Deem, S.L.; Sanchez, C.; Murray, S. Serologic response to a canarypox-vectored canine distemper virus vaccine in the giant panda (*Ailuropoda melanoleuca*). *J. Zoo Wildl. Med.* **2007**, *38*, 363–366. [CrossRef]
33. Geng, Y.; Shen, F.; Wu, W.; Zhang, L.; Luo, L.; Fan, Z.; Hou, R.; Yue, B.; Zhang, X. First demonstration of giant panda's immune response to canine distemper vaccine. *Dev. Comp. Immunol.* **2020**, *102*, 103489. [CrossRef]
34. Dai, Z.; Wang, H.; Feng, Z.; Ma, L.; Yang, S.; Shen, Q.; Wang, X.; Zhou, T.; Zhang, W. Identification of a novel circovirus in blood sample of giant pandas (*Ailuropoda melanoleuca*). *Infect. Genet. Evol.* **2021**, *95*, 105077. [CrossRef]
35. Zhang, J.; Daszak, P.; Huang, H.; Yang, G.; Kilpatrick, A.M.; Zhang, S. Parasite threat to panda conservation. *EcoHealth* **2008**, *5*, 6–9. [CrossRef]
36. Zhu, Y.; Grueber, C.; Li, Y.; He, M.; Hu, L.; He, K.; Liu, H.; Zhang, H.; Wu, H. Mhc-associated baylisascaris schroederi load informs the giant panda reintroduction program. *Int. J. Parasitol. Parasites Wildl.* **2020**, *12*, 113–120. [CrossRef]
37. Qin, Q.; Li, D.; Zhang, H.; Hou, R.; Zhang, Z.; Zhang, C.; Zhang, J.; Wei, F. Serosurvey of selected viruses in captive giant pandas (*Ailuropoda melanoleuca*) in China. *Vet. Microbiol.* **2010**, *142*, 199–204. [CrossRef] [PubMed]
38. IUCN. *Guidelines for Reintroductions and Other Conservation Translocations*; IUCN Species Survival Commission: Gland, Switzerland, 2013; 57p.
39. Zhan, X.J.; Zhang, Z.J.; Wu, H.; Goossens, B.; Li, M.; Jiang, S.W.; Bruford, M.W.; Wei, F.W. Molecular analysis of dispersal in giant pandas. *Mol. Ecol.* **2007**, *16*, 3792–3800. [CrossRef] [PubMed]
40. Zhang, Z.; Swaisgood, R.R.; Wu, H.; Li, M.; Yong, Y.; Hu, J.; Wei, F. Factors predicting den use by maternal giant pandas. *J. Wildl. Manag.* **2007**, *71*, 2694–2698. [CrossRef]
41. Howe, R.W.; Davis, G.J.; Mosca, V. The demographic significance of 'sink' populations. *Biol. Conserv.* **1991**, *57*, 239–255. [CrossRef]
42. Hong, M.S.; Wei, W.; Zhou, H.; Tang, J.F.; Han, H.; Zhang, Z.J. Creative conservation in china: Releasing captive giant pandas into the wild. *Environ. Sci. Pollut. Res.* **2019**, *26*, 31548–31549. [CrossRef]
43. Swaisgood, R.R.; Lindburg, D.G.; White, A.M.; Hemin, Z.; Xiaoping, Z. Chemical communication in giant pandas. *Giant Pandas Biol. Conserv.* **2004**, *106*. [CrossRef]
44. Nie, Y.; Swaisgood, R.R.; Zhang, Z.; Hu, Y.; Ma, Y.; Wei, F. Giant panda scent-marking strategies in the wild: Role of season, sex and marking surface. *Anim. Behav.* **2012**, *84*, 39–44. [CrossRef]

45. Zhou, W.; Nie, Y.; Hu, Y.; Swaisgood, R.R.; Zhang, Y.; Liu, D.; Wei, F. Seasonal and reproductive variation in chemical constituents of scent signals in wild giant pandas. *Sci. China Life Sci.* **2019**, *62*, 648–660. [CrossRef]
46. Li, S.; Mcshea, W.J.; Wang, D.J.; Gu, X.D.; Zhang, X.F.; Zhang, L.; Shen, X.L. Retreat of large carnivores across the giant panda distribution range. *Nat. Ecol. Evol.* **2020**, *4*, 1327. [CrossRef] [PubMed]
47. Griffin, A.S.; Blumstein, D.T.; Evans, C.S. Training captive-bred or translocated animals to avoid predators. *Conserv. Biol.* **2000**, *14*, 1317–1326. [CrossRef]
48. Du, Y.; Huang, Y.; Zhang, H.; Li, D.; Yang, B.; Wei, M.; Zhou, Y.; Liu, Y. Innate predator recognition in giant pandas. *Zool. Sci.* **2012**, *29*, 67–70. [CrossRef]
49. van Heezik, Y.; Seddon, P.J.; Maloney, R.F. Helping reintroduced houbara bustards avoid predation: Effective anti-predator training and the predictive value of pre-release behaviour. *Anim. Conserv. Forum* **1999**, *2*, 155–163. [CrossRef]
50. Wei, F.; Hu, Y.; Zhu, L.; Bruford, M.W.; Zhan, X.; Zhang, L. Black and white and read all over: The past, present and future of giant panda genetics. *Mol. Ecol.* **2012**, *21*, 5660–5674. [CrossRef] [PubMed]
51. Shen, F.; Zhang, Z.; He, W.; Yue, B.; Zhang, A.; Zhang, L.; Hou, R.; Wang, C.; Watanabe, T. Microsatellite variability reveals the necessity for genetic input from wild giant pandas (*Ailuropoda melanoleuca*) into the captive population. *Mol. Ecol.* **2009**, *18*, 1061–1070. [CrossRef]
52. Frankham, R.; Ballou, S.E.J.D.; Briscoe, D.A.; Ballou, J.D. *Introduction to Conservation Genetics*; Cambridge University Press: Cambridge, UK, 2002.
53. Tallmon, D.A.; Luikart, G.; Waples, R.S. The alluring simplicity and complex reality of genetic rescue. *Trends Ecol. Evol.* **2004**, *19*, 489–496. [CrossRef]
54. Shan, L.; Hu, Y.; Zhu, L.; Yan, L.; Wang, C.; Li, D.; Jin, X.; Zhang, C.; Wei, F. Large-scale genetic survey provides insights into the captive management and reintroduction of giant pandas. *Mol. Biol. Evol.* **2014**, *31*, 2663–2671. [CrossRef]
55. Zhao, S.C.; Zheng, P.P.; Dong, S.S.; Zhan, X.J.; Wu, Q.; Guo, X.S.; Hu, Y.B.; He, W.M.; Zhang, S.N.; Fan, W.; et al. Whole-genome sequencing of giant pandas provides insights into demographic history and local adaptation. *Nat. Genet.* **2013**, *45*, 67–99. [CrossRef]
56. Kleiman, D.G. Reintroduction of captive mammals for conservation. *Bioscience* **1989**, *39*, 152–161. [CrossRef]
57. Yao, R.; Xu, L.; Hu, T.; Chen, H.; Qi, D.; Gu, X.; Yang, X.; Yang, Z.; Zhu, L. The "wildness" of the giant panda gut microbiome and its relevance to effective translocation. *Glob. Ecol. Conserv.* **2019**, *18*, e644. [CrossRef]
58. Zhang, M.C.; Huang, Y.; Hong, M.S.; Zhou, S.Q.; Huang, J.Y.; Li, D.S.; Li, R.G.; Liu, D.; Zhou, X.P.; Zhang, H.M. Impacts of man-made provisioned food on learned cub behaviours of giant pandas in pre-release reintroduction training. *Folia Zool.* **2017**, *66*, 58–66. [CrossRef]
59. Wolf, C.M.; Garland, T., Jr.; Griffith, B. Predictors of avian and mammalian translocation success: Reanalysis with phylogenetically independent contrasts. *Biol. Conserv.* **1998**, *86*, 243–255. [CrossRef]
60. Wang, L.; Ding, R.Z.; Zhai, Y.H.; Zhang, Q.L.; Tang, W.; Zheng, N.N.; Hua, G. Giant panda identification. *IEEE Trans. Image Process.* **2021**, *30*, 2837–2849. [CrossRef]
61. Zhu, Z.; Pan, S.; Wei, B.; Liu, H.; Zhou, Z.; Huang, X.; Luo, Y.; Zhou, L.; Zhang, S.; Ma, X.; et al. High prevalence of multi-drug resistances and diversity of mobile genetic elements in escherichia coli isolates from captive giant pandas. *Ecotoxicol. Environ. Saf.* **2020**, *198*, 110681. [CrossRef]
62. Wei, W.; Swaisgood, R.R.; Dai, Q.; Yang, Z.; Yuan, S.; Owen, M.A.; Pilfold, N.W.; Yang, X.; Gu, X.; Zhou, H.; et al. Giant panda distributional and habitat-use shifts in a changing landscape. *Conserv. Lett.* **2018**, *11*, e12575. [CrossRef]
63. Wang, Y.; Lan, T.; Deng, S.; Zang, Z.; Zhao, Z.; Xie, Z.; Xu, W.; Shen, G. Forest-cover change rather than climate change determined giant panda's population persistence. *Biol. Conserv.* **2022**, *265*, 109436. [CrossRef]
64. Wei, W.; Nie, Y.; Zhang, Z.; Hu, Y.; Yan, L.; Qi, D.; Li, X.; Wei, F. Hunting bamboo: Foraging patch selection and utilization by giant pandas and implications for conservation. *Biol. Conserv.* **2015**, *186*, 260–267. [CrossRef]
65. Wei, W.; Swaisgood, R.R.; Owen, M.A.; Pilfold, N.W.; Han, H.; Hong, M.; Zhou, H.; Wei, F.; Nie, Y.; Zhang, Z. The role of den quality in giant panda conservation. *Biol. Conserv.* **2019**, *231*, 189–196. [CrossRef]
66. Sun, X.; Long, Z.; Jia, J. A multi-scale maxent approach to model habitat suitability for the giant pandas in the qionglai mountain, china. *Glob. Ecol. Conserv.* **2021**, *30*, e1766. [CrossRef]
67. Morris, S.D.; Brook, B.W.; Moseby, K.E.; Johnson, C.N. Factors affecting success of conservation translocations of terrestrial vertebrates: A global systematic review. *Glob. Ecol. Conserv.* **2021**, *28*, e1630. [CrossRef]
68. Mccarthy, M.A.; Armstrong, D.P.; Runge, M.C. Adaptive management of reintroduction. *Reintroduction Biol. Integr. Sci. Manag.* **2012**, *12*, 256.
69. Allen, C.R.; Fontaine, J.J.; Pope, K.L.; Garmestani, A.S. Adaptive management for a turbulent future. *J. Environ. Manag.* **2011**, *92*, 1339–1345. [CrossRef] [PubMed]

animals

MDPI

Review

Lessons Learned for Pre-Emptive Capture Management as a Tool for Wildlife Conservation during Oil Spills and Eradication Events

B. Louise Chilvers [1,*] and Pete J. McClelland [2]

1 Wildbase, School of Veterinary Science, Massey University, Palmerston North 4442, New Zealand
2 Independent Researcher, 237 Kennington-Roslyn Bush Rd, RD2, Invercargill 9872, New Zealand
* Correspondence: b.l.chilvers@massey.ac.nz

Simple Summary: Pre-emptive capture or translocation of wildlife during oil spills and prior to pest eradication poison applications have very specific conservation goals to ensure the survival of a threatened regional population or species. This paper reviews reports from pre-emptive captures and translocations of threatened wildlife undertaken during past oil spills and island pest eradications. Species captured, techniques used, outcomes of responses, and lessons learned were assessed and recommendations for the planning and trials needed for future pre-emptive capture operations are described. This paper aims to learn from the past to encourage better use and preparedness for pre-emptive capture as a preventative wildlife conservation tool in the future.

Abstract: Pre-emptive capture or translocation of wildlife during oil spills and prior to pest eradication poison applications are very specific conservation goals within the field of conservation translocation/reintroduction. Protection of wildlife from contamination events occurs during either planned operations such as pest eradication poison applications, or unplanned events such as pollution or oil spills. The aim in both incidences is to protect at-risk wildlife species, ensuring the survival of a threatened regional population or entire species, by excluding wildlife from entering affected areas and therefore preventing impacts on the protected wildlife. If pre-emptive capture does not occur, wildlife may unintentionally be affected and could either die or will need capture, cleaning, and/or medical care and rehabilitation before being released back into a cleared environment. This paper reviews information from pre-emptive captures and translocations of threatened wildlife undertaken during past oil spills and island pest eradications, to assess criteria for species captured, techniques used, outcomes of responses, and lessons learned. From these case studies, the considerations and planning needs for pre-emptive capture are described and recommendations made to allow better use and preparedness for pre-emptive capture as a preventative wildlife conservation tool.

Keywords: pre-emptive capture; translocation; conservation; island eradication; oil spill; wildlife

Citation: Chilvers, B.L.; McClelland, P.J. Lessons Learned for Pre-Emptive Capture Management as a Tool for Wildlife Conservation during Oil Spills and Eradication Events. *Animals* **2023**, *13*, 833. https://doi.org/10.3390/ani13050833

Academic Editor: Chunwang Li

Received: 26 January 2023
Revised: 10 February 2023
Accepted: 13 February 2023
Published: 24 February 2023

1. Introduction

As a wildlife management tool, the use of pre-emptive capture and translocations has risen rapidly in the past two decades [1–3]. In 2020, a review of 145 studies on wildlife capture and translocations indicated that 77% had been carried out for conservation purposes, predominantly to reintroduce or increase species' presence within their indigenous range [3]. Here, we review pre-emptive capture and hold or translocation as techniques for preventing wildlife from entering contaminated areas or removing wildlife from areas before they are oiled or before the use of poisons for pest eradication purposes. The pre-emptive capture, holding, and translocation process undertaken for wildlife during oil spills and pest eradication poison applications is a specific conservation goal within the field of conservation translocation and reintroduction techniques [2]. The aim is to protect a significant proportion of a range-restricted species or significant regional population to reintroduce individuals back to their original range after an impact (oils spill or poison) has been removed to repopulate the area.

The first priority of wildlife protection during contamination events is to minimise the impact of the contaminant on wildlife through prevention. If a contamination event occurs, either planned (poisoning events for pest eradication) or unplanned (oil spills), the protection of wildlife can occur by: (1) stopping the contaminant from reaching and affecting non-target wildlife by containing poison in equipment that does not allow wildlife access or not spreading poison in critical wildlife areas, or containing the oil spill at or close to the source or stopping oil entering the wildlife habitat; (2) stopping wildlife being affected through the removal of wildlife from an affected or about to be affected area, or preventing wildlife from entering affected areas through hazing, deterrence, or pre-emptive capture. If these preventative measures are not undertaken, wildlife may die or need capture, cleaning, and/or medical treatment and rehabilitation before being released back into a clean environment. Avoiding wildlife from being impacted is always the highest priority as it prevents the duress, injury, and possible death of wildlife. Additionally, while it does have its own risks, it will significantly lessen the cost of a wildlife response if wildlife does become affected, and reduces the negative public and media reactions to reports and visual images of impacted wildlife [4]. This review concentrates on the priority of preventing wildlife from entering impacted areas or removing wildlife from areas before the area is impacted, and specifically focuses on pre-emptive capture either to translocate wildlife (move to another location) or to hold wildlife in captivity until release into a clean environment can occur.

The first step needed in all protective and preventative processes is developing a plan, based on analysis of areas at risk of impact (either oiling or where poison will be spread), the vulnerability of species (to both the contaminant and any proposed action), and potential response options for species at risk [5–7]. Each protective and preventative technique is species- and area-specific, and is usually initially based on a species population size and distribution, with species that have a high threat classification (i.e., listed as endangered), are range-restricted, and/or have high cultural importance/or public profile being most likely to be considered for pre-emptive capture. Threatened species usually have a small population size or restricted distribution or endemism, meaning an impact on their habitat could mean the extinction of that species or the local population. Other factors to consider are habitat use, therefore exposure risk, season, and biological factors, such as if the species is breeding at the time of impact [7]. For example, for pest eradication/poisoning events, how species forage is important, i.e., nectar-feeding birds are unlikely to be impacted by an aerial application of cereal poison bait; however, herbivores or omnivores may be vulnerable to primary poisoning as they could eat the bait directly, or omnivores, carnivores, or scavengers that could get secondary poisoning from scavenging poisoned individuals. In oil spills, any species that contacts, digests, or inhales fumes from oil can be affected, and, like poison operations, carnivores or scavengers can get secondary poisoning or oiling from predating or scavenging on other oiled wildlife. Undertaking pre-emptive capture of some species may not be practical or viable, i.e., large animals such as whales cannot be pre-emptively captured; therefore, hazing or deterrence are better options to undertake. For all species, the different stages in life cycles, such as breeding or moulting, can prevent other techniques such as deterrence or hazing from working effectively, and pre-emptive capture may be the only technique that could be successful. This was the case for New Zealand dotterels (*Charadrius obscurus*) in 2011 during the MV Rena spill in New Zealand as the dotterels were breeding when the oil spill occurred, making individuals very territorial, and animals would not have left their nesting sites, eggs, or chicks regardless if disturbance techniques were used.

This manuscript uses past oil spills and pest eradications using toxicants on islands as case studies of pre-emptive capture and holding, or translocation, to highlight lessons learned and considerations of species-specific response option restrictions, and outlines recommendations to allow better use, preparedness, and planning for pre-emptive capture as a conservation tool for threatened wildlife during contamination events.

2. Materials and Methods

An online literature search was undertaken aligned with the PRISMA 2020 guidelines [8] with the aim of creating a list of publicly available articles or reports on the use of pre-emptive capture during oil spill response or island eradication, from 1970 to 2022. Primary sources of information were sourced from scientific journal articles, conference proceedings, and any other grey literature through searches on Google, Google Scholar, or the Web of Science database (search terms were in English and included singular words or combinations of pre-emptive, pre-emptive, capture, wildlife, oil spill, oiled wildlife, and eradication). Additionally, searches were made through the Oil Spill conference websites for Interspill, IOSC, and translocation information from IUCN, including Proceedings of the International Conference on Eradication of Island Invasives. Experts in both fields were also contacted for any additional grey literature that was available but not yet published.

3. Results

The most striking result from this research is how few of the undertakings of pre-emptive capture of wildlife for prevention from contamination have been written into publicly available reports, journal articles, conference proceedings, or grey literature (Table 1). There have been over 600 island eradications of invasive rodents, many of which were multi-species eradications [9–12], and 1000s of oil spills that have affected wildlife [13]. There are multiple articles that highlight the need, advantages, and brief outlines on why pre-emptive capture should be undertaken but not many examples of when it has been undertaken or recommendations for what species should be considered, planning considerations needed, or factors to be taken into account before attempting pre-emptive capture [9–14]. However, even from the articles that do mention pre-emptive capture or translocation being undertaken, most only mention that it occurred, and there are few reports on how wildlife was captured or held, with what methodology, what proportion of the population was captured, processes during captivity, or short- or long-term survival or reproduction results after their release. Outlined below and in Table 1 are summaries of the 11 documented case studies of pre-emptive capture of wildlife during oil spill responses or island pest eradications that were assessed. Locations of case studies are shown in Figure 1.

Table 1. Case study summaries highlighting species protected, control measures, outcome and lessons.

Case Study (Reference)	Species Protected	Control	Outcome	Lessons
Oil Spill Response				
MV Iron Barron, Tasmaina, 1995 [15,16]	Little blue penguins (*Eudyptula minor*)	Translocation 480 km from spill after wildlife cleaned and rehabilitated to allow time for area to be cleaned before wildlife returns	863 translocated, 56% reported returned within 4 months, no difference in survival rates recorded between translocated and non-translocated wildlife	Translocation considered effective, recommend trialing distances before being implemented

Table 1. *Cont.*

Case Study (Reference)	Species Protected	Control	Outcome	Lessons
MV Treasure, South Africa, 2000 [17–19]	African Penguins (*Spheniscus demersus*)	19,506 penguins were pre-emptive captured and translocated ~700 km away to allow time for area to be cleaned before wildlife returns. 3350 orphaned chicks captured and hand reared	One year after the spill, 84% of the translocated birds were re-sighted, compared with 55% of the captured, cleaned, and released birds. Of the 3350 chicks collected approximately 2300 were fledged and released	Translocations considered effective however greater consideration of conditions prior to and during transport needed. Preemptive capture and hand rearing of chicks was a successful conversation practice which can be used for oil spills, droughts and other human and natural impacts.
Deepwater Horizons, USA 2010 [20]	Brown Pelicans (*Pelecanus occidentalis)*	Translocation and supplementary feeding away from spill area after wildlife cleaned and rehabilitated to allow time for area to be cleaned	No morality of translocated birds reported, birds mixed with local flock and stayed for 4 to 6 weeks	Translocations and supplementary feeding considered successful. Shorter time period of feeding suggested and tracking of translocated individuals
MV Rena, New Zealand, 2011 [21]	Northern New Zealand dotterels (*Charadrius obscurus aquilonius)*	60 dotterels pre-emptively caught and held for 60 days	90% survival to release	Critical to have a dedicated captive management team. Strong recommendation that if shorebirds are preemptively captured, that the clean-up of their habitat is prioritised to enable as early release as possible.
Eradication Operation				
Kapiti Island, New Zealand 1996 [22]	North Island weka (*Gallirallus australis grey)*	Capture and transfer of 243 weka to mainland NZ	Some Weka not transferred survived the aerial poisoning and no reintroduction back to the island was made. Weka now breed prolifically on the island and are fully recovered	Species at risk should be identified through both non-toxic bait trials and knowledge from species at risk from previous operations

Table 1. *Cont.*

Case Study (Reference)	Species Protected	Control	Outcome	Lessons
Whenua Hou/Codfish Island 1998 [23]	Fernbirds (*Bowdleria punctata wilsoni*), Short-tailed bats (*Mystacina tuberculatus tuberculatus*)	Fernbirds—21 birds transferred to a nearby rat-free island and poison placed in bait stations in highest density fernbird habitat instead of aerial spread Bats—captured and translocated onto another island and 386 held in captivity on island for ~90 days	Fernbirds—transferred birds successfully translocated, established, and bred and have not been transferred back. Most fernbirds on the island were thought to be killed. However enough survived or naturally reintroduced to recover and expanded their range without rats. Bats—capture and release unsuccessful, none know to survive. Capture and hold on the island was considered successful	Dedicated husbandry teams are needed for the pre-emptive capture of species during eradication projects. The additional cost of an additional rat eradication and transfer of a security population to another island was considered warranted even though not needed in the end.
Seychelles 2000 [24,25]	Seychelles magpie-robins (*Copsychus sechellarum*), Seychelles fodys (*Foudia sechellarum*), Aldabran giant tortoises (*Geochelone gigantea*)	590 individuals from the 3 species were held in captivity on the island for up to 90 days during eradication	All individuals survived capture and were released. Magpie robins breed in captivity	Dedicated husbandry teams are essential for success and allow for increased knowledge and capability for the aviculture of species
Anacapa Islands, California 2001 and 2002 [26]	Anacapa deer mouse (*Peromyscus maniculatus anacapae*), Peregrine falcons (*Falco peregrinus*)	Aerial poisoning was conducted over two years. Pior to each drop deer mice were live captured and held in captivity or before the second application mice (from the soon to be poisons island) were transferred into the wild on the now rat-free island Raptors were live captured prior to rodenticide applications (peregrine falcons, red-tailed hawks, barn owls, and burrowing owls). Most were released on the mainland in suitable habitat; peregrine falcons were held and released back onto Anacapa 3 weeks after rodenticide applications	There were no signs of rats or wild deer mice on the islands after poison applications. Deer mice that had been captured were released back onto rat-free islands 5 months after applications. In both years, >90% of the deer mice taken into captivity were released. Captive holding and translocation significantly reduced raptor mortality. One granivorous bird species, rufous-crowned sparrow, *Aimophila ruficeps Obscura* ,showed an unexpected significant decline	This was the first recorded rodent eradication that ensured a native endemic rodent, which showed to be equally susceptible to the bait as the rats, to survive. Eradication showed the importance of learning from previous operations, particularly based on species similar to raptors, as some granivorous birds may require captive-holding efforts or no-drop zones to minimize risk for non-target impacts as seen on Codfish Is, NZ. Demonstrates the need for well-designed data-driven mitigations.

Table 1. *Cont.*

Case Study (Reference)	Species Protected	Control	Outcome	Lessons
Galapagos 2012 [27]	Pinzón giant tortoise (*Chelonoidis duncanensis*), Pinzón lava lizards (*Microlophus duncanensis*), Galapagos hawks (*Buteo galapagoensis*)	15 tortoises captured and held on another island for 2 years 40 lizards held in captivity until 10 days after second bait spread 60 hawks were captured and held in captivity until 12–14 days after second bait spread	All tortoises survived, were released, and have since bred 87% survival rate of lizards in captivity Unfortunately, 22 hawks died 12 to 170 days after the release of secondary poisoning therefore 10 were recapture treated with Vit K and not released until poison levels known to reduce	Rodenticide lasted longer in the environment than expected. Lizards did not eat bait in laboratory trials, but did in the field, emphasising both laboratory and field trials should be undertaken to determine species at risk
Lord Howe Island, Australia 2019 [28,29]	Lord Howe woodhen (*Gallirallus sylvestris*) and pied currawong (*Strepera graculina crissalis*)	Trial preemptive capture of both species prior to poison spread. 85% of woodhen population and 50% of currawong population captured before eradication operation and held until one month after.	All woodhen and currawong survived capitivity and woodhen population now quadruple pre-eradication population size	Importance of pre-emptive capture trials to understand how to manage wildlife in captivity Eradication also showed the importance of learning from previous operations particularly based on similar species
Gough Island, Tristan da Cunha, UK 2021 [30–34]	Gough bunting (*Rowettia goughensis*) and moorhen (*Gallinula comeri*)	Bunting and moorhens were trial preemptively captured and held before poison spread 84 moorhens and 100 buntings captured and held during poisoning	80 moorhens and 103 bunting released Buntings continue to do well; however, the status of moorhens is unknown Unfortunately, the rodent eradication was not a success	Recommended that the avicultural project be run separately but parallel to the eradication operation A dedicated husbandry team with a comprehensive plan was essential

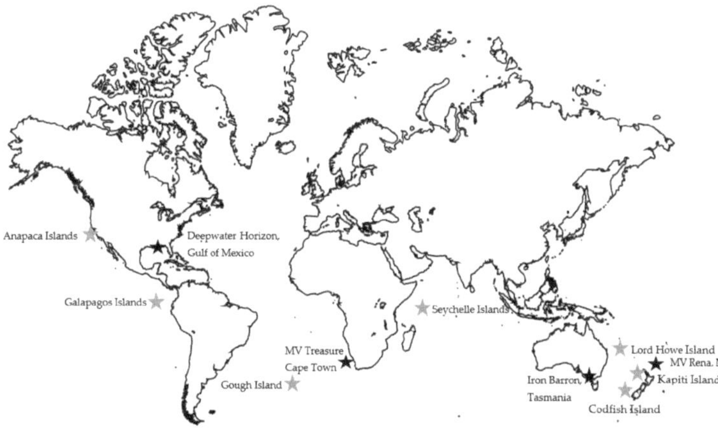

Figure 1. Locations of case studies. Oiled wildlife responses are shown as a black star. Eradication operations are shown as a grey star.

3.1. Case Studies—Oil Spills

3.1.1. Australia MV Iron Barron Oil Spill 1995

On 10 July 1995, the MV Iron Barron encountered bad weather coming into the port of Launceston in northern Tasmania, Australia, grounding on Low Head, Hebe Reef, leaking an estimated 325 tonnes of heavy bunker fuel oil [15]. Little blue penguins (*Eudyptula minor*) were significantly impacted by the spill with an estimated 10,000 to 20,000 killed and 1894 oiled birds captured, cleaned, and rehabilitated in an improvised rehabilitation facility [16]. The penguins were ready for release before their habitat had been cleaned, particularly as it was a large complex area with many islands over which the oil had spread. Rather than prolonging captivity, which increases the risk of disease and stress, and as breeding was imminent, a translocation strategy to release cleaned and rehabilitated penguins at different distances from the oiled site was trialled. This was undertaken to determine the optimal distance to release rehabilitated penguins so that they returned to their habitat after it had been cleaned. Twenty-five VHF-tagged penguins were translocated 360 km from the spill site on the east coast of Tasmania, and their movements were tracked from the air. Two birds returned to their original capture site within 3 days, not enough time to clean up the area, prompting a new release site 120 km further south (480 km in total). After the first trial, it was decided that the translocation site 480 km away was appropriate for the circumstances, and a further 863 penguins were translocated. At least 56% of the birds released further south returned to Low Head in four months, after their habitat had been cleaned. Monitoring found no differences in the survival rate of translocated and non-translocated birds.

Lessons learned: While translocation was considered effective in this situation, it is recommended that translocation protocols should be trialled before being implemented [15,16].

3.1.2. South Africa MV Treasure Oil Spill 2000

The MV Treasure spilled approximately 400 tonnes of heavy fuel oil onto the coast of South Africa near Cape Town on 23 June 2000. The spill occurred near the two major breeding colonies, Robben and Dassen Islands, of the endangered African Penguins (*Spheniscus demersus*). A total of 19,000 oiled penguins were caught, cleaned, rehabilitated, and returned to a clean environment. Over 1660 birds died during captivity, most from the negative impacts of the oil [17]. To prevent even more penguins from being oiled, a further 19,506 penguins were captured, relocated, and released at Cape Recife near Port Elizabeth, ~700 km to the east of Cape Town [18]. These penguins, whether oiled or pre-emptively captured, represented over half of the known, endangered, declining population of African penguins at the time of the spill [19].

Relocated birds returned quickly to their breeding islands, with the faster returning in 11 days and most returned within two to four weeks [17]. This indicated that Cape Recife was an appropriate location for release because it was a suitable distance to allow time for the oil to be cleared before the birds returned, but close enough for birds to return within a month, thereby minimising any disruption to breeding and moulting. Of the 19,506 penguins translocated, 241 died between being captured and release at Cape Recife due to some being transported in closed trucks causing CO_2 poisoning. Additionally, before transport, those kept on Dassen Island were kept fenced in an area on the island with limited access to drinking water and no areas to swim. Both factors contributed to the higher mortality of those pre-emptive captured birds [19]. Additional to the adults, 3350 orphaned chicks were also pre-emptively captured and reared in captivity and released back into their clean environment when they had fledged. Of the 3350 chicks collected, approximately 2300 were fledged and released [17].

Prior to the MV Treasure spill, South Africans' seabird oil spill rescue plans focused on catching and treating oiled birds as soon as possible, before releasing them back into the wild; preventing birds from becoming oiled was not part of any plan [19]. This wildlife response is still the largest relocation response for oiled wildlife globally and, due to its success, the implementation of relocating birds before they became oiled has been implemented as a response option and documented to have been an effective conservation measure [17]. One year after the MV Treasure spill, 84% of the evacuated birds had been re-sighted, compared with 55% of the captured, cleaned, and released birds.

Lessons learned: The two overall lessons from the pre-emptive capture of African penguins were greater consideration of conditions prior to and during transport to translocation sites to prevent deaths, and consideration of distance transported so that the wildlife's return allowed enough time for the oiled areas to be cleaned, but the distance was not too far to cause individuals to get disorientated or lost, or to cause major disruption to breeding or moulting cycles. A second conclusion is that pre-emptive capture and raising of penguin chicks is a successful conservation practice that continues today for African Penguins, not only during oil spills, but also droughts, colony disturbances, and other human and natural impacts on this endangered species (https://sanccob.co.za; accessed on 15 February 2023).

3.1.3. USA Deepwater Horizons Oil Spill 2010

On 20 April 2010, the Deepwater Horizon well exploded 66 km off the coast of Louisiana, in the Gulf of Mexico, and before being capped, three months later, more than 780,000 tonnes of crude oil were spilled. There were numerous impacts on the environment and wildlife, and because of the length of time oil continued to be spilled, some wildlife that had been cleaned and rehabilitated were ready for release long before their environment was cleaned. Brown Pelicans (*Pelecanus occidentalis)* were one of the species impacted, with more than 700 rehabilitated in south-eastern Louisiana alone [24]. To overcome the lack of a clean habitat for their release, 182 oil-rehabilitated pelicans were translocated from south-eastern Louisiana to Rabbit Island in south-western Louisiana, an island that was not impacted by the spill and had non-impacted pelicans breeding on it. The aims of this translocation were to enable monitoring of movements of translocated groups and determine if translocation would delay pelicans returning to their habitat, and therefore getting re-oiled, and to be able to monitor mortality, determine the integration of translocated pelicans with local pelican groups, and determine if supplemental feeding of translocated birds prolonged occupation on the island, therefore again reducing the likelihood of re-oiling [20]. Daily surveys were undertaken at the island for six weeks from the date of translocations, with supplementary feeding occurring twice a day for four weeks. There was no mortality of rehabilitated birds recorded and it was observed that translocated pelicans mixed readily with local pelican flocks. Many of the local and translocated pelicans moved away from the island within 4 to 6 weeks, likely due to natural and human-induced factors.

Lessons learned: The translocations and supplementary feeding program of the brown pelican were considered successful at reducing the movement of pelicans back into oiled areas. However, habituation to the feeding vessel and supplementary feeding were observed both from the rehabilitated and local pelicans. For future translocations, it is suggested shorter time periods of supplemental feedings should occur, using alternative feeding strategies such as blinds or remote feeders due to the easy habituation of pelicans to humans. For tracking of movement of rehabilitated birds, it is recommended a subset of individuals be radio/satellite tagged for documentation of movements and mortality.

3.1.4. New Zealand MV Rena Oil Spill 2011

On 5 October 2011, the container vessel MV Rena ran aground on Astrolabe Reef, Bay of Plenty, New Zealand, and within days spilled approximately 350 tonnes of heavy fuel oil. The endangered Northern New Zealand dotterels (*Charadrius obscurus aquilonius*—a small ~140 g shorebird) were pre-emptively captured as part of the oiled wildlife response to ensure the survival of a regional population. The pre-emptive capture occurred as it was considered that if these small birds became significantly oiled their chances of survival were minimal despite cleaning/rehabilitation [21]. Sixty dotterels were caught, with over half the birds already having some level of oil contamination. This population of dotterels represented ~6% of the global population of this species and the majority of the local population within the area of the spill. Many pairs were already breeding and nesting at the time of the spill, so other deterrence or hazing activities would not have worked as the birds are territorial and would not have moved away from their nests. This was the first time wild adult New Zealand dotterels pairs had been held in captivity for a prolonged period. Birds were caught in their breeding pairs, with each pair held in individual enclosures, blocked from view of other pairs to prevent territorial and fighting behaviours which would have been normal during breeding. There was a 90% survival rate of the New Zealand dotterels held in captivity during the MV Rena oil spill response over a ~2-month period [21]. Dotterels took 1–15 days (median 5 days) to convert to the captive diet. Sixty-one percent of birds obtained minor abrasions from contact with enclosure netting during captivity due to their flighty behaviour which did not affect survival; however, seven birds (11.7%) developed respiratory disease, with six of these dying from aspergillosis causing pneumonia-type deaths [21]. Intensive captive husbandry was needed to convert the birds to a captive diet, minimise injuries, and manage pododermatitis/foot sores.

Lessons learned: It was critical to have a dedicated captive management team for these birds. The challenges that come with managing wild adult shorebirds in captivity and converting to captive diets are well recognised within the wildlife rehabilitation community. Additionally, shorebirds are species considered to respond poorly to the stresses of capture and captivity [35]. Therefore, although the pre-emptive capture and management of shorebirds during an oil spill to minimise the effects of oil spills carries significant costs and risks to the birds, it is considered essential in emergency management situations for high-priority/at-risk species. Additional to normal capture stressors, clinical signs of respiratory disease were not observed until the last half of the time the birds were in captivity. Therefore, a strong recommendation for the management of shorebirds that are pre-emptively captured is that the clean-up of their habitat is prioritised to enable the early return of birds to the wild.

3.2. Case Studies—Island Eradications

3.2.1. New Zealand—Mice and Rat Eradication/Poisoning, Kapiti Island 1996

After the eradication of cats (*Felis catus*), deer (*Cervidae* spp.), pigs (*Sus* spp.), goats (*Capra* spp.), and possums (*Trichosurus vulpecula*), by hunting and trapping, from the rugged 19.65 km^2 Kapiti Island off the south-west coast of the North Island, New Zealand, the Department of Conservation of New Zealand also successfully eradicated Norway and Pacific rats (*Rattus norvegicus* and *R. exulans*) in 1996 using helicopter broadcast of brodifacoum cereal baits [22]. Trials with non-toxic baits were carried out on North Island weka (*Gallirallus australis grey*) and little spotted kiwi (*Apteryx owenii*), both flightless birds found on the island, to help determine the risks of poisoning for these non-target species [27]. North Island weka at the time were classified as endangered and were expected to be affected by the eradication activities both from primary and secondary poisoning, particularly as weka are omnivores and will scavenge and kill other species. From the non-toxic trials, measures to minimise the effects of the poison application on fauna at risk were put in place, which included the capture and holding in captivity or translocation to reserves on mainland New Zealand of 243 weka, and the transfer of 66 New Zealand robins (*Petroica australis*), which had previously been identified as being at risk to nearby Mana Island. Post-poisoning call rate monitoring indicated that weka call rates were significantly lower after poisoning; however, it could not be determined if that was caused by the removal of weka from the island (not yet returned or released at the time of the call counts) and/or the poisoning operation, because no call rate monitoring was undertaken in the period between the removal and the poisoning for comparison. However, the fact that weka calls were heard meant that some survived the poisoning operation, and together with the birds released after the operation, they are now distributed throughout Kapiti Island and breeding prolifically [22].

Lessons learned: Species at risk should be identified through both non-toxic bait trials and knowledge from species at risk from previous operations. Monitoring between pre-emptive capture and poison applications should be undertaken to allow the determination of the impacts of both.

3.2.2. New Zealand—Rat Eradication/Poisoning, Whenua Hou Nature Reserve/Codfish Island 1998

Whenua Hou Nature Reserve/Codfish Island is located 3 km NW of Stewart Island, New Zealand, and is the protected island home to the largest population of the endangered Kākāpō (*Strigops habroptilus*), a large flightless native parrot. Following the removal of possums and South Island weka (*Gallirallus australis australia*), eradication of the Pacific rat was undertaken on Codfish in August 1998, using a combination of aerial applications and bait station cereal pellets containing brodifacoum. In preparation for the eradication, a smaller island, Putauhinu (96 ha), was eradicated of Pacific rats the year before in 1997, so that a population of fernbirds (*Bowdleria punctata wilsoni*), endemic to Codfish Island, could be established [23]. Additional to the transfer, a 37 ha block of the best fernbird habitat known on Codfish, containing the densest population of fernbirds, was poisoned using bait stations at 25 m intervals instead of using aerial baiting, which had been shown during field trials elsewhere to cause a high death rate in fernbirds. All Kākāpō (except one that could not be found) were removed from the island prior to the poison application and temporarily held on a separate island. Short-tailed bats (*Mystacina tuberculatus tuberculatus*) were also managed, with 50 being captured and released onto Ulva Island, a predator-free island off Stewart Island; however, this was unsuccessful. Additionally, during the poison applications, four purpose-designed "batteries" were constructed on Codfish Island with 386 short-tailed bats held for nearly three months. There was no observable loss to the bat population linked with the bait application although individuals are likely to have been lost. Nine bats were lost up until the last week of the capture program, when 42 died in one event due to heat stress in one of the roost boxes. Despite this sad event, the operation was still considered a success given how difficult bat husbandry can be. The bat protection and monitoring was undertaken by a team of 5–7 people and this investment of single-task personnel is one of the main reasons for its success. The 21 fernbirds that were transferred to Putauhinu were confirmed to have bred, and follow-up checks on Putauhinu have shown that the fernbird population has continued to increase and expand its range on Putauhinu. It appeared most fernbirds were lost on Codfish due to the bait application, despite the management, with very few recorded for 2 years after. However, enough survived to rebuild and recover not only to the population's original range, but to also expand into a variety of habitats in the absence of rats [23]. This meant the planned reintroduction from Putauhinu was not required.

Lessons learned: Dedicated husbandry teams are needed for the pre-emptive capture of species during eradication projects. Although the bait stations in the areas of the fernbirds achieved the goal of reducing fernbird mortality, it was thought that as fernbirds outside the area affected by the aerial bait died, fernbirds within the bait station area expanded their range and therefore became more exposed to aerially laid bait. Therefore, it was thought that the impact may have been lessened by expanding the size of the core area in which only bait stations were laid, thus increasing the percentage of birds within the core area. This result also led to the conclusion that the additional cost of rat eradication and transfer of a security population to Putauhinu was warranted even though it proved to not be necessary. This eradication also proved that field trials are important for poison eradication, as fernbirds were thought to be insectivores mainly preying on spiders and hence at little risk from the baiting operation. However, field trials showed that fernbirds when presented with brodifacoum bait would eat it, and indicated that the species would be heavily impacted by aerial bait, therefore leading to the mitigation work of bait stations in the area where the fernbirds were in high abundance (Pete McClelland pers comm).

3.2.3. Seychelles—Cat, Rabbit, Rats and Mice Eradication 1996–2000

Between 1996 and 2000, attempts were made to eradicate five introduced mammal species, feral cat, rabbit (*Oryctolagus cuniculus*), ship rat (*Rattus rattus*), Norway rat, and house mouse (*Mus domesticus*), on four inhabited Seychelle islands. As there were no rat-free islands in close proximity for the transfer of species at risk, 590 individuals from three threatened native species, the Seychelles magpie-robins (*Copsychus sechellarum*, $n = 39$), Seychelles fodys (*Foudia sechellarum*, $n = 330$, 50% of the known population), and the Aldabran giant tortoises (*Geochelone gigantea*, $n = 218$), thought to be at risk from primary and/or secondary poisoning, or for public goodwill in the case of tortoises, were held in captivity for the three months of the eradication program [24]. During the captivity of these species across the islands, the avicultural knowledge and capability of staff increased enormously. The captivity of these species during eradication was very successful, with magpie-robins breeding successfully during three months in captivity [25]. All tortoises, Seychelles fodys, and magpie-robins were successfully released within 3 months after bait application.

Lessons learned: Dedicated husbandry teams are essential for success and allow for increased knowledge and capability for the aviculture of species and in the region. This was one of the first major human-occupied island eradication programs and its success led to the planning of eradications on the likes of Galapagos and Lord Howe Islands (see below). It was an important conclusion at the time that land held privately, human habitation, or tourism activities need not be seen as barriers to eradication projects, as island-based tourism activities can provide the financial and human resources to restore and maintain threatened endemic biodiversity.

3.2.4. California, USA—Rat Eradication/Poisoning, Anacapa Islands 2001–2002

Eradication of black/ship rats from Anacapa Islands, US Channel Islands National Park, California, was undertaken in 2001 and 2002 [26]. This was the first aerial application of a rodenticide in North America and the first attempt in the world to eradicate a rodent from islands while preserving a native endemic rodent on the same islands. There are three islands in this group and, to ensure the presence of the native deer mouse (*Peromyscus maniculatus anacapae*), the rodenticide application was staggered over two years so that a wild population was always present on one or more islands [26]. Concurrently, mice populations from each island were held in captivity during poison applications. Additional to the mice, to avoid as much as possible birds being affected by the application, bait was made using colouring and sizing that deterred gulls and granivorous birds, resident raptors were captured and held or translocated, and a 15 ha no-drop zone was established on West Anacapa to create a refuge for granivorous birds, particularly the Santa Cruz Island rufous-crowned sparrow *Aimophila ruficeps obscura*. In the no-drop zone, rats were poisoned using bait stations that were inaccessible to granivorous birds.

Prior to the first poisoning in 2001, 185 deer mice were live captured from East Anacapa and held for five months. Of these, 174 were released after poisoning. Prior to the second poisoning in 2002, 373 and 365 deer mice were captured from Middle and West Anacapa, respectively, and held in captivity, while concurrently 715 and 308 mice from Middle and West Anacapa were captured and translocated to rat-free East Anacapa. Five months after the second eradication, 358 and 360 captive mice were reintroduced to Middle and West Anacapa, respectively. Raptors were live captured prior to rodenticide applications (including eight peregrine falcons *Falco peregrinus*, nine red-tailed hawks *Buteo jamaicensis*, four barn owls *Tyto alba*, and six burrowing owls *Athene cunicularia*). Most were released onto suitable habitat on mainland California, except peregrine falcons, which were held and released back onto Anacapa 3 weeks after rodenticide applications. A total of 94 birds (16 species) were identified from carcass searches following rodenticide applications. Of the 63 birds tested for brodifacoum, 59 (94%) tested positive [26].

Lessons learned: The successful recovery of the Anacapa deer mouse following the eradication demonstrates that it is feasible to eradicate invasive rodents from islands when native rodents or other susceptible native animals can be held in captivity and kept away from poison. Captive holding and translocation significantly reduced raptor mortality.

However, captive holding or other mitigation measures (no-drop zones) may be necessary for sedentary granivorous passerines, as previously used for fernbirds during the eradication of rats from Codfish Island.

3.2.5. Galapagos—Rat Eradication/Poisoning, Pinzón Island 2012

In December 2012, brodifacoum bait was spread on Pinzón Island (1815 hectares), Galapagos, to eradicate black rats which had prevented the Pinzón giant tortoise (*Chelonoidis duncanensis*) from breeding successfully for nearly a century. Two years prior to the poisoning, 15 adult Pinzón tortoises were brought into captivity and housed on Santa Cruz Island for release after the eradication; all survived, and breeding has been recorded since the eradication. The two other species of concern were Pinzón lava lizards (*Microlophus duncanensis*) and Galapagos hawks (*Buteo galapagoensis*). Forty Pinzón lava lizards were taken into captivity prior to baiting and maintained in enclosures on Pinzón Island, and were released 10 days after the second bait application as it was determined that, due to bait degradation, the risk of poisoning would by then be minimal [32]. Two lava lizards escaped captivity and five captive lizards died during captivity, resulting in a survival rate of 87%. Sixty Galapagos hawks were taken into captivity and held in purpose-built aviaries on Pinzón Island. All survived captivity and were released 12–14 days after the second aerial bait application. However, within 12 to 170 days after release, 22 mortalities of tracked Galapagos hawks were recorded [27]. Unfortunately, reported to be due to the arid conditions of the island, residual poison persisted in lava lizards. The remaining Pinzón Island Galapagos hawk population (n = 10) was recaptured, returned to captivity, and treated with Vitamin K1, while the toxicological levels of Pinzón lava lizards were monitored [27]. These captive Galapagos hawks represented 15% of the original population and were released when risk was considered acceptable, in July and August 2016. As of 2018, eight hawk nests had been observed on Pinzón with chicks and fledglings confirmed.

Lessons learned: The rodenticide used in this eradication remained in the ecosystem much longer than in any previous rodent eradication project worldwide. This resulted in the secondary poisoning of predatory hawks long after expected; therefore, understanding the longevity of poisons in the local environment and possible pathways into at-risk species is essential to ensure captive wildlife are held for an appropriate length of time so as not to be impacted. Similar to the Whenua Hou fernbird experience, this eradication also highlighted the importance of field base trials, as laboratory trials do not always reflect the response of wildlife in the field. Lava lizards did not eat the rodent bait in the laboratory; however, they did in the field, leading to a greater impact on themselves and the hawks than expected.

3.2.6. Australia—Rat Eradication/Poisoning, Lord Howe Island 2019

Lord Howe is a permanently human-inhabited island group approximately 1455 hectares in size and having a diverse landscape, where rats have already been implicated in the extinction of five endemic bird species and at least 13 species of endemic invertebrates. After the successful eradication of cats, pigs, and feral goats from the Lord Howe Island group, ship rats and mice were then targeted. In 2019, brodifacoum baits were distributed across the island depending on habitat type and land use using aerial distribution in the uninhabited areas, and hand broadcast and locked bait stations in the inhabited areas. Following field observations on a range of species on the island in which they were presented with non-toxic bait, two species were thought to be at risk from the bait, Lord Howe woodhen (*Gallirallus sylvestris)* and Lord Howe pied currawong (*Strepera graculina crissalis*), and successful pre-emptive captive trials were undertaken for these species in 2013, prior to baiting. Twenty-two woodhens and ten currawongs were captured and held in captivity, with all individuals subsequently released successfully back into the wild. The woodhens were captured in family groups or pairs and held together in pens, and initial trials showed the need to be careful with the species' diet as they put on weight quickly on the captive diet [28,29]. From this trial, despite the woodhens normally being very territorial, they were held in groups of 20–30 with great success. The idea to hold the woodhens together was undertaken from the experience with weka (a similar bird to the woodhen), on Kapiti Island, New Zealand (see above). For the main poison application, to minimise any potential impact, at least 85% of the woodhen population and 50% of the pied currawong population were placed into captivity. Birds were held for at least one month before baiting, and until risks of primary or secondary poisoning were considered no longer present. From ongoing surveys of the island, by the second autumn woodhen survey following the rodent eradication, 778 woodhens were recorded over a two-week period. This number nearly quadruples the population survey results prior to rodent eradication.

Lessons learned: The Lord Howe Island eradication showed how important it was to build on the learning from previous operations based on similar species, i.e., woodhens vs. weka, not only to decide which species need managing but how they can be managed. The undertaking of pre-emptive capture trials before the poison application allowed a greater understanding of how animals would react to captivity, including understanding that they can put on weight easily with captive diets and can be held together in larger numbers than normal when needed, and assuring the local community of its success. This understanding allowed for better-conditioned individuals to be released back into their environment, with current surveys showing woodhen are thriving on the rodent-free island.

3.2.7. United Kingdom/South Atlantic—Mouse Eradication, Gough Island 2021

An attempt was made by the Royal Society for the Protection of Birds (RSPB) and Tristan da Cunha to eradicate mice from the rugged 6500 ha Gough Island between June and August 2021. Gough Island is part of a World Heritage Site in the southern Atlantic and is one of the world's most important seabird breeding areas, with 22 species of seabird species breeding on the island, many of which are globally threatened, as well as two endemic threatened land bird species. Invasive non-native mice have been responsible for demographically unsustainable levels of chick mortality in seabirds [30,31]. However, it was the two endemic land birds, the Gough bunting (*Rowettia goughensis*) and Gough moorhen (*Gallinula comeri*), for which primary and secondary poisoning was of greatest concern during the eradication, as many of the seabirds would not be present on the island at the time of the poison application. Trials on the capture and holding of these two land bird species began early in the programme planning with 25 buntings and 30 moorhens captured and held for 6 weeks between April and September 2010. Over March to May 2021, 84 moorhens (pre-eradication population estimate of 3500–4250 pairs) [32] and 100 buntings (from a population of 1041–1889 individuals; RSPB unpublished data) were captured and held in captivity during the mouse eradication poison application. Eighty moorhens and 103 buntings were subsequently released back into the wild after the completion of the bait application. Follow-up surveys had shown that there was still a significant wild population of buntings after the bait applications that was then joined by the safeguard bunting population. However, as expected, the wild moorhen population was significantly impacted, hence the importance of the aviculture operation. In late September 2021, once any sign of bait on the island and especially in the lowland moorhen habitat had disappeared (extensive searches were undertaken to validate this), the captive moorhen population was released into their preferred habitat. At the time of writing the status of the moorhen population is still not known: monitoring has proved difficult (e.g., few birds calling, larger potential habitat available, dense vegetation), and while moorhens remain on Gough, no breeding has as yet been recorded to show that the population is beginning to rebuild (A. Callender pers comm.).

Lessons learned: It is highly recommended that the avicultural project be run as a separate parallel operation so as not to be overshadowed in its importance by the "high-profile" baiting operation. A dedicated husbandry team was essential for the capture, care, and survival of these two species. A comprehensive plan for all stages of the aviculture operation is required and it should be followed unless there is good justification to do otherwise. Adequate resourcing is crucial, especially considering sufficient capacity over the holding period to allow members of the team to have downtime, particularly on remote islands such as Gough. The documentation and recommendations from the pre-poisoning capture and holding of moorhens and buntings from Gough Island was an important tactic in the success of the survival of the species for release after the eradication attempt [33]. Trials need to be carried out early and critically accessed, and final design undertaken by a combination of aviculturists and eradication specialists where appropriate so, if necessary, teams can work together. Trials should aim to hold the birds for as long as they are likely to be held for the operation when possible, as the initial Gough trials were not of sufficient length to test how issues such as pododermatitis might affect the birds.

Unfortunately, the eradication attempt was not successful, possibly due to slug consumption of the aerial spread bait, which reduced the amount available for rodents and meant not all mice accessed a lethal dose [34]. An independent review panel is currently assessing the Gough eradication attempt and will report its findings in 2023.

4. Discussion

The conservation goal of pre-emptive capture/translocation of threatened wildlife during oil spills or eradication operations is to protect a biologically significant proportion of a range-restricted species or significant regional population to reintroduce individuals back to their original range to re-establish the population after an impact has been removed. There are examples both in oiled wildlife response and, particularly, eradications where wildlife was considered to be at risk, however, the at-risk population was not a biologically significant proportion of the species, range-restricted species, or significant regional population (i.e., could be reintroduced from other regions if impacted). Examples of these include eradications that impacted Giant Petrels (*Procellariiformes* spp.) and Skuas (*Stercorariidae* spp.) on Macquarie Island, Australian subantarctic [36], Antipodes Island, New Zealand subantarctic [37], and South Georgia, South Atlantic [38]. It should also be noted that these species would also be difficult to hold in captivity, and particularly to hold sufficient numbers of individuals for the risk period to re-establish the population.

The main result from this review is how few of the pre-emptive capture/translocations have occurred for oiled wildlife response and how few have been documented for island eradication operations considering the number of both that have occurred in the last three decades [10–13]. Despite this, there are still valuable lessons to be learned from what has been documented.

The most important lessons learned from both responses is the importance of planning and a specific, dedicated team for the capture and care of captive wildlife. In the case of eradications, planning can be very specific as the site and species present are known well in advance. Whereas planning for oiled spills is more likely to be generic because the specifics of the event, e.g., timing, location, season, etc., are uncertain. However, there will be known endangered or range-restricted species that can be identified within a region or country that can have plans developed for them in advance in the case of a spill in their area. In general, eradication operations have significantly more time and ability to learn from both laboratory and field-based trials, including non-toxic bait trials, to determine species likely to be at risk, and to be able to trial the capture and holding of species prior to poisoning event, as seen for Gough and Lord Howe [28,33]. However, it is only recently that the capture and captive care of protected wildlife has been undertaken by specialist rehabilitators/zoological carers (Lord Howe, Taronga Zoo, and Gough, Royal Society for the Protection of Birds) for eradication operations, and that there has been greater documentation and reporting of the methodology of pre-emptive capture and care techniques, successes, and recommendations.

Conversely, since the Exxon Valdez oil spill in 1989, oiled wildlife response has almost always been undertaken by professional and/or experienced wildlife veterinarians or rehabilitation centres, and has involved greater monitoring and documentation of events.

Unfortunately for wildlife, due to the random, unexpected, and usually instantaneous nature of oil spills, and lack of planning, pre-emptive capture has not been undertaken frequently in oiled wildlife response. The speed at which the oil covers and impacts the environment and wildlife is often too fast to allow pre-emptive activities to occur; however, those that have occurred have been reasonably documented (i.e., [15–17,19,21,39]). Additionally, this review has highlighted that it is not only pre-emptive capture and translocation prior to wildlife being oiled that is a successful, useful management tool for oiled wildlife response. The translocation of cleaned rehabilitated birds outside the area of the oil, to lengthen the time before wildlife return to their habitat, has been shown not only to reduce their chance of being re-oiled, but also reduced the time spent in captivity, therefore reducing secondary problems that can occur in captivity such as pododermatitis [21].

For both oil spills and eradications, an essential lesson is understanding and using the knowledge learned from previous operations to improve current operations. These learnings can be everything from understanding species likely to be impacted, to how to manage and care for species, and for how long to keep them in captivity. However, this type of information needs to be written up and made publicly available from past responses for these lessons to be learned and used in the future.

All reports and articles on pre-emptive capture outline how logistically challenging it can be and, depending on preparation time prior to an oil spill or eradication poison application, pre-emptive capture can seem almost unfeasible. However, the case studies above outline how, with consideration and planning, particularly for planned eradication poison applications, pre-emptive capture or translocations can be successful and save significant proportions of populations or range-restricted species from potential extinction events. For an oil spill event, decisions for pre-emptive capture must be made in a time-critical window, meaning delays could result in wildlife being oiled or injured, or dying. Therefore, understanding the requirements as suggested below prior to a spill occurring, and therefore the activation of a predetermined wildlife plan and personnel immediately after an oil spill, is needed to ensure protective and preventive actions can be undertaken if the situation allows [13].

Primary requirements for pre-emptive captures include:

- Before a spill or eradication—determine potential species at risk: consider the numbers and species of wildlife, their threat classification and geographic extent, the animals' behaviour (seasonal, feeding, breeding), response options available for each species, and whether it is practical for the species to be kept in captivity or if capture and translocation are more appropriate [13]. For eradication activities, this includes developing an inventory of non-target species, including bait-competitors, and a simple food web model to try and understand all possible primary and secondary poison routes pathways (e.g., [40]). Both laboratory and field trials are recommended.
- Conduct applicable capture planning (techniques and personnel) to ensure animal welfare, i.e., conduct site assessment for capture and housing: consider site accessibility and the prioritisation of locations (accessibility, tide, weather), and have knowledge of species behaviour and the geographical area, and lists of experts and pertinent contacts.
- Plan for appropriate captive care arrangements (housing, husbandry, personnel expertise, etc.).
- Plan and possibly trial relocation solutions (release location, transport, site fidelity, predicted time to return, energetic costs of return, etc.).
- Ensure the plans for aviculture can logistically be undertaken given the species, scale of operation, and numbers of individuals or species that need to be held.
- Critically, gain approvals from relevant government agencies and first nations groups, where applicable, for the capture, handling, and holding and transfer/release of wildlife.

5. Conclusions

The difficulty of capturing wildlife safely and providing for their health in a captive environment or during relocation must never be underestimated. The risk of impacts from oiling or primary or secondary poisoning must be weighed against the risks of injury, disease, or death of wildlife during the entire pre-emptive capture and holding process.

To determine the effectiveness of a wildlife pre-emptive capture process, it is critical to monitor and quantify the short- and long-term success or failure of the project. Relative to the number of oil spills and eradication operations that have occurred, direct counts of mortality and pre- and post-event wildlife monitoring studies are still rare. These types of research are critical to fully understand the total and long-term benefits of pre-emptive capture operations of wildlife [41–43]. One of the strongest recommendations from this review is that, once species are identified that are suitable and likely to require pre-emptive capture and holding or translocation, the development of prospective techniques for them should be undertaken by a dedicated and experienced team, and fully documented, with outcomes made publicly available to inform future conservation planning.

Author Contributions: Conceptualization, B.L.C.; Methodology, B.L.C.; Formal Analysis, B.L.C.; Investigation, B.L.C. and P.J.M.; Resources, B.L.C. and P.J.M.; Writing—Original Draft Preparation, B.L.C.; Writing—Review and Editing, B.L.C. and P.J.M.; Visualization, B.L.C.; Supervision, B.L.C.; Project Administration, B.L.C.; Funding Acquisition, B.L.C. All authors have read and agreed to the publication version of the manuscript.

Funding: This research received no external funding.

Institutional Review Board Statement: Not applicable.

Informed Consent Statement: Not applicable.

Data Availability Statement: All data used in this review is already publicly available.

Acknowledgments: Thank you to Andrew Callender for his input into the section on Gough Island in this paper. Thank you also to Bridey White and the anonymous reviewers who contributed to the paper. The Gough Island Restoration Programme was undertaken by the Royal Society for the Protection of Birds and Tristan da Cunha. Thank you to Massey University.

Conflicts of Interest: The authors declare no conflict of interest.

References

1. Seddon, P.J.; Armstrong, D.P.; Maloney, R.F. Developing the science of reintroduction biology. *Conserv. Biol.* **2007**, *21*, 303–312. [CrossRef] [PubMed]
2. IUCN/SSC (International Union for Conservation of Nature/Species Survival Commission). *Guidelines for Reintroductions and Other Conservation Translocations. Version 1.0*; IUCN.SSC: Gland, Switzerland, 2013; Available online: https://portals.iucn.org/library/efiles/documents/2013-009.pdf (accessed on 15 February 2023).
3. Resende, P.S.; Viana–Junior, A.B.; Young, R.J.; de Azevedo, C.S. A global review of animal translocation programs. *Anim. Biodivers. Conserv.* **2020**, *43.2*, 221–232. [CrossRef]
4. Chilvers, B.L.; Finlayson, G.; Ashwell, D.; Low, S.I.; Morgan, K.J.; Pearson, H.E. Is the way an oil spill response is reported in the media important for the final perception of the clean-up? *Mar. Pollut. Bull.* **2016**, *104*, 257–261. [CrossRef] [PubMed]
5. Ott, G.L. Fish and Wildlife Contingency Planning for Oil Spills: The Need for A National Wildlife Plan. *Int. Oil Spill Conf. Proc.* **2001**, *1*, 609–615. [CrossRef]
6. Clumpner, C. Wildlife Response: Part of the Plan. *Int. Oil Spill Conf. Proc.* **2005**, *1*, 349–352. [CrossRef]
7. Chilvers, B.L.; Battley, P.F. Species prioritization index for oiled wildlife response planning in New Zealand. *Mar. Pollut. Bull.* **2016**, *149*, 110529. [CrossRef]
8. Page, M.J.; McKenzie, J.E.; Bossuyt, P.M. The PRISMA 2020 statement: An updated guideline for reporting systematic reviews. *Syst. Rev.* **2021**, *10*, 89. [CrossRef]
9. DIISE. *The Database of Island Invasive Species Eradications, Developed by Island Conservation, Coastal Conservation Action Laboratory UCSC, IUCN SSC Invasive Species Specialist Group*; University of Auckland and Landcare Research New Zealand: Auckland, New Zealand, 2019; Available online: http://diise.islandconservation.org/ (accessed on 15 February 2023).
10. Veitch, C.R.; Clout, M.N. *Turning the Tide: The Eradication of Invasive Species*; IUCN SSC Invasive Species Specialist Group: Gland, Switzerland; IUCN SSC Invasive Species Specialist Group: Cambridge, UK, 2002; pp. Viii + 414.
11. Veitch, C.R.; Clout, M.N.; Towns, D.R. (Eds.) Island Invasives: Eradication and Management. In Proceedings of the International Conference on Island Invasives, Gland, Switzerland, 8–12 February 2010; IUCN: Gland, Switzerland; CBB: Auckland, New Zealand, 2011; pp. xii + 542.
12. Veitch, C.R.; Clout, M.N.; Martin, A.R.; Russell, J.C.; West, C.J. (Eds.) *Island Invasives: Scaling Up to Meet the Challenge*; 2019 Occasional Paper SSC no. 62; IUCN: Gland, Switzerland, 2019; pp. xiv + 734.

13. Chilvers, B.L.; Morgan, K.J.; White, B.J. Sources and reporting of oil spills and impacts on wildlife 1970–2018. *Environ. Sci. Pollut. Res.* **2021**, *28*, 754–762. [CrossRef]
14. Pitt, W.C.; Berentsen, A.R.; Shiels, A.B.; Volker, S.F.; Eisemann, J.D.; Wegmann, A.; Howald, G. Non-target species mortality and the measurement of brodifacoum rodenticide residues after a rat (*Rattus rattus*) eradication on Palmyra Atoll, tropical Pacific. *Biol. Conserv.* **2015**, *185*, 36–46. [CrossRef]
15. Hull, C.L.; Hindell, M.A.; Gales, R.P.; Meggs, R.A.; Moyle, D.I.; Brothers, N.P. The efficacy of translocating little penguins Eudyptula minor during an oil spill. *Biol. Conserv.* **1998**, *86*, 393–400. [CrossRef]
16. Goldsworthy, S.D.; Giese, M.; Gales, R.P.; Brothers, N.; Hamill, J. Effects of the Iron Baron oil spill on little penguins (*Eudyptula minor*). II. Post-release survival of rehabilitated oiled birds. *Wildl. Res.* **2000**, *27*, 573–582. [CrossRef]
17. Barham, P.J.; Crawford, R.J.M.; Underhill, L.G.; Wolfaardt, A.C.; Barham, B.J.; Dyer, B.M.; Leshoro, T.M.; Meÿer, M.A.; Navarro, R.A.; Oschadleus, D. Return to Robben Island of African Penguins that were rehabilitated, relocated or reared in captivity following the Treasure oil spill of 2000. *Ostrich* **2006**, *77*, 202–209. [CrossRef]
18. Crawford, R.J.M.; Davis, S.A.; Harding, R.T.; Jackson, L.F.; Leshoro, T.M.; Meÿer, M.A.; Randall, R.M.; Underhill, L.G.; Upfold, L.; van Dalsen, A.P.; et al. Initial impact of the Treasure oil spill on seabirds off western South Africa. *S. Afr. J. Mar. Sci.* **2000**, *22*, 157–176. [CrossRef]
19. Wolfaardt, A.C.; Underhill, L.G.; Altwegg, R.; Visagie, J.; Williams, A.J. Impact of the Treasure oil spill on African penguins Spheniscus demersus at Dassen Island: Case study of a rescue operation. *Afr. J. Mar. Sci.* **2008**, *30*, 405–419. [CrossRef]
20. Selman, W.; Hess, T.J.; Salyers, B.; Salyers, C. Short-Term Response of Brown Pelicans (*Pelecanus occidentalis*) to Oil Spill Rehabilitation and Translocation. *Southeast. Nat.* **2012**, *11*. [CrossRef]
21. Gartrell, B.D.; Battley, P.F.; Clumpner, C.; Dwyer, W.; Hunter, S.; Jensen, M.; McConnell, H.M.; Michael, S.; Morgan, K.J.; Nijman, P.; et al. Captive husbandry and veterinary care of seabirds during the MV Rena oil spill response. *Wildl. Res.* **2019**, *46*, 610–621. [CrossRef]
22. Empson, R.A.; Miskelly, C.M. The risks, costs and benefits of using brodifacoum to eradicate rats from Kapiti Island, New Zealand. *New Zealand J. Ecol.* **1999**, *23*, 241–254.
23. McClelland, P.J. Eradication of Pacific rats (*Rattus exulans*) from Whenua Hou nature Reserve (Codfish Island), Putauhinu and Rarotoka Islands, New Zealand. In *Turning the Tide: The Eradication of Invasive Species*; Veitch, C.R., Clout, M.N., Eds.; IUCN SSC Invasive Species Specialist Group: Gland, Switzerland; IUCN: Cambridge, UK, 2002; pp. 173–181.
24. Merton, D.; Climo, G.; Laboudallan, V.; Robert, S.; Mander, C. Alien mammal eradication and quarantine on inhabited islands in the Seychelles. In *Turning the Tide: The Eradication of Invasive Species*; Veitch, C.R., Clout, M.N., Eds.; IUCN SSC Invasive Species Specialist Group, IUCN: Gland, Switzerland; Cambridge, UK, 2002; pp. 182–198.
25. Millett, J.E.; Parr, S.J.; Shah, N.J. *Seychelles Magpie-Robin Recovery Programme Quarterly Report No. 39 and 40*; Birdlife Seychelles: Victoria, Seychelles, 2000.
26. Howald, G.; Donlan, C.; Faulkner, K.; Ortega, S.; Gellerman, H.; Croll, D.; Tershy, B. Eradication of black rats Rattus rattus from Anacapa Island. *Oryx* **2010**, *44*, 30–40. [CrossRef]
27. Rueda, D.; Campbell, K.J.; Fisher, P.; Cunninghame, F.; Ponder, J.B. Biologically significant residual persistence of brodifacoum in reptiles following invasive rodent eradication, Galapagos Islands, Ecuador. *Conserv. Evid.* **2016**, *13*, 38.
28. Wilkinson, I.S.; Priddel, D. Rodent eradication on Lord Howe Island: Challenges posed by people, livestock, and threatened endemics. In *Island Invasives: Eradication and Management*; Veitch, C.R., Clout, M.N., Towns, D.R., Eds.; IUCN: Gland, Switzerland, 2011; pp. 508–514.
29. Taronga Conservation Society Australia. Captive Management for Woodhen and LHI Currawong Associated with the Lord Howe Island Rodent Eradication Project. 2014. Available online: https://lhirodenteradicationproject.org/wp-content/uploads/2017/12/2.-Taronga-Zoo-Captive-Trial.pdf (accessed on 15 February 2023).
30. Cuthbert, R.; Hilton, G. Introduced house mice Mus musculus: A significant predator of threatened and endemic birds on Gough Island, South Atlantic Ocean? *Biol. Conserv.* **2004**, *117*, 483–489. [CrossRef]
31. Caravaggi, A.; Cuthbert, R.J.; Ryan, P.G.; Cooper, J.; Bond, A.L. The impacts of introduced house mice on the breeding success of nesting seabirds on Gough Island. *Ibis* **2019**, *161*, 648–661. [CrossRef]
32. Cuthbert, R.J.; Sommer, E. *Gough Island Bird Monitoring Manual*; RSPB Research Report No 5; Royal Society for the Protection of Birds: Sandy, UK, 2004.
33. Rexer-Huber, K.; Parker, G. *Captive Husbandry of the Gough Island Bunting and Moorhen*; RSPB Research Report; Royal Society for the Protection of Birds: Sandy, UK, 2011; p. 42. ISBN 978-1-905601-29-5.
34. Samaniego, A.; Jolley, W.; McClelland, P. A lesson for planning rodent eradications: Interference of invasive slugs during the Gough Island mouse eradication attempt in 2021. *Wildl. Res.* **2022**. [CrossRef]
35. Piersma, T.; Ramenofsky, M. Long-Term Decreases of Corticosterone in Captive Migrant Shorebirds That Maintain Seasonal Mass and Moult Cycles. *J. Avian Biol.* **1998**, *29*, 97–104. [CrossRef]
36. Springer, K. Eradication of invasive species on Macquarie Island to restore the natural ecosystem. In *Recovering Australian Threatened Species: A Book of Hope*; Garnett, S., Latch, P., Lindenmayer, D., Woinarski, J., Eds.; CSIRO Publishing: Melbourne, Australia, 2018; pp. 13–22.
37. Horn, S.R.; Cox, F.S.; Elliott, G.P.; Walker, K.J.; Russell, J.C.; Sagar, R.L.; Greene, T.C. Eradication confirmation of mice from Antipodes Island and subsequent terrestrial bird recovery. *New Zealand J. Ecol.* **2022**, *46*, 3488. [CrossRef]

38. Martin, A.; Richardson, M. Rodent eradication scaled up: Clearing rats and mice from South Georgia. *Oryx* **2019**, *53*, 27–35. [CrossRef]
39. Gartrell, B.D.; Collen, R.; Dowding, J.E.; Gummer, H.; Hunter, S.; King, E.J.; Laurenson, L.; Lilley, C.D.; Morgan, K.J.; Mcconnell, H.M. Captive husbandry and veterinary care of northern New Zealand dotterels (*Charadrius obscurus aquilonius*) during the CV Rena oil-spill response. *Wildl. Res.* **2013**, *40*, 624–632. [CrossRef]
40. Innes, J.; Barker, G. Ecological consequences of toxin use for mammalian pest control in New Zealand—An overview. *New Zealand J. Ecol.* **1999**, *23*, 111–127.
41. IPECA. *A Guide to Oiled Wildlife Response Planning*; IPIECA Report Series V 13; IPIECA: London, UK, 2014; pp. 1–52. Available online: http://www.oiledwildlife.eu/publications/ipieca-guide-oiled-wildlife-response-planning (accessed on 15 February 2023).
42. PIECA. *Key Principles for the Protection and Care of Animals in an Oiled Wildlife Response*; IOGP Report 583; IOGP: London, UK, 2017; pp. 1–74. Available online: https://www.ipieca.org/resources/awareness-briefing/key-principles-for-the-protection-care-and-rehabilitation-of-oiled-wildlife/ (accessed on 15 February 2023).
43. Hunter, S.A.; Tennyson, A.J.D.; Bartle, J.A.S.; Miskelly, C.M.; Waugh, S.M.; McConnell, H.M.; Morgan, K.J.; Finlayson, S.T.; Baylis, S.M.; Chilvers, B.L.; et al. Assessing avian mortality during oil spills: A case study of the New Zealand MV 'Rena' oil spill, 2011. *ESR* **2019**, *39*, 303–314. [CrossRef]

MDPI AG
Grosspeteranlage 5
4052 Basel
Switzerland
Tel.: +41 61 683 77 34

Animals Editorial Office
E-mail: animals@mdpi.com
www.mdpi.com/journal/animals

www.ingramcontent.com/pod-product-compliance
Lightning Source LLC
Chambersburg PA
CBHW042013260926
51850CB00020B/3276
* 9 7 8 3 7 2 5 8 2 3 9 5 6 *